P. CLAY SHERROD, President of the Midsouth Astronomical Research Society, Little Rock, Arkansas, has been aiding the Smithsonian Astrophysical Observatory, Cambridge, Massachusetts, since 1974. He has made over 1000 observations for the United States Naval Observatory in Washington, D.C., and he is a frequent contributor to numerous scientific periodicals and publications.

PHalarope Books are designed specifically for the amateur naturalist. These volumes represent excellence in natural history publishing. Each book in the PHalarope series is based on a nature course or program at the college or adult education level or is sponsored by a museum or nature center. Each PHalarope Book reflects the author's teaching ability as well as writing ability.

BOOKS IN THE SERIES

THE CURIOUS NATURALIST
John Mitchell and the Massachusetts Audubon Society

THE AMATEUR NATURALIST'S HANDBOOK
Vinson Brown

NATURE DRAWING: A TOOL FOR LEARNING
Clare Walker Leslie
(also in the Art & Design Series)

OUTDOOR EDUCATION: A MANUAL FOR TEACHING IN NATURE'S CLASSROOM
Michael Link, Director, Northwoods Audubon Center, Minnesota

NATURE WITH CHILDREN OF ALL AGES: ACTIVITES AND ADVENTURES FOR EXPLORING, LEARNING, AND ENJOYING THE WORLD AROUND US
Edith Sisson, The Massachusetts Audubon Society

WILDLIFE WATCHING: A COMPLETE GUIDEBOOK
Charles Roth, Massachusetts Audubon Society

NATURE PHOTOGRAPHY: A GUIDE TO BETTER OUTDOOR PICTURES
Stan Osolinski
(also in the Art & Design Series)

A COMPLETE MANUAL OF AMATEUR ASTRONOMY: TOOLS AND TECHNIQUES FOR ASTRONOMICAL OBSERVATIONS
P. Clay Sherrod with Thomas L. Koed

A COMPLETE MANUAL OF AMATEUR ASTRONOMY

TOOLS AND TECHNIQUES FOR ASTRONOMICAL OBSERVATIONS

P. CLAY SHERROD
WITH THOMAS L. KOED

Foreword by Leif Robinson, Editor, *Sky & Telescope* Magazine

PRENTICE-HALL, INC., Englewood Cliffs, New Jersey 07632

Library of Congress Cataloging in Publication Data

Sherrod, P. Clay.
A complete manual of amateur astronomy.

(A Spectrum Book)
Bibliography: p.
Includes index.
1. Astronomy—Observers' manuals. I. Koed, Thomas L. II. Title.
QB64.S47 522 81-2441
ISBN 0-13-162115-7 AACR2
ISBN 0-13-162107-6 (pbk.)

This Spectrum Book is available to businesses and organizations at a special discount when ordered in large quantities. For more information, contact: Prentice-Hall, Inc., General Book Marketing, Special Sales Division, Englewood Cliffs, New Jersey 07632.

A SPECTRUM BOOK

10 9 8 7 6 5 4 3

Printed in the United States of America

Prentice-Hall International, Inc., *London*
Prentice-Hall of Australia Pty. Limited, *Sydney*
Prentice-Hall of Canada, Ltd., *Toronto*
Prentice-Hall of India Private Limited, *New Delhi*
Prentice-Hall of Japan, Inc., *Tokyo*
Prentice-hall of Southeast Asia Pte. Ltd., *Singapore*
Whitehall Books Limited, Wellington, *New Zealand*

Surely there is something in the unruffled calm of Nature that over-awes our little anxieties and doubts: the sight of the deep-blue sky and the clustering stars above seem to impart a quiet to the mind.

JONATHAN EDWARDS
American theologian, 1703–1758

I deliberately interrupt my sleep and my vast speculation, crying out with the Psalmist King that great sentiment:

Great is the Lord, and great is His power; of His wisdom there is no limit. Praise Him, O heavens, praise Him, O sun, moon and planets; use every sense to perceive your Creator, every tongue to declare Him. Praise Him, O judges of harmonies discovered. Praise the Lord your Creator, you my soul, as long as I shall live. For from Him, and through Him and in Him are all things—both things known by the senses and things understood by the mind—as well as those things of which we are entirely ignorant. Because there is still more beyond. To Him be praise, honor, and glory forever!

JOHANNES KEPLER
German astronomer, 1571–1630

To my patient and beautiful wife,
HARRIETT

CONTENTS

PREFACE

This guide began as a pile of tiny scraps of paper, each with an important message that only the amateur astronomer could interpret.

Thousands of information-hungry amateur astronomers have had their first look around in the sky and have seen pretty much all there was to see. The next step is either to give up the hobby, thinking that changes in space cannot be seen unless one uses large observatory instruments, or to turn to an astronomy of a different sort by adding a scientific element to what has become a possessive avocation. After changing telescopes many times, always looking for that bigger and better instrument to satiate the thirst for some sort of scientific endeavor, the nonprofessional astronomer soon realizes that it is not the instrument but the person who uses it who unravels the mysteries of the skies. And the mysteries are there, as well as their answers, for those *prepared* to discover them.

The library, it would seem, is the logical source for all the information needed to begin some systematic research with the equipment the amateur astronomer has. Not so . . . there are plenty of textbooks on theoretical and practical astronomy and lots of popular reading on the newly found black holes and other items of cosmological fascination. There are all sorts of books to boggle the imagination—but we need something to *clear* our imagination and to set us on the right path to consistent studies with our telescopes.

Perhaps the most help to me in my infancy of astronomy, just prior to giving up hope of contributing to the science of astronomy, was the journal *Sky and Telescope,* which combines a pleasing mixture of astronomical bewilderment and interesting monthly projects that any amateur astronomer can do. My first copy of that magazine still sits on my library shelf—with no covers left. Bit by bit, the knowledge jelled, and the enthusiasm to contribute something meaningful to the science of astronomy grew. There were projects that needed to be done, and today the projects are even more significant in number and importance. Many things in the sky still baffle man, and there are just not enough professional astronomers around to carry the load.

I began with Jupiter in 1969, knowing enough that I should perhaps draw on paper what the telescope was showing me. As soon as the cloud belts of that great planet began to show signs of change I wanted to know *why* they were changing. I scribbled and gathered little tidbits of information on proper observing procedures for Jupiter and stuck them in the back of a notebook so that I would not forget the techniques. Several books that I had been lucky enough to find, particularly B.M. Peek's *The Planet Jupiter,* were earmarked for reference.

I logically assumed in that day of revived interest in space that a book would soon appear that would merge all the references and procedures necessary for scientific studies suitable for amateur astronomers and that it would serve as a guide for such research. It never appeared.

Then, my interests diverged, as will yours, going from comets and meteors to variable stars, to Saturn and Mars, and lunar occultations. My collection of equipment grew. Soon I had at my disposal a fine observatory, and I was convinced that the efforts I could make in the science were important. And all the notebooks grew . . . each having in its pocket those scraps of paper with their tidbits of knowledge to aid me in my studies.

Other astronomers saw what I was doing in my studies and wanted equally to contribute to the science. So I organized the little bits of paper into monographs that demonstrated either the techniques I had taught myself for 12 years or techniques that I had learned by not being too shy to ask the persons I considered most highly qualified to teach me.

And so the collection of scraps and notes finally became a book.

The amateur astronomer's equipment is now becoming more and more sophisticated, with such tools as filar micrometers, photoelectric photometers, chilled emulsion cameras, and the like. It is becoming increasingly possible for the amateur to contribute some information to the science of astronomy and make some mark in the world of science.

Too many amateur astronomers buy a fine telescope, examine the sky for about one year, and then assume that the universe is static and unchanging except as seen through the glass giants of the world's professional observatories. Then the enthusiasm wanes and the telescope is put away. Conversely, other amateurs have a thirst for learning more about the universe. Astronomy is a unique science in that as we learn more and more, the universe becomes even less known and more mysterious. In the theoretical end of things, astronomy allows the average person to think away as far as the mind will allow. And what I had assumed all along has turned out to be correct: now more than ever before the professional astronomer cannot monitor all the fantastic objects in space, and many go virtually ignored except through the effort of amateurs.

I hope that what is said in the following pages—the guidelines for your contribution to such areas of astronomy as comets, meteors, planets, the moon and sun, occultations, and so much more—will fill that void that I experienced 12 years ago and provide you with the guide for serious channeling of your efforts in astronomy. Good luck to all who begin. I am sure that in some—perhaps many—archives of astronomy and science *your* name will appear for having participated in one of the many research projects that follow.

ACKNOWLEDGMENTS

This book is somewhat different from most astronomy books that are available today. Rather than defining the basic elements of the science, this book presents many of the research projects suitable and useful to the nonprofessional astronomer. And because I never intended it to be a broad overview of astronomy, many subjects have been purposely omitted. It is not within the scope of this book to teach the reader the physics of astronomy. That can be acquired from the many excellent texts already on the library shelves. Nonetheless, some basic data regarding the subject of each chapter were required to enable the amateur to understand the methods to be used in the research projects. My familiarity with the subject and the manuscript became a handicap, and I often assumed that my readers would have the same familiarity. It is also easy for the initial enthusiasm with which a manuscript is begun to wane when the author is faced with the necessary but tedious processes of review and revision. Then it is necessary that he be spurred over his occasional writer's blocks and blind spots by those who can view his manuscript in an objective way.

On this project I had the good fortune that many technical writers do not have. Shirley Covington, Managing Editor of Spectrum Books at Prentice-Hall, Inc., who sparkplugged this book from manuscript form to final publication, has in residence an enthusiastic amateur astronomer—her son, Paul Covington, to whom she could turn for technical assistance. Paul's help was entirely voluntary. He read the entire manuscript, challenging, correcting, suggesting changes, and citing sources. There were times when I was absolutely positive regarding some fact or figure, yet I was wrong, and Paul diplomatically brought these things to my attention. Having been a scientific editor in the field of carcinogenic research for many years, I can appreciate the kind of objectivity (as well as devotion to the subject) that results in this kind of insistence on correctness and clarity. It is not often that we thank persons for noting our mistakes and oversights, but in this case the gratitude is justified. Throughout this book are many sentences and paragraphs that were rewritten by Paul, as well as an entire section on the rings of Saturn that incorporates the 1980 Voyager spacecraft findings. In addition, his (and Shirley's) indefatigable photographic forays into the cold twilight of last winter with the Newtonian telescope that he had recently built led to the photograph that appears on the cover of this book. Paul's contributions to this book abound.

Several sections of this book have benefitted from contributions from other sources. The observing forms in the Appendixes were compiled by the American Association of Variable

Star Observers (AAVSO) and by the various sections of the Association of Lunar and Planetary Observers and are used with their permission.

The Henry Exposure Graphs appear by the courtesy of the Midsouth Astronomical Research Society, Inc. of Arkansas, but they were originally prepared by James M. Henry of North Little Rock. These predicted exposure times were not merely calculated or estimated; they were diligently derived by months of searching through old issues of *Sky & Telescope* magazine for the best photographs of the moon, planets, and sun. By carefully noting the exposure times and film combinations, Mr. Henry derived the graphs. The first of the graphs was published in 1973.

Chapter 14, "Astrophotography for the Amateur Astronomer," was written by Thomas Koed, an active amateur astronomer with an impressive astrophotography laboratory in his home in Maumelle, Arkansas. His photographs have appeared in many of the popular astronomy journals during the past 10 years, the latest having been taken using his 12.5-inch (32-cm) Newtonian telescope with a fine 8-inch (20-cm) Schmidt camera mounted alongside. Mr. Koed also wrote the section on polar alignment of portable and permanently mounted telescopes in Chapter 1. Thomas Koed and I are partners in an astronomical retail business in Little Rock. Besides his help in the writing of these sections, I am indebted to him for his understanding of my devotion to completing this book during working hours while he absorbed much of the work load of the business.

Other friends were called on for help. Linda Derden survived deadline after deadline for three months, until the final manuscript was typed. Brian Sherrod aided in some last-minute photography and copy work, which would not have been possible had it not been for his assistance.

But before all the busy work began, there were those who took an interest in the project, not knowing if it would ever be published but knowing that it should be. Mary Kennan, Spectrum Books editor, lobbied to have the original manuscript accepted for publication, no small feat when one considers the many repetitious books on astronomy that are introduced each year. Mary Kennan enabled this book to make it to publication and helped to make its distribution a success even *before* the first copies were printed.

Certainly not the least—but I feel it should be last—is the "thank you" that I can never say to a man who was a dear friend to me and to every other astronomer who knew him: Dr. Joseph Ashbrook. For many years editor of the prominent journal *Sky & Telescope,* Joe was perhaps the greatest helper and advisor an amateur astronomer ever had. He was never too busy to talk to someone with a question, nor busy as he was, was he ever too busy to personally answer a letter from an astronomy enthusiast. This is the way I came to know Joe in 1971; from then until the summer of 1980 many letters and telephone calls were exchanged between us. It was Dr. Ashbrook who initially encouraged me to write this book, and he gave me understanding in many areas that were not clear to me. In every chapter of this book, there is some personal input by Dr. Joseph Ashbrook. For that reason, the book should have an added dimension of meaning to every amateur astronomer. Joe was unique in his contribution to the world of astronomy, which he made more fascinating to all of us. Joe Ashbrook died August 4, 1980, only minutes after congratulating me on the acceptance for publication of the manuscript of this book and promising to write the Foreword. It seems only fitting that the Foreword be written by Leif Robinson, a close friend and associate of Dr. Ashbrook and his successor as editor of *Sky & Telescope* magazine, because Leif began as long ago as the early 1960s to urge that such a book as this be written to give aid to those amateurs who wished to contribute to the science of astronomy. As the final touches were being put on *A Complete Manual of Amateur Astronomy,* Leif Robinson's enthusiasm was a lift to all of us.

FOREWORD

On reading Clay Sherrod's manuscript, my first reaction was: "Too bad a book like this wasn't available when I first became seriously interested in astronomy!" In format and presentation it lays bare many secrets of an epicure who has feasted on things celestial and who now wishes to pass along some well-tried recipes. This effort is all the more appreciated at a time when most books on astronomy appear to be either sorry collections of typescript for professionals or warmed-over homilies for the public.

Any amateur—whether using unaided eyes for constellation study or large sophisticated telescopes for research—will find a potpourri of valuable ideas in *A Complete Manual of Amateur Astronomy.* To be frank, any title that contains the word *complete* scares me. This one may fulfill such a promise. Some unorthodox minutiae include a table for converting Universal Time to decimals of a day, report forms of national associations, and information on how to start an astronomy club.

Yet this is not a book for only the amateur astronomer. I can imagine joy in the eyes of instructors of observational astronomy courses and of directors of public nights at planetariums and museums. In fact, this book could not have come at a more propitious time. The 1980s opened with a rebirth of public interest in science after the psychological and federal letdown following the Apollo moon landings (which all too rapidly became routine mirror images) and the Vikings' exploration of Mars (which failed to detect heralded life forms or other goodies for the media mill). Touted less but historically equal were the fantastic space robots that voyaged to Mercury, Venus, Jupiter, and Saturn. Together they yielded renaissance perspectives concerning the evolution of our solar system. Someday those electronic visions may be likened to the human visions of Galileo, Kepler, and Copernicus. But, sadly, such profound discoveries are appreciated at their moment only by scientists, philosophers, and artists. That confederacy, unfortunately, does not attract federal support or a broad audience.

Sherrod's book should modify this passive attitude, simply because it will encourage people to *observe* the sky, thereby raising private questions about the celestial vault, how it works, and what things make it up. His broad survey will also provide the perspective necessary to appreciate better the next bright comet or the long-overdue supernova that might gleam in some future daylit sky.

Who *is* watching Mars tonight? Does anyone know what's happening on or above that enigmatic planet, now that the Vikings have been shut down? After the last close-up pictures were taken, after the last active experiments were performed on the planet's soil and atmosphere, it's all too easy to imagine that we've learned all there is to know. Yet, despite the recent death of the Vikings, Martian clouds continue to come and go, the polar caps expand and recede, and the wave of change still spreads under a summer sun. The professional whose program has ended or whose money has run out will not know. The day of the amateur has returned, not only to keep records as before but to interpret them in the context of space-age revelations. This is exciting; let's hope that Sherrod's book will kindle a timely new fire!

It's widely assumed that the era of amateur contributions to learning is over, the logical consequence of seeing big-time science at work. This is far from the truth, as proven by K. Ikeya and T. Seki who wrote their names in the sky by discovering one of the greatest comets seen in the twentieth century. And, following the tradition of Sir William Herschel—an eighteenth-century one-time amateur and acknowledged father of stellar astronomy—thousands of aficionados continue to monitor the light changes of a myriad of stars—some that brighten and dim over decades, others over minutes. In 1975 hundreds of amateurs "discovered" Nova Cygni, which overnight changed the face of the Northern Cross. Yet only one, Ben Mayer, anticipated such a happening and photographically chronicled the hour-by-hour brightening of this new star with an automatic camera—an unprecedented feat in the curious annals of astronomy.

On a much slower time scale, I also experienced one of those magic moments of discovery. In 1960 a bright spot appeared in the north polar region of Saturn. Although I wasn't the first to see it, I did follow it night by night in an effort to determine the planet's rotation period. (From reading, I knew that Saturn's rotation period at such a high latitude was virtually indeterminate.) This was high amateur excitement, although tempered by hours of study and frustration in attempting to understand the message of the observations. Eventually, at age 18, I published in a professional journal. The windows to the universe still remain open.

It is often said that astronomy, unlike chemistry or physics, is not an experimental science. This is only partly true. Of course, the astronomer—amateur or professional—cannot go into space and mix various combinations of molecules, atoms, and dust to see what might fall out—stars, planets, whatever. Yet it is possible to sort out observations and codify them, and from this, an empirical or mathematical model can be constructed and tested to see if it "fits." (To me, this seems just as valid as playing around in some chemical kitchen.) With the computer revolution at hand, the ancient nightmare of processing data has faded into the azure sky of discovery. It is now possible for amateurs, using nothing more than an affordable home computer, to undertake analyses that professionals would have quailed at 20 to 30 years ago. Orbits and ephemerides of comets and minor planets, period changes of variable stars, artificial satellite predictions, occultation analysis: none should hold terrors for the inquisitive amateur.

The whole of natural science awaits contributions from dedicated and knowing persons who will collect and analyze what they see. Although the two tastes are quite different, they need not be mutually exclusive. Consider, on a terrestrial level, the millions of birdwatchers who annually derive satisfaction from "checking-off" as many species as possible; this is the end to itself. But there is also a relatively minute corps of bird students (call them amateur ornithologists) who examine the ebb and flow of the avian world and attempt to make sense of what they find. From such people, using data gleaned primarily from the dilettanti, the 1950s crash of the peregrine falcon was documented in Europe and North America. Thus, the world was alerted to the potential genocide that might result from continuing the use of DDT.

Professionals do not have the time, and usually not the funds, to monitor continually populations of plants, animals, or stars. Nor do they have the luxury of exploring seemingly barren ground; contemporary careers simply do not permit such bold excursions. Thus, the pursuit of the routine on one hand and the arcane on the other falls into amateur hands. Each can do what he might at his own pace. If knowledge for mankind does not come from such efforts, excitement, pleasure, and learning for the devotée will. This is as assured as the sun rising tomorrow. But will it . . . how do *you* know?

Sky & Telescope

LEIF J. ROBINSON

INTRODUCTION

AMATEUR ASTRONOMY IN AMERICA

Above us, the sparkling stars of the night skies stretch out like thousands of diamonds suspended on the curtain of space. Unfolding through the beauty and the mysteries of this seemingly endless expanse are patterns and answers familiar to those willing to study them—not only the professional astronomer but the casual stargazer as well. Across our country—indeed, the world—millions of amateur astronomers are involved with the study of the skies simply because of their love for the heavens.

There is an affinity for the eternity of space experienced by all of mankind, a kind of motherhood in the stars to those who study space. Since the beginning of humans on earth, there have been stargazers, but never in the history of astronomy has the interest developed at such a pace as we are experiencing now. From the movie screen to the backyard, people are now involved with the night sky, and there is a place for any person wanting to try a hand at amateur astronomy.

In our country's infancy there were no "professional" astronomers. The study of the skies from American soil depended solely on the efforts of the public. Perhaps there is not so wide a gap in astronomy as in other scientific fields between the "amateur" and the "professional." The fact that the amateur receives no pay for his or her efforts seems to be the only dividing line. This dedication and love for the science of astronomy are what spur the amateur into action and cause people with busy lifestyles to make time in the wee hours of night to experience the fascination of space.

Almost 350 years ago, on June 25, 1638, an eclipse of the moon was recorded by a Rhode Island amateur astronomer, marking the first observation in astronomy on American soil for a scientific purpose. I say "for a scientific purpose" simply because this observer bothered to note the length of time required for the earth's shadow to pass entirely across the face of the moon. An observation by Thomas Hariot was made of a bright comet in 1585, but this North Carolina amateur did not record any of his observations, so no scientific data remain.

Telescopes during the late seventeenth century were of little help in bringing the influence of science to America. John Winthrop, Jr., the first governor of Connecticut, owned two of the largest telescopes of that time. However, these small instruments would not be considered good enough for use even by the beginning stargazer of today. Winthrop's telescopes were eventually given to Harvard College in 1672 and were put to excellent use there, in spite of their limited size. The great comet of 1680 was observed with one of these telescopes from Harvard, thus beginning the first observations of the sky be an "established American observatory," the observer being a nonprofessional resident of the College. A severe fire destroyed all but one of Winthrop's instruments, as well as most of Harvard College, late in the seventeenth century. The remaining telescope was, at the time of the fire, on a foreign expedition to observe a transit of Venus across the disk of the sun. However, the fine brass vintage telescope eventually fell into the hands of British collectors and now rests far away in England.

By the eighteenth century two almanacs had been published that aided the ordinary person living in colonial America and called the happenings of the skies to the attention of otherwise unaware readers. Benjamin Franklin's *Poor Richard's Almanac* was the best seller of the time, with as many as 100,000 copies sold each year in a society of only 2 million people. The reason for the success of this and other almanacs was the importance of astronomical knowledge in all walks of life. It was needed to navigate across the countryside, rivers, lakes, and oceans, for surveying the ever-increasing land of America, for charting the wilderness, and even to tell the time of day. Astronomy was also important to astrology because the casting of horoscopes was of much greater importance to the people of the eighteenth century than it is today. Farmers and merchants used the almanac for weather forecasting throughout the year to aid in the growing, harvesting, and transporting of crops.

By the early eighteenth century, American amateur astronomers began to contribute significantly to European science. Aurorae, sunspots, eclipses of the sun and moon, Jupiter's satellites, Mercury, Venus, and particularly meteors and fireballs dominated the scene of American research. Between 1700 and 1776 over 100 papers were contributed by amateur New England astronomers for publication by the Royal Astronomical Society of Britain. In one such paper, John Winthrop, IV, the grandson of the Connecticut governor, discussed the possibility that meteors come from interplanetary space (outside of the earth's atmosphere) rather than from within the clouds, as was the accepted theory at the time. Until 1807, however, Winthrop's theory created very little stir in Britain and Europe, and people did not take his belief seriously. In that year, a huge fireball that was spotted over the skies of Connecticut broke apart, dropping hundreds of pounds of fragments to the ground, thus supporting Winthrop's claim that the meteors must be heavy particles floating in space.

Still, with so much individual effort being put forth, the year 1800 began with nothing in the way of an observatory for public or research use. In 1832 the Astronomer Royal of England noted, "I am not aware that there is any public observatory in America, though there are some very able observers." The dawn of American professional astronomy began midway in the nineteenth century when the Naval Observatory at Washington, D.C., was established in 1844. In 1847 the giant 15-inch refractor at Harvard College was put to use by the father of American astronomy, William Cranch Bond, a clockmaker from Boston. Interestingly, Bond was self-trained in his knowledge of astronomy and was an amateur in the strictest sense until taking the Harvard appointment. His enthusiasm for astronomy set the pace for the future of all American astronomers, both professional and amateur.

By 1850 there were many observatories of professional quality in the United States, and by 1885 this number had grown to at least 144, most of which were supported and funded only through public endowment, with no government funds. But almost as if slapped in the face, the American amateur astronomer—the very element that fought for and supported the study of space for two centuries—began to lose prestige after 1850, primarily because of the great numbers of professional observatories and the large telescopes being put in them. Not that efforts could not be made by the amateur, but the enthusiasm waned. Even into the twentieth century this feeling of subservience to the professionals continued. In the first half of this century those who might have been involved in the study of the sky experienced two wars and a crippling depression. The sky was still there, but the eyes to see it were lacking.

TO THE PRESENT

Perhaps if it had not been for a few die-hard amateurs and telescope makers in New England and a handful of enthusiasts in California, the amateur's contribution to astronomy might have made way entirely for the glass giants of the professional. By 1950 amateur astronomy got that shot of enthusiasm it needed, and a revival began that would sweep the country like wildfire. Just what the sparkplug was, we still do not know. Some reasons why the public may have taken off with such great enthusiasm for the stars are the following:

- ☆ Two magazines, neither of which might have continued on their own, merged to provide one strong voice for the efforts of the amateur and to keep the public abreast of sky events. *The Telescope* and *Sky* became *Sky and Telescope,* the best-read astronomical journal in American history.
- ☆ Telescope making had become an interesting hobby for young and old alike, and even those not interested in watching the stars began to make scopes in their spare time.
- ☆ At the same time, commercial outlets emerged across the country providing factory instruments to the amateur with quality that rivaled that of the professional telescopes.
- ☆ The space age had opened with Sputnik and lighted a fire in the intellect of the public. Moonwatch teams, consisting of amateur astronomers, were established to follow the paths of the new space satellites as they crossed the sky. This activity allowed amateurs to feel once again that they were contributing to the science of astronomy.
- ☆ The interest developed so rapidly that public planetariums were established throughout the United States, increasing the enthusiasm of the people that had been whetted by newspaper reports of our space achievements.
- ☆ Then, using the moonwatch teams and the planetariums as focal points, amateurs soon discovered others with the same interest and astronomy clubs began to be formed, first in the large metropolitan areas, and then spreading even to rural areas in every state. Even the clubs themselves began to organize on a national basis, and organizations supporting amateur research soon gave the amateur astronomer a reason to develop knowledge.

In Cambridge, Massachusetts, the American Association of Variable Star Observers prompted amateurs to follow the light changes of stars that are not constant in brightness. Now, nearly 75 years after its formation, the AAVSO has accumulated over 4 million observations of scientific quality—all from amateur stargazers. Thirty years ago, The Association of Lunar and Planetary Observers was founded in New Mexico to provide an outlet for serious amateur research on the moon and planets, and to date it has accumulated one of the most extensive records of observations of the planets ever amassed—again all from the nonprofessional observer. And even for the not-so-serious person, national organizations support our efforts. The Astronomical League is composed of virtually every major astronomy club in the United States, supporting the amateur's causes, sponsoring publications, and organizing national meetings to promote that exciting feeling of being involved with astronomy.

THE AMATEUR ASTRONOMER'S CONTRIBUTION TO SCIENCE

Can the amateur astronomer still contribute to the science as in the past? It would seem that in this day of sophisticated computerized space ships and telescopes that do not even need an operator for a night's observing, little could be done to supplement the professional's work. Nothing could be further from the truth. Perhaps the best example of just how far an amateur astronomer can take his or her interests was seen at the turn of the twentieth century, just when it seemed that the amateur had lost any influence in a society filled with professionally trained astronomers and giant telescopes.

An American businessman, Percival Lowell, had a theory. It was a unique theory dealing with the planet Mars, perhaps the most mysterious celestial object of his day. Lowell was not an astronomer, nor had he had any formal training in astronomy other than what he had taught himself. Yet he had an idea, and he had the intellect and fortitude to resist all the criticism of the professional-oriented society. Indeed, in 1893–1894 Lowell constructed what was to become the foremost observatory in the world for the studies of planets. Located in Flagstaff, Arizona, the Lowell Observatory shines above all the rest in its uniqueness; it was the home of "Mars fever," the enthusiasm and eagerness to find life on Mars, that swept the country in that period. And it was at Lowell that the only discovery in the United States of a planet—Pluto—was made. The Lowell Observatory continues to be a foremost center for the study of comets, Mars, Jupiter, and more. More uniquely, however, it continues to be like no other major research facility: it still is a private observatory, founded by an amateur astronomer, and supported entirely by his perpetual endowment for its continuance.

It was Lowell's perseverance and his financial ability to support his hobby that enabled him to make his contribution to science. Today, it is for different reasons that the amateur astronomer continues to provide scientific data otherwise unavailable to the professional. And, unlike Lowell, the amateur's efforts are heartily welcomed by the professional, no matter how insignificant the observations might at first seem.

The amateur astronomer is able to watch the sky in a way that the professional cannot. Indeed, many professionals rarely get a chance actually to look at the sky, being restricted to electronic instrumentation within some room beneath the observatory floor while a technician operates the tracking mechanism for the observatory telescope. Amateur astronomers are not restricted in things to observe. If a bright new comet appears, the amateur's scope can be turned to it for valuable observations. On the other hand, the professional usually has waited long and patiently for time on one of the large telescopes to undertake a pet project. Even a bright comet is not allowed to interrupt a project that has been a year in waiting.

In many cases, the professional observatory is too *well* equipped to observe many of the things studied by the amateur. Many times sophisticated instrumentation on large observatory telescopes cannot be removed, nor would anyone wish to remove it if it could be. Consequently, many of the largest telescopes can no longer be looked through. The amateur is thus given the edge for studies of the planets and other sudden transient events. And while the professional telescopes are getting more and more complicated and specialized, the amateur's telescopes are getting larger and larger, rivaling many times the full capability of a neighboring college or university. Indeed, perhaps the best-instrumented, modest sized observatories are those belonging to amateurs. These small private observatories are complete with photoelectric photometers with strip-chart recorders, bifilar micrometers for measurements of double stars, specialized cameras for planetary photography, complete libraries, darkrooms, and much more—all of which are used every clear night.

But what can the stargazer without all this fancy gadgetry do in amateur astronomy? Truly, the sky is the limit. Even a good pair of eyes can be useful for astronomical observation during passages of bright comets and meteor showers, and even observations of many variable stars that can be seen with the naked eye.

Most of the bright comets discovered by the naked eye have been found by amateur astronomers using crude home-built, wide-field telescopes or good binoculars. Searching for comets right after sunset in the western evening sky, or just before dawn in the eastern sky, can be a relaxing way to learn the sky and an exciting way to literally spell your name across the sky. Japanese amateur astronomers—notably, Iwata, Ikeya, Kosato, and Seki—have found considerable numbers of bright comets and have received national fame in their native country. Likewise, Australian amateur astronomers can hold their own in comet discoveries. The most successful comet searcher of the twentieth century, William Bradfield of Australia, has discovered as of now 13 comets—all with a home-built 6-inch telescope assembled from scrap lumber and burlap twine.

In addition to discovering comets, amateurs devote much time to monitoring these objects as they approach and pass close to the sun. Again, the professional does not have the time for such studies, and if it were not for the amateur astronomer, much valuable data would be lost. Even the naked eye or binoculars can be useful in such observations to note the development of the curving tail of the comet, sudden increases in brightness, and other remarkable details when the comet approaches the intense radiation of the sun.

Another naked-eye project for the amateur is the surveillance of meteor showers as they come from the *radiant,* that point in the sky from which a shower of meteors appears to originate. Meteors belonging to shower groups can be followed night after night to determine when the maximum number per hour can be seen. It is possible through naked-eye observation to pinpoint the time of maximum to within a half hour, giving the professional data that give clues as to the motion of the meteor cloud, its density, and the part of space in which the meteors originate.

The photographically minded amateur can provide some useful information from meteor watching as well. A simple 35 mm camera mounted on a tripod and shooting with color slide film can record the tracks of bright meteors as they pass, thus providing astronomers with a rule as to the point of origin of the meteor.

Almost all the data being systematically recorded on the planets other than that sent back by spacecraft are supplied by the amateur astronomer who diligently draws and photographs the visible faces of Mars, Jupiter, and Saturn. The professional astronomer's attention has turned away from the planets in an effort to explain celestial phenomena of greater cosmological significance; but they have turned their attention from the planets before the answers were at hand. So again, the amateur astronomer, with the volunteer labor, provides the necessary data.

Amateurs study the clouds, early morning mists and fogs, violent dust storms that occur in Martian summers, and the changes in the polar caps of Mars and its dark regions, known as maria, or seas. As in Percival Lowell's time, the amateur continues to provide most of the continuous observations of the mysterious red planet.

Jupiter offers a realm of research for the amateur astronomer that is not studied by the professional. Monitoring of the circulation patterns of Jupiter's clouds, the rotational patterns and changes of the Great Red Spot, and the development of massive storm centers on Jupiter are all best studied by the amateur astronomer's equipment. The solar system, forgotten by the professional, becomes a field of badly needed research for the nonprofessional astronomer.

Perhaps the area above all others in which the professional is admittedly indebted to the amateur's efforts is the study of the stars. Stars that vary in light are numbered in the millions throughout our Milky Way galaxy; some take years to make a complete light cycle, others only take minutes. Because of the great numbers of these stars and, many times, because of the unpredictable nature of their light changes, professionals rely almost exclusively on the observations of amateurs to plot the nature of each star's light. From these plottings, the very evolution of the star can be speculated on. In many cases, stars not seen by the naked eye as anything but a point of light are discovered to be double stars, and their masses, the size of their orbits, and even their distance from earth can all be determined through the dedication of thousands of amateur observers of variable stars.

The possibilities are endless. Whether a person is a casual stargazer or an advanced amateur astronomer involved with supplemental research, the sky continues to provide a new show every night.

THE FUTURE OF AMATEUR ASTRONOMY

A lot of people are becoming involved with astronomy. Indeed, it appears that almost everyone has that latent interest in the cosmos evolving from the subconscious to the conscious. People are *excited* about the sky, and the awareness is changing into a curiosity. As the answers to the curiosities begin to unfold, the fascination of astronomy becomes a habit—an overwhelming obsession that relaxes, teaches, tranquilizes, yet stimulates the imagination of all who pursue its nature.

Astronomy has been called "the fastest growing hobby of the American culture." Astronomers are appearing almost monthly on the "Tonight Show," and even Johnny Carson is an avid amateur astronomer. Newspaper articles appear daily announcing the discovery of some new stranger-than-science phenomenon of deep space, and weekly columns are now syndicated in the country's largest newspapers to keep readers abreast of the happenings in the night skies.

The fact is that astronomy involves every one of us—we are made of the stars themselves. And people can seek their roots in the stars just as they can in libraries and genealogical records.

And as long as the sky can be seen, the trend will continue—to search the skies with eyes eager to learn is a natural for man. We can look into the dark night sky and peer into the past as it was millions of years ago; we can speculate, theorize, imagine as far as our minds want to go.

The sky belongs to us all; it is there for the looking. And amateur astronomy is *free* for the curious to explore. Tonight, begin exploring the night sky. Begin with a comfortable lawn chair or a blanket spread on the ground. Invite a friend or involve the whole family. Soon, before your eyes, the fascination with the night sky, the constellations, the planets as they trek slowly through the unchanging heavens, and the darkness of space itself begins to unfold. The more we learn, the more complex the mysteries become. Welcome to the world of amateur astronomy. It is like heaven on earth!

1

THE SELECTION OF A TELESCOPE

Getting that first telescope or the instrument you have been planning as your final investment is an exciting adventure. A new telescope can bring hours—even years—of enjoyment to the whole family. Moving up to a larger, more efficient instrument allows the serious amateur astronomer to engage in research, astrophotography, and make a contribution to the scientific field of astronomy.

However, in either case—for the beginner or the semiprofessional astronomer—the choice of a telescope is a serious matter and one that should be thoroughly planned prior to the final investment. As in many other endeavors, there is no one perfect telescope for general all-purpose, all-interest viewing. Many telescopes are well suited for studies of the planets, exhibiting very fine detail of Mars, Jupiter, and Saturn's magnificent ring system. Others show these objects too glaringly without enough built-in magnification to do much good, yet these instruments excel in astrophotography or in views of the star clouds stretched through the summer Milky Way. Picking up some popular astronomy magazine with its dazzling array of advertisements from commercial sources merely compounds the difficult question: "Which telescope is best suited for my interests?"

THE MANY FUNCTIONS OF AN ASTRONOMICAL TELESCOPE

The beginning astronomer is usually baffled by claims for "power," that is, magnification that will seemingly bring anything one wishes into view. Small department-store types of telescopes are often advertised as having such excessive magnifications as 454, 600, or even 1000. Such magnifications are not practical, even in the largest amateur telescopes. What, then, is the purpose of the telescope? It would seem at first that magnification is the primary purpose of the instrument, but there are more important functions to look for in choosing an instrument.

Light Grasp

Opinion varies, and much is already in print about the quality and performance of astronomical telescopes, but there is no substitution for *size,* that is, the diameter of the aperture of the primary optics. A larger telescope will exhibit more subtle detail in nebulous objects and show fainter stars because it simply has more surface area on which light is collected. Basically, before all other considerations, a telescope must be a light collector. This does not mean that instruments of smaller aperture are not suitable for astronomical studies, but only that in most cases added light grasp is an asset. The brightness of all celestial objects is designated on a somewhat arbitrary scale, known as the *apparent magnitude scale,* in which we assume that a star near the naked-eye limit—approximately magnitude 6—is about 100 times fainter than a star of the magnitude 1. The basis for the assumption arose early in the second century BC when the astronomer Hipparchus compiled a star catalog containing about 1000 stars; each of the brightest stars of the major constellations were assigned *first magnitude,* whereas those just at the limit of naked-eye vision were said to be *sixth magnitude.* Consequently, a difference of five magnitudes corresponds to a ratio of 100:1, or—as suggested by astronomer Norman R. Pogson in 1856—each magnitude corresponds to the fifth root of 100, or 2.512. Put another way, a star of magnitude 1 is simply 2.512 times brighter than one of magnitude 2. It is somewhat confusing to those new to astronomy that the brigher stars have lower numbers than do the fainter ones. Exceptionally bright objects even have negative numbers to allow for the continuation of the Hipparchus/Pogson concept.

Such magnitudes are known as *apparent,* or *relative,* because they serve only to intercompare stars and celestial objects as we perceive their brightnesses from earth. Obviously, a star appears quite bright to us if it is very close, and it appears fainter as it moves farther away. It is important to astronomers to understand the "real" or *absolute magnitude* of objects. How much light does a star actually emit into space? All stars emit light in relation to their size and temperature. By knowing the distance of the celestial object, we can compute the real brightness because the light we receive from the object varies inversely to the square of the distance that the object is from us.

Comparing the absolute magnitudes of stars allows astronomers to evaluate in more depth the physical nature of the processes of those stars. If we were able to establish just how bright each celestial object would appear if all were positioned at exactly the same distance from earth, the evaluations are simplified. Thus, the absolute magnitude of any celestial object is defined as the magnitude that that object would have if it were positioned at a distance of 10 parsecs in space, or about 32.6 light years. Our own star, the sun, shines with an apparent magnitude of -26.5, yet at that distance its absolute magnitude is a mere +4.85, near the limit of the naked eye, which is magnitude +6.

In most studies by amateur astronomers, references made to magnitude concern *apparent* magnitude—the brightness of the object as we see it in our instruments or with our naked eyes. The professional astronomer and theorist are more concerned with the *absolute* magnitude. Yet the apparent magnitude and the absolute magnitude can be derived each from the other. The first advantage of light grasp is that the *limiting magnitude* of a telescope increases with increased aperture. We think of limiting magnitude in terms of the faintest star that is visible under the best conditions in any given aperture. Most tables that compare aperture to limiting magnitude seem somewhat conservative in my estimation. There is no doubt that the most acute eye is capable of perceiving a faint star just beyond another's view, but the values should be near constant for the majority of observers (See Table 1-1).

In very exacting situations, with trained observers, the magnitude limit of any telescope can be increased by adjusting the magnification upward slightly, thus increasing the threshold contrast of the object and the background sky.

The figures given in Table 1-1 can be misleading for persons not prepared to interpret them first. For example, even though a 25 cm (10-inch) reflector might be capable of showing stars to magnitude 14.2, there is no way the same telescope could be used visually to survey for galaxies of that same magnitude. When magnitudes of celestial objects are expressed, it is in terms of *integrated light,* that is, light compacted to a point source, like a star, and then measured. Thus, for example, the bright galaxy Messier 33 in Triangulum is a quite difficult object to view in a 25-cm telescope, even though its total brightness is greater than magnitude 6. Such *extended*

TABLE 1-1. Limiting visual magnitudes of telescopes of selected aperture.[a]

Aperture	*Magnitude*	*Aperture*	*Magnitude*
6 cm (2.4 in.)	10.5	20 cm (8.0 in.)	13.9
8 cm (3.1 in.)	11.0	25 cm (10.0 in.)	14.2
10 cm (4.0 in.)	12.0	32 cm (12.5 in.)	14.7
15 cm (6.0 in.)	13.4	40 cm (16.0 in.)	15.5

[a]Limiting photographic magnitudes are 3.5 magnitudes fainter.

objects will appear considerably brighter if their sizes are relatively small by comparison to Messier 33, whose light is spread out over an area greater than 1° of sky.

Increased aperture will likewise increase the resolution of an instrument, but atmospheric turbulence and differential cooling and heating of optics of larger telescopes will often cancel the resolution gained by an increase of objective diameter.

Resolution

Another important function of a telescope, *resolution*, is related not only to the size of the objective lens or mirror, but to the quality of the optics and the conditions under which they are used. According to Dawes' criteria of telescope resolution, the larger the diameter of a telescope's optics, the finer the detail that can be seen, as expressed in the formula:

$$sep'' = \frac{4.56}{d}$$

where *sep''* represents the minimum separation in seconds of arc of two stars of equal magnitude, at which each star is clearly seen to be distinct from the other in a given aperture. Dawes' criterion is 4.56, and *d* is the diameter of the telescope lens or mirror, in inches. Table 1-2 lists telescopes of various apertures and their resolving power, according to Dawes' limit.

The limits set forth by the foregoing equation seem to hold fairly well but only in instances of excellent visibility and steady air. If the air is quite turbulent, due to the rapid mixing of warm and cool air, no telescope can expect to achieve the performances given in Table 1-2. Likewise, only the finest optical surfaces can attain resolution close to the figures in Table 1-2. In addition, instruments of short focal length (f/6 or faster) generally cannot resolve nearly as well as telescopes of longer focal length. Consequently, we would expect the long-focus, clear-aperture refractors to excel in resolution—and they do, in cases reaching or exceeding Dawes' limit.

Resolution of an optical system is also affected by the bending and scattering of light that must happen through most reflector telescopes. In classical Newtonian and Cassegrainian telescopes the light as it enters the tube must pass across not only the secondary mirror but also the supporting mechanism (the "spider")

TABLE 1-2. Resolution of telescopes according to Dawes' criteria.

Aperture (cm)[a]	*Resolution (sec of arc)*	*Aperture (cm)*	*Resolution (sec of arc)*
6	2.2	20	0.6
8	1.5	25	0.5
10	1.2	32	0.4
15	0.8	40	0.3

[a]See Table 1-1 for equivalents in inches.

for that mirror. In the catadioptic, or compound, telescopes, in which a front correcting lens supports the secondary mirror, the scattering of light is much less severe.

Magnification

If a telescope has great light-gathering power (large aperture) and good resolution (aperture + quality optics + air steadiness), it also should be well suited for magnification. As a matter of fact, in very steady conditions a magnification of 50 power per inch of aperture must be used to achieve optimum resolution. Anything more, and the image degrades rapidly. Even under the steadiest conditions, the eye is a limiting factor in maximum, and even minimum, magnification.

For example, under dark sky conditions, with the human eye totally "dark-adapted," the pupil should measure about 8 mm. A beam of parallel light wider than the eye's ability to receive it emerging from an eyepiece of very low power results in *vignetting* (i.e., loss of perception of the entire field of view). Thus, every telescope is ruled by the low-power law for minimum magnification. This law is expressed by the formula:

$$m_{\ell} = \frac{\text{aperture of telescope in mm}}{8 \text{ mm}}$$

where m_{ℓ} represents the lowest usable magnification for a particular instrument, and the aperture of the telescope is expressed in millimeters divided by 8 mm (the diameter of the pupil of the human eye when fully dark-adapted).

It has probably occurred to you at this point that there is most likely some criterion for determining the maximum power that can be utilized. The human eye achieves greatest resolution when the pupil is constricted to only 3 mm. This is because constriction causes the focal ratio of the eye to change from about f/2.7 to about f/7. Yet, when substituting 3 mm for the 8 mm of the fully dilated eye, the resulting values are still less than the highest useful magnifications. I have found, through practice, that the most practical upper limit of magnification is about five times the lower limit. Table 1-3 gives recommended lowest, optimum, and highest magnifications of telescopes of various aperture.

The upper limits of magnification on any given night, of course, will depend on the state of the atmosphere for that night. On the crisp cool nights when the sky is amazingly transparent, the steadiness of the air is usually quite poor. Conversely, when there is a heat inversion in the summer months, with stagnant air overhead, the air is usually quite steady, allowing for excellent views of the planets and double stars. Like resolution, the magnification potential of a telescope increases on such nights because the boiling of the earth's air is at minimum and definition is steady and uninterrupted.

Telescopes that have a minimal number of light obstructions in the optical path are potentially higher magnifiers than other types. Refractors, with their clear unobstructed aperture, usually can hold more magnification while maintaining steady images than can a reflector of equivalent size.

TABLE 1-3. Recommended lowest and highest magnifications for given telescopes.

Aperture (cm)	*Lowest Practical*[a]	*Optimum*	*Highest*	*Aperture (mm)*	*Lowest Practical*[a]	*Optimum*	*Highest*
60	15	25	100	200	50	70	200
80	20	30	150	250	60	85	250
100	25	35	175	320	80	105	320
150	40	50	180	400	100	135	400

[a]Optimum values are about three times minimum values by formula.

TELESCOPE DESIGN AND RECOMMENDED USE

Three basic types of telescopes are of interest to beginning and advanced amateur astronomers. All three types are capable—at least to some degree—of performing the functions of light grasp, resolution, and magnification. Any type of telescope can be used for almost every project described in the following chapters, but some are better suited for particular uses than others. The *refractor* telescope is that which is commonly brought to mind when someone mentions "astronomy." The refractor uses a lens as its light collector at one end, and an observer at the other. Its counterpart is the *reflector,* which uses a mirror, as its name implies. The third common telescope in amateur hands is the compound, or *catadioptic* design, which uses both mirrors and lenses to provide long focal lengths and quality image formation. Each of the three types has certain advantages and disadvantages, but when properly used each gives endless enjoyment and is a tool whereby the projects described in the following chapters can be begun with scientific vigor.

The Refractor

The refractor telescope is what the layman envisions as a telescope: a long tube with a lens at one end and a human eyeball at the other. The front element, or *objective,* is a convex lens of long focal length (see Figure 1-1a), followed by a second lens of equal size. Together the pair of lenses form an *achromatic lens,* one that should provide images in their true colors.

However, any piece of glass has prismatic capability, refracting various wavelengths of light in varying degrees, and even achromatic lenses are not free from some spurious violet coloration. The longer the focal ratio of the system (f/12 and longer), the less detrimental the color becomes. If it were not for the second, or *flint lens,* behind the convex glass (the *crown lens*), the spurious blue introduced through the aberration of light would render the instrument almost useless.

To produce an achromatic objective for a refractor requires an optician to grind, polish, and figure four surfaces, not just one, as in the case of a reflector telescope. Such work is time consuming and expensive; likewise, the pure glass, free of defects and bubbles, must be carefully and individually selected from expensive stock. The heart of the refractor—its objective lens—is therefore a very costly item.

Refractors become quite massive in the larger sizes. Even a 4-inch refractor has a tube 60 inches long and must be mounted high enough to allow the observer easy access beneath the eyepiece when the instrument is aimed overhead. Because of this requirement for rigidity to accommodate the long tube, the height of the instrument, and the size of the optical path, refractors must be mounted on much more massive mountings than a reflector of normal (f/8) focal ratio with an equivalent aperture. Figure 1-2 shows a 4-inch refractor in use. Transportation of refractors with apertures of 4 inches and greater also becomes quite a problem. If one does not have some type of observatory shelter in which to house the instrument, it must be disassembled each night following observing and taken back indoors. If one wishes to take the telescope to the mountains where the skies are darkest, the refractor is usually of prohibitive size. Nonetheless, the refracting telescope is still the favorite among many critical observers because it provides very dark background sky, large image scale, and superb resolution. I have even heard the comment, "I would surely rather look *at* the sky than to have to watch a *reflection* of the same thing." Perhaps there are still some aesthetics involved in amateur astronomy.

The first refractor used for astronomical observations to any degree was made by the famous Galileo Galilei in 1609. Through it he discovered the four major satellites of Jupiter, craters on the moon, sunspots, and the phases of Venus.

The refractor is an excellent choice for the discriminating observer of the moon, planets, and sun (instrumental requirements for those projects are discussed in Chapters 6, 8-10, and 5, respec-

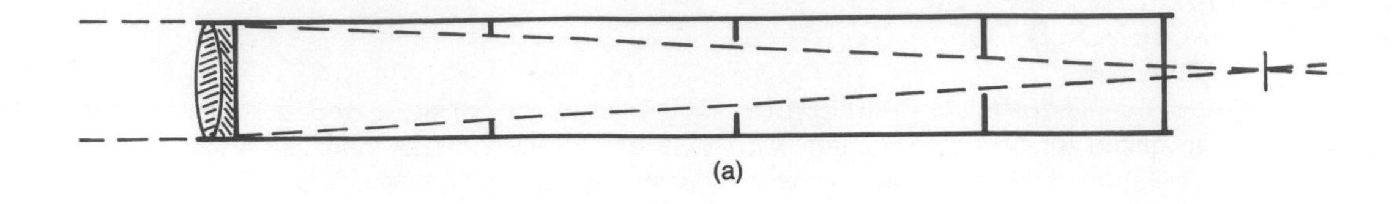

(a)

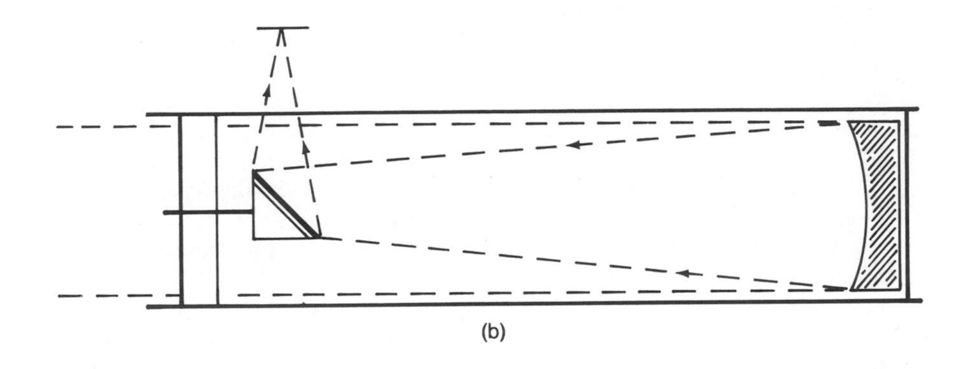

(b)

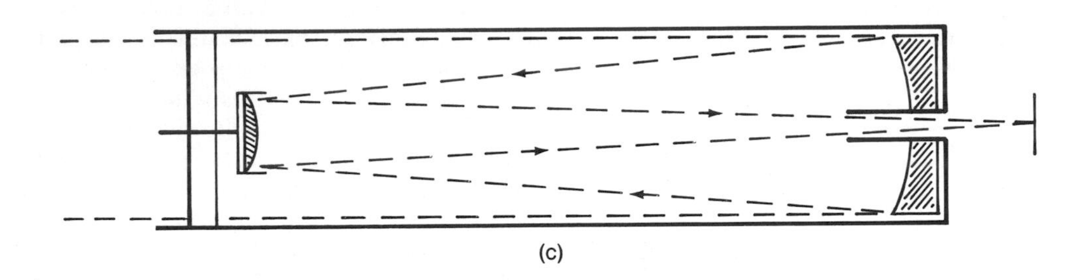

(c)

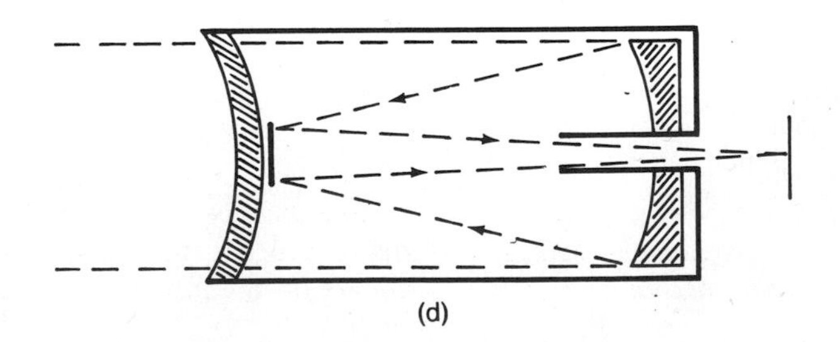

(d)

FIGURE 1-1. Basic designs of (a) the refractor, (b) the Newtonian reflector, (c) the classical Cassegrain, and (d) the Maksutov catadioptic.

FIGURE 1-2. A commercially made 10-cm (4-inch) refracting telescope.

tively). The instrument in sizes 4 inches and larger provides crisp, sharp images of such objects, and is less subject to internal heating currents than open-tube instruments. The long focal length of the refractor is advantageous for planetary study, resulting in a large image scale with minimal eyepiece magnification.

However, the refractor telescope is quite expensive compared to the other types, and it is not particularly outstanding for observations of variable stars, faint comets, or galaxies. A refractor with at 2.5-inch aperture costs $200 or more, depending on the mounting that holds the instrument. One with a 6-inch lens can cost up to $18,000. The largest refracting telescope in use in the world today remains the 40-inch Yerkes telescope in Wisconsin. No larger lens can ever be used effectively because the lenses must be supported around the edge (so as not to obstruct the light path). Sag would occur in the thick middle portion of lenses larger than 40 inches, distorting the clarity of the images produced. Table 1-4 demonstrates the optimum applications for the refractor telescope.

TABLE 1-4. Uses of the refractor telescope.

Use	*Chapter*	*Description*
Solar observing	5	Excellent instruments for sunspot observations; good for solar photography.
Comet searching	4	Best by far for lowest-power, wide-field searches is a 6-inch, f/5 optimum instrument.
Comet study	4	Excellent for wide-field observation, provided lens is of f/5 to f/8; dark background sky.
Moon observing	6	Highest resolution; excellent for photography.
Planet observing	8-10	High resolution, potential magnification; clear and sharp images.

The Reflector

A well-made reflecting telescope is perhaps the best buy on the telescope market, offering maximum performance, portability, adequate light grasp, and a very comfortable price tag. For amateur use there is one ever-popular type of reflecting telescope—the Newtonian reflector devised by Isaac Newton in 1672. This style uses a highly polished parabolic mirror, finished to the curve of a parabola, which focuses light in the same manner as a refractor lens. Unlike the refractor, however, the reflector allows the light to be diverted rather than pass through the optics, and thus the light is free of chromatic aberration.

The main parabolic mirror, called the *objective mirror,* reflects and converges the light that falls on its surface. A second, smaller mirror at the top end of the telescope (see Figure 1-1b) intercepts the light and diverts it at a right angle out of the tube and to the eyepiece, where the observer is positioned. So, unlike the refractor, the observer is positioned at the top of the telescope, and the objective rests at the bottom.

The Newtonian reflector is the best all-round instrument for both the beginning and advanced amateur astronomer, primarily because telescopes of large aperture are priced considerably less than refractor or compound instruments and are thus affordable to those wishing to pursue scientific studies in astronomy. A good 6-inch Newtonian reflector, which will nearly rival any telescope of equal size, can be bought for around $300.

For the amateur specializing in some aspect of astronomical research, the Newtonian telescope offers many chances for custom modification. Mirrors can be produced in a variety of focal ratios, from super-fast f/3 paraboloids for deep-sky astrophotography to long-focus f/12 spherical curves for high resolution planetary and lunar studies.

FIGURE 1-3. A Newtonian reflecting telescope of 32-cm (1.25-inch) diameter. (Courtesy, *Arkansas Gazette,* photograph by Don Jones)

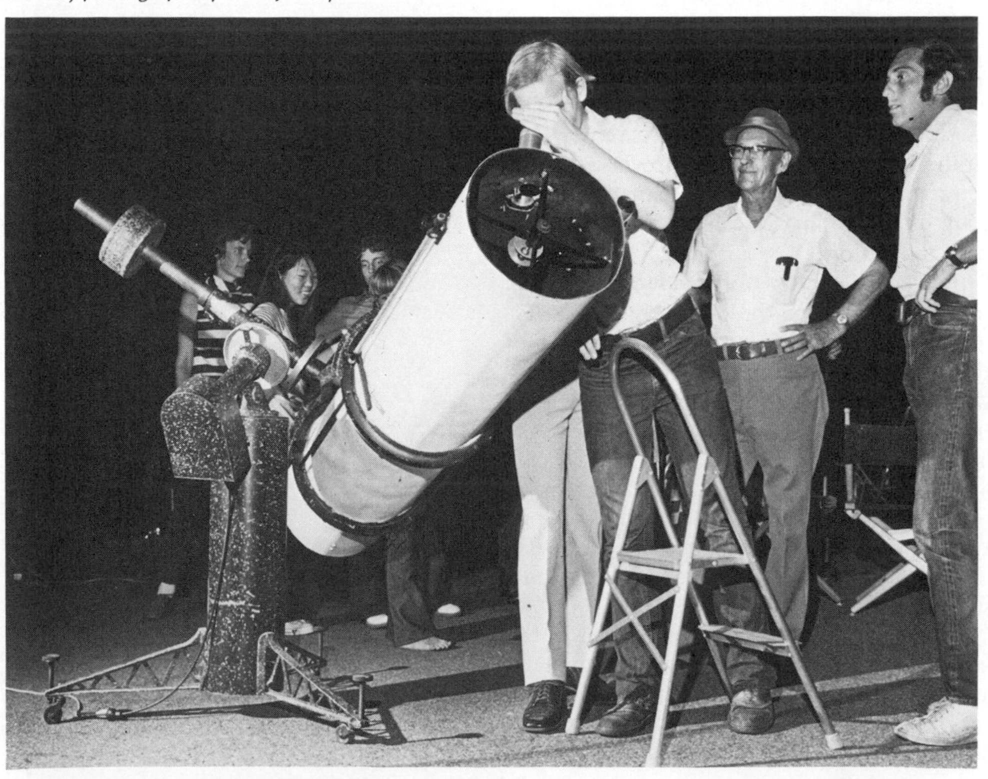

TABLE 1-5. Uses of the Newtonian reflecting telescope.

Use	*Chapter*	*Description*
Comets	4	Excellent in richest-field (f/5 or faster) instruments; wide, bright field plus high magnification and resolution for details near nucleus.
Lunar	6	Very good photographically; excellent for occultations of faint stars.
Planets	8–10	If optics are very good, high resolution/high magnification work is possible; colors very true.
Asteroids	11	In moderate (12½-inch or 32-cm +) aperture sizes, excellent for photometry work, extending observer's range to possibly a dozen variable asteroids.
Variables	12	Perhaps the best instrument of choice, offer good low-power wide fields for quick star acquisition and best limiting magnitude per cost; many faint irregular stars are within the limits of a 20-cm telescope.
Photoelectric photometry	13	The Newtonian and Cassegrain telescopes are preferred for their lack of chromatic aberration.
Photography	14	Newtonian is best all-round telescope for astrophotography, with relatively fast focal ratios, wide field, and large aperture.

Because the reflector is virtually aberration free, it is the ideal choice for photoelectric photometry (see Chapter 13). It is difficult to discuss the applications of reflecting telescopes without listing the projects covered in almost every chapter of this book. Ideal uses of the Newtonian telescope are given in Table 1-5.

The Cassegrain Reflector

If one chooses a reflecting telescope but desires a somewhat larger image scale than can be provided by the Newtonian design, a Cassegrain telescope can be considered. Commercial sources for such instruments are limited, and the performance of these instruments varies from telescope to telescope.

A pure Cassegrain telescope has a primary mirror with a short focal ratio—say, f/3 or f/4. The light cone reflected off this is reflected by a secondary, which is not flat but a convex hyperboloid. On reflection from the secondary, the light cone is now much narrower and subtends a much smaller angle. Hence, its rays now converge much more slowly and behave as if the light were reflected by a primary of longer focal length—f/10 to f/20. A secondary with an amplification of 4x will cause the light from an f/4 primary to act as if the primary were f/16. Because the effective focal length is increased, so is the available magnification. It is possible to construct a Newtonian–Cassegrain telescope, in which the secondary mirror can be replaced as desired, to make the scope vary from a f/4 Newtonian with an optical flat to divert the light at a right angle to the light path entering the tube, to an f/12 to f/20 Cassegrain in which the light is amplified by the curved surface of the other secondary mirror and reflected in line with the optical axis through a hole in the primary mirror. This arrangement provides the capability of low-power, wide-field work with the Newtonian while still maintaining the possibility of high resolution studies of planets in the Cassegrain mode.

I have used many Cassegrain telescopes and, although there have been exceptions, have found that most are inferior by far to the simpler

Newtonian design and perform poorly by comparison, even at high magnifications. This is particularly true in sizes less than 40 cm (16 inches). In many cases, problems inherent in Cassegrain systems arise from poor optical alignment on the part of the observer. Such problems can be overcome once the person achieves perfect collimation of the optics. Such alignment is not nearly so straightforward as with the Newtonian design (see Chapter 2), and many amateur astronomers give up trying to align the optics of the Cassegrain reflector after some time. Even after alignment is achieved, *maintaining* the collimation is likewise difficult with the long focal lengths involved in this type of instrument. Other considerations that contribute to my misgivings about small Cassegrain telescopes arise from problems that cannot be controlled or eliminated by the astronomer. For example, the secondary mirror of Cassegrain telescopes is quite large in relation to the primary mirror which results in considerable light loss when the path of the light from the celestial object being viewed crosses the obstruction.

On the other hand, a well-adjusted, quality Cassegrain telescope can be a powerful research tool, provided that the projects for which it is used are those best suited for that instrument. For example, because the eyepiece end of the telescope is easily accessible, the Cassegrain telescope is preferred for photoelectric photometry. The heavy detector/telescope coupler (see Chap-

FIGURE 1-4. A commercially made Schmidt-Cassegrain telescope.

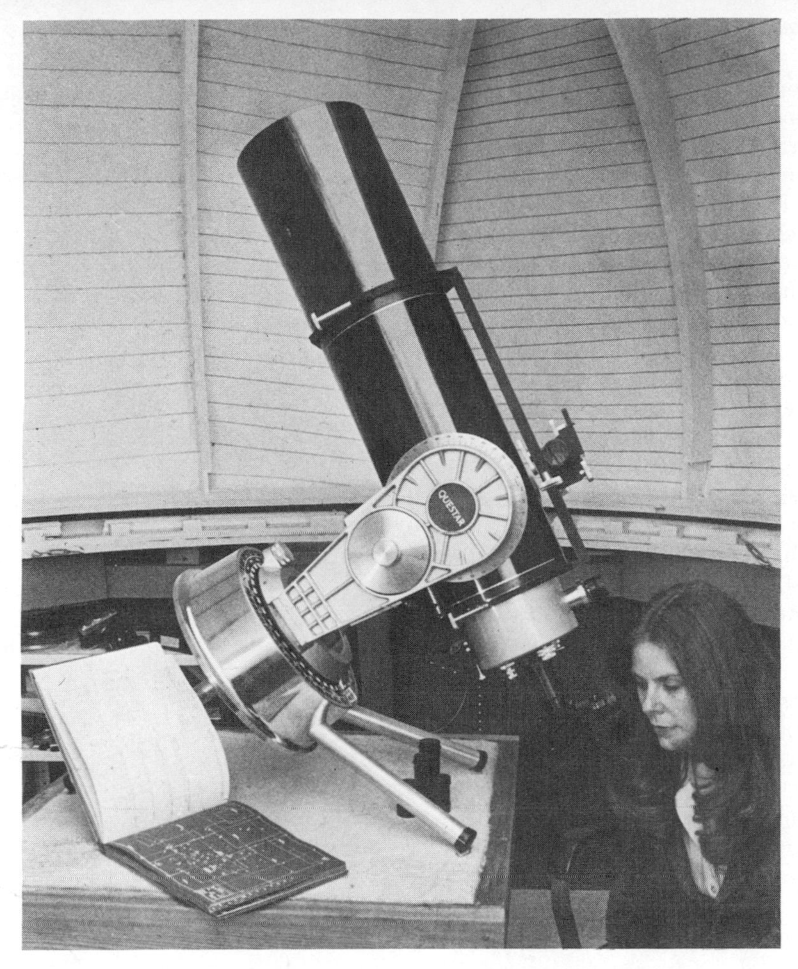

FIGURE 1-5. A commercially made 18-cm (7-inch) Makustov telescope.

ter 13) requires balancing when it is attached to the telescope. On the Newtonian such an adjustment is difficult if the weight is excessive because the coupler must be attached at a right angle to the mechanical axis in right ascension. Likewise, the large image scale and accessibility to the eyepiece allow precision in micrometer measurements, as discussed in Chapters 4, 8, 9, and 10.

The Compound Telescopes

The compound, or catadioptic, telescopes, which are derivations of the Cassegrain design, offer such advantages as compactness and portability, excellent optical quality, large image scale, and good buys when compared to refractor telescopes. As a general rule, the catadioptic telescopes, primarily Schmidt–Cassegrain and Maksutov designs, are good performers in almost every respect and combine all the advantages of the other types of telescopes with few disadvantages. They are priced between Newtonian reflectors and refractors in cost per size of aperture.

The most popular telescope in amateur hands today is the compound instrument, primarily because of its adaptability to photography and its portability. Most amateur astronomers are suburban residents who must travel miles to the nearest dark site. Large refractors and Newtonian telescopes are usually too cumbersome to transport. The compound telescopes fill a great need, therefore.

The Schmidt–Cassegrain telescopes use a spherical primary mirror. The problems inherent in such a mirror (say, compared to the paraboloid of the Newtonian type) are reduced to negligible amounts by having the light pass through the *correcting lens* before reaching the primary. By having such a glass "window" the telescope is also less subject to moving air within the tube, which can spoil otherwise good definition. The Maksutov telescopes are generally more expensive than the Schmidt–Cassegrain of comparable size, and they are difficult to obtain in apertures greater than 32 cm (12 inches). Because both types of compound telescope function similarly to the Cassegrain design, optical folding must occur, and there must be a secondary mirror in the light path, as in the other reflector types. In the Schmidt–Cassegrain telescopes, the secondary mirror is somewhat larger than in the equivalent Maksutov, resulting in slightly better performance in the Maksutov. However, because the Schmidt–Cassegrain design allows for effective use at f/10 or faster compared to the seemingly slow f/15 of the Maksutov, a somewhat better field of view is realized, brighter images result, and there is a distinct advantage for astrophotography in the Schmidt–Cassegrain. The compound telescope can be used ideally for the studies shown in Table 1–6.

TABLE 1–6. Uses of the compound (catadioptic) telescope.

Use	*Chapter*	*Description*
Comets	4	Schmidt–Cassegrain good for photography of coma, observations of center and head.
Moon	6	Both excellent, providing maximum resolution and good photographic capability.
Planets	8–10	Performance of Maksutov equal to refractor quality in most respects; Schmidt–Cassegrain good for photography.
Asteroids	11	Both are excellent for photometry studies, but cost-prohibitive in sizes large enough to study fainter objects.
Variables	12	Restricted to brighter members.
Astrophotography	14	Schmidt–Cassegrain excellent in all respects.

SOME RECOMMENDATIONS

So what telescope does one choose to get started in scientific research projects as described in the following chapters? Actually a 6-cm (2.4-inch) refractor could keep an enthusiastic viewer busy for a long time, but perhaps at least a 15-cm (6-inch) Newtonian should be the bare minimum to recommend as a telescope capable of diversified scientific study in astronomy.

Equally important as the size of the telescope is the manner in which the telescope is mounted. An equatorial mounting, preferably with clock drive, is always preferred over an altazimuth mounting. Only in the area of comet searching is the latter preferred for methodical sweeping of the sky. If I were to make a choice between a 25-cm (10-inch) telescope on a flimsy pipe mounting with no clock drive and a fully equipped 15-cm (6-inch) telescope with equatorial mounting, setting circles, and clock drive, the choice would be easy—aperture is not *that* important. What good is large aperture if the wind blows the telescope so violently that the image cannot be viewed steadily? Astrophotography is virtually impossible without a good mounting and clock drive. Planets are difficult to measure and draw if they are drifting out of the field of view so rapidly that one hand is occupied with making adjustments while the other attempts to draw the planet and hold the clipboard simultaneously.

So, there is more to a telescope than size. Added aperture is important, but it is not so important as to outweigh all other factors. Small instruments have their uses, too. The quality of the optics and mechanics is equally important. Likewise, the "usability" of the instrument should be considered: if I buy a great big telescope, how often will I be willing to dismantle the instrument and take it outdoors for an evening's observing? At first, quite a bit perhaps . . . but later on?

A telescope in the 15-cm to 25-cm range, with a full complement of necessary accessories for detailed study of celestial objects, and one that will be used more often because of its portability, is a much better investment than some enormous telescope that requires several persons to move it with no auxiliary equipment whatsoever.

ATMOSPHERIC CONSIDERATIONS

One final word concerning small, medium, and large instruments. The state of the earth's air is usually the determining factor for the performance of any given telescope. As telescope aperture increases, so increases its resolution—not just the resolution of celestial objects but of atmospheric turbulence as well.

For this very reason, smaller instruments frequently exhibit finer detail than 32-cm reflectors. Closer doubles can be resolved, and the image appears steady. In the 25-cm and above class of instruments, air currents and the unsteadiness of the nighttime skies become increasingly annoying. Yet on the same nights the observer with somewhat smaller instruments keeps on plugging away at the data. No observer has perfect skies. If transparency and dark skies are needed for low-power work, the larger apertures usually perform tops. But on the other hand, if high resolution/high magnification steady images are required, the smaller instruments can be used more often than larger ones. It is that simple.

More is said in the ensuing chapters on the selection of the proper telescope for each research project. Obviously no one observer will possess every telescope noted, so some compromise is necessary. It is best to find your special interest before you buy the final instrument. If you are interested only in planets, you must consider a different telescope from that of the observer wanting to photograph the sky or examine faint galaxies for possible supernovae.

THE SELECTION OF APPROPRIATE STAR ATLASES

Perhaps equally important to the selection of the proper observing instrument is the supplementing of a good star atlas for reference, plotting, and the extrapolation of data. In several following chapters many appropriate atlases suited for such use are discussed, yet there are some guidelines for the matching of one's projects to the choice of atlas. Some considerations for choosing a good atlas are the following:

- ☆ Is the atlas of a scale large enough to be easily readable and not confusing in the dark of night?
- ☆ Are objects other than stars plotted for quick reference so that the observer can differentiate, for example, between a cluster and a galaxy on the atlas?
- ☆ Is a grid provided so that accurate measurements or placements of some transient interloper (e.g., a comet or asteroid) can be easily determined?
- ☆ Is the atlas impervious to dewing and the effects of nighttime use?
- ☆ Can the atlas be used conveniently at the telescope?
- ☆ Is the atlas accurate to the epoch in which it is being used? Are the stars positioned properly?
- ☆ What is the approximate limiting magnitude of the atlas?
- ☆ Is a standard reference catalog available for the atlas in which the positions, magnitudes, and other pertinent data are listed?

All the considerations listed are important but in ways that might first elude the beginning observer. For example, it might appear at the outset that a star atlas with a limiting magnitude of 13 is perhaps better suited for detailed studies than on reaches on reaches only magnitude 9.0. Granted, more stars appear in the first atlas, but will the atlas be easy to use for plotting the path of a comet? In many cases, I find that a simple atlas with some good reference stars to magnitude 8.0 or 9.0 is much more suited for such plotting than one with many stars. It is quite easy to "get lost" on a chart that shows thousands of tiny stars which obliterate stars that otherwise could be used as convenient steppingstones at the telescope.

Some Recommendations

General Use

For learning the sky, or for the plotting of major meteor showers, bright comets, and other naked-eye phenomena, a simple atlas with a limiting magnitude near that of the eye should be selected. The atlas should have north at the top, with indexing in the margins for right ascension and declination, should a telescope be used. Some general-use atlases are the following:

- ☆ *Popular Star Atlas,* Inglis. This is a good booklet for beginners, listing stars to magnitude 5.5, and including 50 nonstellar objects (galaxies, clusters, etc.). Inexpensive and well suited for a "first" star chart. Bound in cloth and durable. Published by Sky Publishing Corp., Cambridge, Massachusetts.
- ☆ *Field Guide to the Stars and Planets,* Menzel. This is perhaps the finest sky guide in pocket form. The book contains not only sky charts for each month of the year but a beautiful set of photographs covering the entire sky. Each photograph is printed both positively and negatively. The negative print shows the exact positions and designations of stars, objects, and constellation outlines. This is excellent for binocular viewing and is a good reference for even the most serious observers. Available in both cloth and paper bindings. Available through Sky Publishing Corporation, Cambridge, Massachusetts.
- ☆ *Norton's Star Atlas,* Norton. An atlas now in its seventeenth edition, *Norton's* is perhaps the most overrated of all sky charts, because most of this expensive book consists of explanations of sky phenomena and little of it is charts. However, the atlas portion of the book shows stars to less than magnitude 6.0 and nonstellar objects numbering over 600. Published by Sky Publishing Corporation, Cambridge, Massachusetts.

Positional Work for Comets, Minor Planets, And Other Faint Objects

Atlases that are easily marked for plotting the changing positions of minor planets and comets, and for the planning of nova searches are of a character different from those for general observing. Positional work for very faint objects necessitates a chart with a faint limiting magnitude. As well, the atlases in the following list allow for the exact measurement of magnitude and position, either through transparent grids supplied with the atlases or through accompanying catalogs.

- ☆ *Skalnate Pleso Atlas of the Heavens,* Becvar. An excellent choice for nearly every application. This atlas is available in a deluxe edition, bound in Lexitone plastic, with fold-out multicolor charts, or in "desk" (black stars on white background) and "field" (white stars on a black background for use at the telescope) editions. The latter two editions are very modestly priced and excellent for plotting brighter phenomena. Over 32,000 stars and 1,850 nonstellar objects are plotted. A catalog is available for reference. Published by Sky Publishing Corporation, Cambridge, Massachusetts.
- ☆ *Handbook of the Constellations,* Vehrenberg and Blank. This atlas features indexed constellations for quick referral, each constellation isolated on a separate chart. The facing page of each chart gives coordinates and physical data for nonstellar objects (1,050 in all), lists of prominent double and variable stars, and the names and positions of bright key stars in the constellations, very useful

TABLE 1-7. Instruments required or preferred for astronomical research.

Subject	*Chapter*	*Naked Eye*	*Binocular*	*Refractor*	*Newtonian*	*Cassegrain*	*Schmidt–Cassegrain*	*Maksutov*
Aurorae	2	Excellent[a]	Good	—	—	—	—	—
Meteorology	2	Excellent	—	—	—	—	—	—
Meteors	3	Excellent[a]	Fair	Poor	Poor	Poor	Poor	Poor
Comet search	4	Poor	Fair	Excellent[a]	Good	Poor	Poor	Poor
Comets	4	Fair	Good	Good	Excellent[a]	Good	Good	Fair
Solar	5	Poor	Poor	Excellent[a]	Poor	Poor	Good	Good
Lunar	6	Poor	Poor	Excellent[a]	Excellent	Good	Excellent	Excellent
Occultations	7	—	Poor	Good	Excellent[a]	Good	Good	Good
Mars	8	—	—	Excellent (15 cm +)	Good	Good	Good	Excellent[a]
Jupiter	9	—	—	Excellent (15 cm +)	Excellent	Good	Good	Excellent[a]
Saturn	10	—	—	Excellent (15 cm +)	Excellent	Good	Good	Excellent[a]
Minor planets	11	—	Possible	Fair	Excellent[a]	Good	Good	Fair
Variable stars	12	Fair	Good	Good	Excellent[a]	Excellent	Excellent	Fair
Novae search	12	Good	Good	Fair	Excellent[a]	Excellent	Good	Fair
Photoelectric Photometry	13	—	—	Poor	Fair	Excellent[a]	Good	Fair
Astrophotography	14							
piggyback[b]		—	—	Good	Good	Good	Excellent[a]	Excellent
lunar		—	—	Excellent	Excellent[a]	Good	Good	Excellent
planetary		—	—	Good	Excellent[a]	Excellent	Good	Good
deep sky		—	—	Poor	Excellent[a]	Poor	Excellent	Poor

[a]Best choice, considering size requirements, performance, and cost.
[b]See Chapter 14.

for setting circle use. The book is bound in heavy plastic for durability in use at the telescope. Published by Sky Publishing Corporation, Cambridge, Massachusetts.

☆ *AAVSO Variable Star Atlas*. Combined with the *Sky and Telescope Guide to the Heavens,* this atlas is excellent for variable star studies, giving the positions and nearby comparison stars for many variables. In addition, the atlas can be used for asteroid and comet studies, providing both visual and photoelectric magnitudes of many stars across which some transient object may pass, allowing accurate determinations of magnitude. The atlas is adapted from the 1967 *Smithsonian Astrophysical Observatory Star Atlas,* which is now out of print. Because of a convenient printed grid on each chart, measurement of position is easily facilitated, thus the chart is excellent for comet measurement and the plotting of minor planets for visual or photoelectric photometry. Reference to many of the stars plotted on the atlas can be obtained through the *Smithsonian Astrophysical Observatory Star Catalog,* which is currently available. The atlas is published by Sky Publishing Corporation, Cambridge, Massachusetts.

☆ *Photographic Star Atlas,* Vehrenberg. For the studies of very faint comets, minor planet photometry, or even verification of a suspected faint nova, this atlas is perhaps the finest available. The atlas consists of loose sheets, each centered on a designated position of the sky, with some overlap between charts. All are boxed in two containers, and transparent grids are provided in scales set to various declinations for easy measurement. No reference catalog is available, although the *SAO Star Catalog* provides an excellent supplement. The atlas is published privately by Hans Vehrenberg, and can be obtained through Sky Publishing Corporation, Cambridge, Massachusetts.

Certainly, no observer can be expected to participate every clear night in every research project, so that eliminates many problems in the choice of the proper instrument. Keep in mind that *whatever instrument you might now have,* it is usable for some form of research in astronomical science. Even your naked eyes can be useful. And, indeed, all instruments are welcome in the research. Do not let the size or the type of telescope discourage you from participating in the exciting world of discovery. Give it a try! Learn the limitations of your instrument (even the glass giants of the world are limited in some particular aspect of astronomy), and you can devote the rest of your life to filling the astronomical void that only the amateur astronomer can fill.

REFERENCES

Ingalls, A.G., *Amateur Telescope Making: Vols. I, II, III*. New York: Scientific American, 1953.

Johnson, B.K., *Optics and Optical Instruments*. New York: Dover Publications, 1960.

Micaika, G.R., and W.M. Sinton, *Tools of the Astronomer.* Cambridge: Harvard University Press, 1961.

Moore, Patrick, *Astronomical Telescopes and Observatories for Amateurs.* New York: Norton, 1973.

Muirden, James, *The Amateur Astronomer's Handbook*. London: Cassell, 1969.

Naval Personnel, Bureau of, *Basic Optics and Optical Instruments*. New York: Dover Publications, 1969.

Page, Thornton, and Lou William Page, "*Telescopes: How to Make Them and Use Them,*" Vol. 4. *Sky and Telescope Library of Astronomy*. New York: Macmillan, 1966.

Paul, Henry, *Telescopes for Skygazing.* New York: Amphoto, 1970.

Roth, Gunter D., *Astronomy: A Handbook*. Cambridge, MA: Sky Publishing Corp., 1975.

Sidgwick, J.B., *Amateur Astronomer's Handbook*. London: Faber and Faber, 1958.

Thompson, Allyn J., *Making Your Own Telescope*. Cambridge, MA: Sky Publishing Corp., 1947.

2

INSTRUMENT SETUP AND MAINTENANCE

DETERMINATION OF LATITUDE AND LONGITUDE OF AN OBSERVING SITE

For exacting studies in astronomy, particularly for the timing of lunar occultations, the exact coordinates in latitude and longitude of one's observing site on the earth must be determined to within 10′ to 20′, and the altitude above sea level must be known to an accuracy of better than 10 feet (3 m). All reports issued by the advanced amateur astronomer should include these values, so that further credibility will be given to facts reported from that person's station.

The amateur astronomer may choose from several ways to determine exact latitude and longitude. Of these, perhaps the most impressive and interesting is "surveying in" the observatory, that is, meticulously timing meridian transits of stars that have known positions. If the observatory is located a great distance from any notable landmark, this is probably the only method available.

Some observers, just for the curiosity of doing it, have determined their positions to within a mile, using stars measured with small marine sextants and a cup of water held by hand to serve as an artificial horizon. Using such instrumentation, accuracy within a mile is quite remarkable. However, for surveying-in practices to be efficient enough for accurate reduction of data (certainly one mile is much too great a margin of error), either a photographic zenith tube or a transit instrument would have to be erected.

However, even the sophisticated devices designed primarily for such determinations require procedures that can lead to substantial errors that result from inaccurate leveling of the instrument, small mathematical miscalculations, and unknown precession of stars from their reported positions.

So, perhaps the easiest and most reliable method available to the amateur by which very precise determinations can be made is by directly graphically measuring your positions on the topographic maps supplied by the United States Geological Survey. The maps are available for any observer's location within the United States. Many local libraries have copies of these maps available for the observer's general area. The original survey from which these maps were compiled was completed in 1960, and the results were field checked in 1963 and again in later

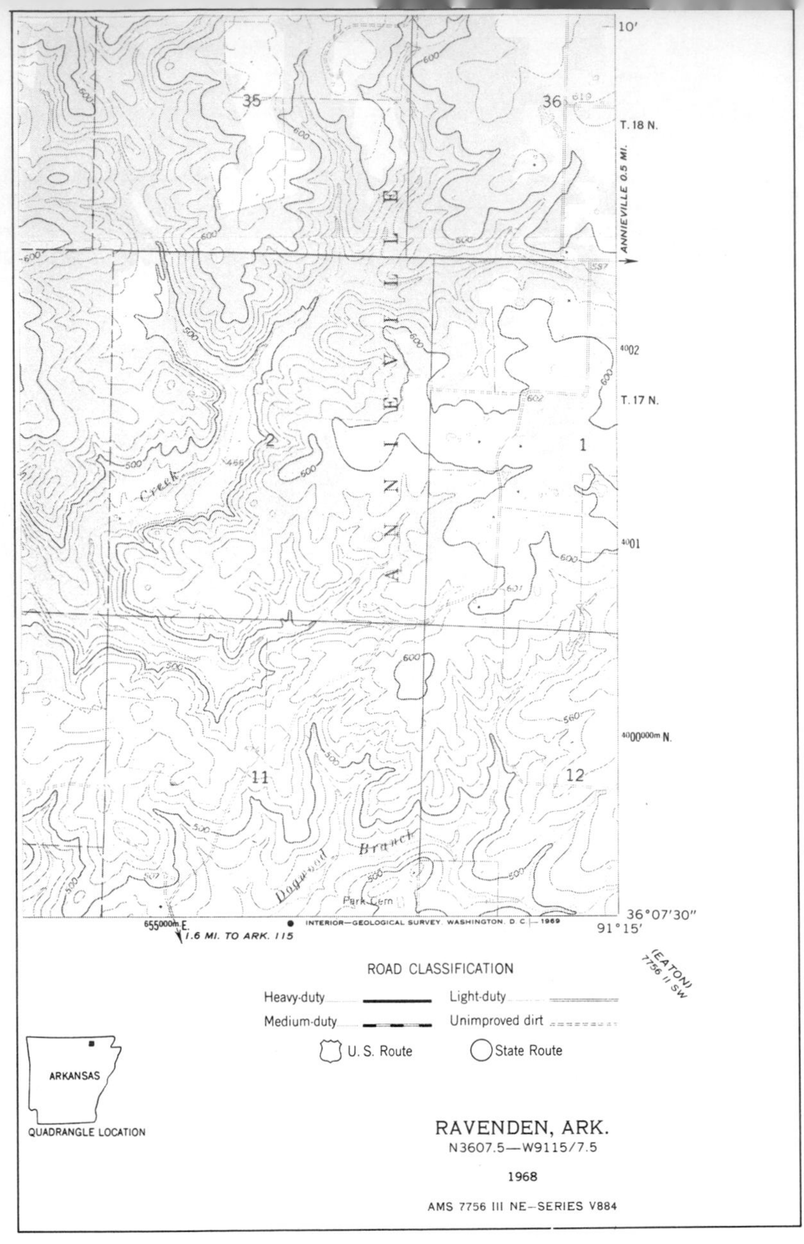

FIGURE 2–1. A sample section of a topographic survey map.

years. Various additions to these maps (new roads, permanent buildings, churches, graveyards, etc.) have appeared in recent years, all these additions appearing in purple on the maps. These landmarks appearing in purple have not been field checked, so their accuracy is somewhat less than that of the remainder of the map. The topographic maps have a scale of 1:24,000, so they are quite precise for such determinations, perhaps more precise than transit observations done with amateur instruments.

Because the maps are already scaled in increments of latitude and longitude, no conversion is necessary. You merely measure the position in relation to the standard starting latitude and longitude for each particular map. These standards appear at each corner of each map, and thus all measurements are made in relation to one of the four corners. In addition, marks appear for every 0° 02′ 30″ of latitude and longitude from each standard. For example, on the maps near 35° north latitude, 6.25 inches correspond to each mark in longitude, whereas 7.56 inches are given for each mark in latitude. The latitude increments are constant no matter what the continental location, but the longitude increments (inches or millimeters between marks) vary considerably depending on the latitude of the observer.

If you have any doubts as to the accuracy of the measurements, the wisest thing is to have a local surveying company measure the position accurately. Many such companies are equipped with large map tables capable of measuring to an accuracy of about 0.1 arc sec, which is much more accurate than most of us can determine with simple rulers. Because streets are shown, *but not labeled,* you must first, before measuring, determine without a doubt which street or road corresponds closest to the observatory. Much care should be exercised when determining the final position to which the measurement is to be made. It is quite easy to mistake a position by as much as 200 yards, much more than can be allowed for these determinations. Checking the maps for various landmarks, such as railroad tracks, churches, crossroads, cemeteries, large buildings, and hills, aids in zeroing in on the correct position. Small streams and ponds are also shown on the maps, which aids greatly if the location is in a great field or other isolated location.

Another reason for using the topographic maps to determine location is that the contours of all hills and valleys are provided in grids for every 10 feet of altitude. The grids enable you to determine to within 5 to 10 feet the correct elevation (above sea level) of your observing site. To participate in the observations of lunar occultations, you must know the correct elevation as well as the latitude and longitude. The maps have been compiled through aerial photography, and they have subsequently been field-checked. Only mathematical miscalculations, therefore, will result in poor readings. I always make a point to measure three times, and if all three measurements are in fair agreement, I determine a mean value. If the three measurements differ more than 0.04 arc sec, I assume that I am a bit rusty in using the method or that perhaps the measuring rule is not divided accurately, which often is the case.

Maps for your area can be obtained for $2 each from:

United States Geological Survey Commission
Denver, Colorado 80225

FIGURE 2-2. Elevation grids on topographic maps.

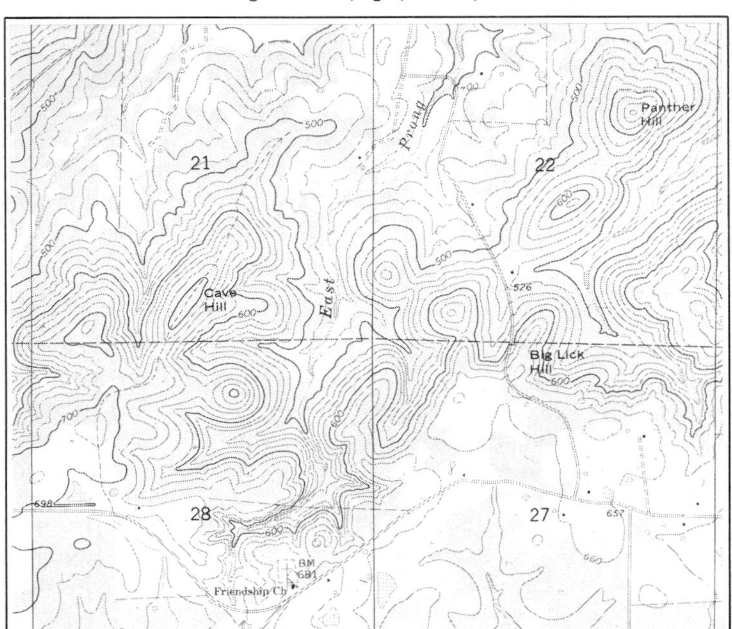

However, your state geological commission usually will have the correct map, and it will aid in the proper selection and possibly in the measurement. A folder describing topographic maps and the symbols used on these maps is also available through these agencies on request.

POLAR ALIGNMENT FOR TELESCOPES

For an equatorially mounted telescope to perform the way it was designed to operate, proper polar alignment must be achieved. This alignment is accomplished when the polar axis of the mounting parallels the earth's axis of rotation. The accuracy of this adjustment determines the amount of declination drift that will be noticed in the telescope over an extended period of time.

Before we discuss methods of alignment, we should say a few words about the degree of accuracy that must be achieved in that alignment. A vast difference exists between the alignment requirements of the visual observer using low magnification and those of the astrophotographer making a long-exposure photograph. Also, many degrees of perfection exist in alignment between these two extremes. For example, observing Jupiter at 300x to make detailed drawings or micrometer measurements can be quite a hassle if the observer's hand must be on the declination control, rather than on the filar micrometer or on the pencil.

The permanent telescope should be aligned as accurately as possible. However the time needed to set up and accurately align a portable scope is taken away from what could be observing time. So you must consider how much time is needed to set up when you plan the night's observing schedule.

ALIGNMENT OF PORTABLE TELESCOPES

The portable telescope usually has one of two types of mountings. Most Newtonian reflectors and almost all refractors are commonly mounted on the *German equatorial* mounting, whereas short-focus Newtonians and Cassegrains are usually mounted on the *yoke* or *fork* mounting. The latter type is the quickest and easiest to set up, so our discussion starts there.

Aligning the Fork Mounting

As with any telescope, the fork-mounted instrument can be set up for casual observing simply by pointing the polar axis toward Polaris. Simply put the telescope at a declination of 90° (assuming that the setting circle is properly set), and adjust the mounting (with the "wedge" or tripod) until the star is centered in the field of the finderscope. This is sufficient for setting circle use, and accuracies of about 1° can be expected.

Because the fork mounting enables the optical path and the polar axis to be coincident, two unique methods of alignment can be used with this type of mounting. In both, the finderscope is the primary tool, and no fancy gadgets are needed. However, a few check points should be made before the actual alignment procedure. First, the finderscope must be accurately aligned to the optical path of the primary instrument (centered, field to field). A quick check with the telescope on a bright star should be made. Next, make sure the cross hairs in the finderscope are aligned with the two axes of the telescope, with one hairline parallel to the mounting's declination axis.

Also check to make sure that the declination setting circle is reading accurately. The reading of the circle can easily be checked on a fork mounting as follows:

1. Point the telescope away from the base and set it to a declination of 90°.
2. Place a low-power eyepiece in the telescope and focus on a star field.
3. Leaving the declination clamp locked, rotate the telescope around the right ascension axis while observing the star field. If the declination is true to 90°, the star field will appear to rotate in concentric circles around a point in the center of the eyepiece. If

FIGURE 2-3a. Fork-mounted telescope.

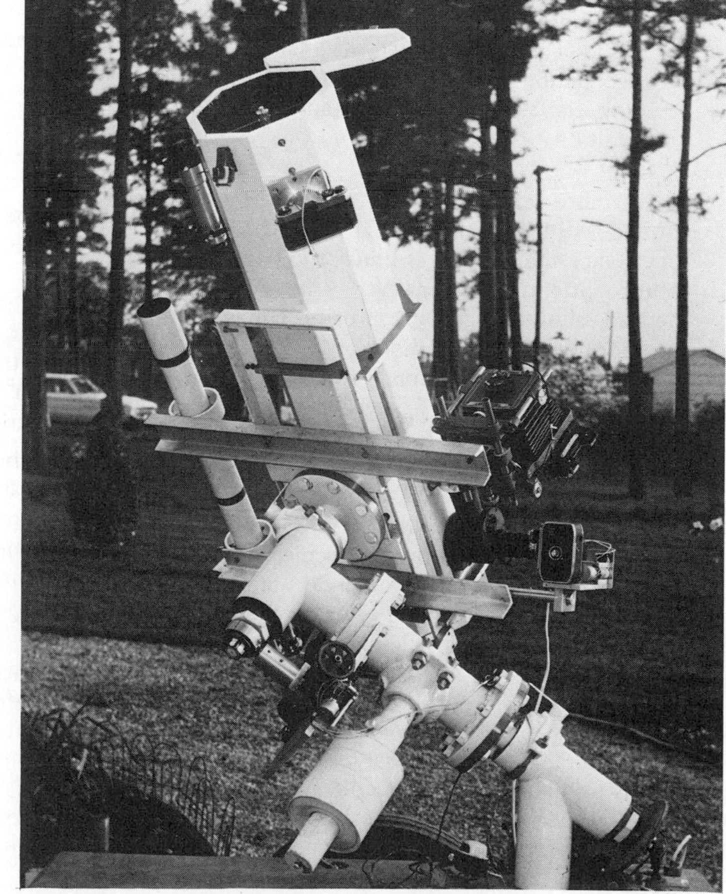

FIGURE 2-3b. German equatorially mounted telescope. (Courtesy, Midsouth Astronomical Research Society, Inc., photograph by Dr. J. P. Prideaux)

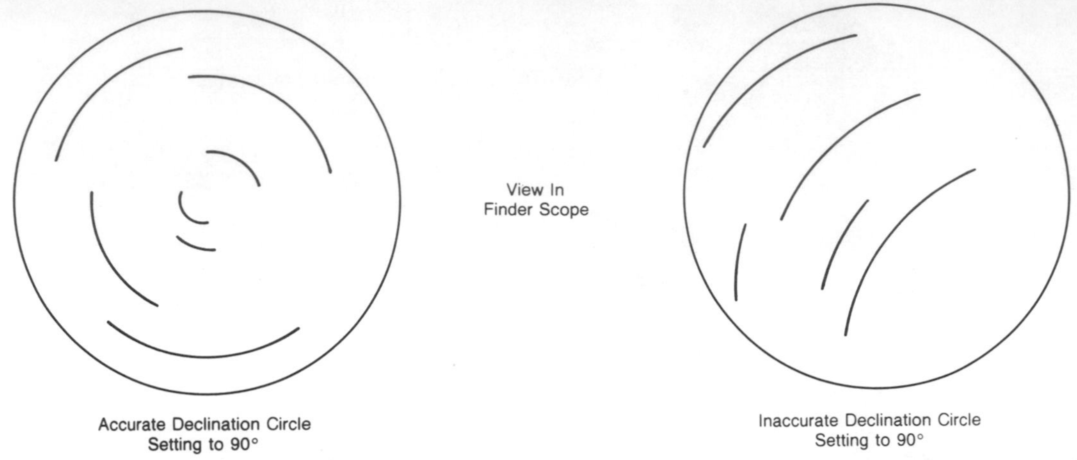

FIGURE 2-4. Appearances of star movement in accurately aligned and improperly aligned declination set at 90°.

the star images are moving in arcs, rather than circles, around the center of the eyepiece field, the declination circle is not set accurately.

4. Move the telescope in declination slightly and repeat the procedure until the star field circles about the center. At that point, the declination is exactly on 90°, and the setting circle should be adjusted to read that position.

A more refined method, which can be performed within five minutes, allows greater accuracy. Once the concept is understood and a few setups attempted, however, the foregoing method allows alignment precise enough for astrophotography. The more refined concept is based on using the telescope's finderscope to offset the proper distance from Polaris to the north celestial pole. This pole is offset almost 1°–52′ (min) arc—from Polaris toward the last star on the Big Dipper's handle. A standard 6 X 30 finderscope usually has a field of view of about 3.5°, or 210′ arc. This is a very convenient scale for this offset, and even if larger finders are used, it is good practice to have such a finderscope mounted just for polar alignment. From the center of the crosshair to the edge of the field of view, then, is 105′ arc. *Half* of this distance (a fourth of the way from one edge to the other edge) is 52.5′ arc, precisely the offset from Polaris to the north celestial pole (NCP).

Now that the offset scale is known, the proper *direction* of the offset is needed. This is where the alignment of the cross hairs to the axis of the telescope is important. With the declination of the telescope set to 90°, rotate the telescope about the right ascension axis until one cross hair, parallel to the declination axis, is lined up on an imaginary line drawn between Polaris and the end star in the Dipper's handle. The right ascension axis is then locked to maintain this alignment. The observer now moves the tripod, or adjusts the altitude and azimuth of the equatorial fork mounting, so that Polaris is located on that cross hair, half way from the center of the finderscope to the edge of the field of view. The telescope is moved *toward* the Dipper star for the alignment. This offset is demonstrated in Figure 2-5, and the procedure is as follows:

1. Turn the telescope to the declination reading of 90° (assuming that the setting circle is properly set (Figure 2-5a).
2. For the fork mounting, make sure that the mechanical axis of the telescope is aligned (Figure 2-5b) with the line extending from the end star of Ursa Major through Polaris and into Cassiopeia.
3. In the standard 6 X 30 finderscope (field = 3.5°), celestial north is only one-fourth the distance of the field offset from Polaris. Therefore, one must merely move the mounting of the telescope toward the end star in the handle of the Big Dipper until Polaris is halfway from center to edge in the finderscope (Figure 2-5d). Remember that the finderscope will show a reversed (left-to-

right) field from what is seen with the eye. The finderscope must be aligned with true north, south, east, and west coordinates with the telescope.

Always remember that the finderscope inverts (i.e., reverses left to right) the image. Consequently, make sure that the telescope is physically moving *away* from Polaris and *toward* the last star of the Dipper handle. If the tail star of the Dipper is not seen (lower culmination), the easternmost star of Cassiopeia (Epsilon) can also be used for the alignment, being opposite the Dipper star from Polaris. Using the same line, extending from Epsilon Cassiopeiae through Polaris, the offset would be on the opposite side of Polaris from the distance spanning those two stars.

This alignment procedure can be used with finderscopes other than 6 X 30, although the convenient field of the 6 X 30 finderscope allows quicker alignment. The 5 X 24 finder common on many commercial telescopes usually has a field of about 4.0°, and 8 X 50 wide-field

FIGURE 2–5. Using a finder scope to align to celestial north.

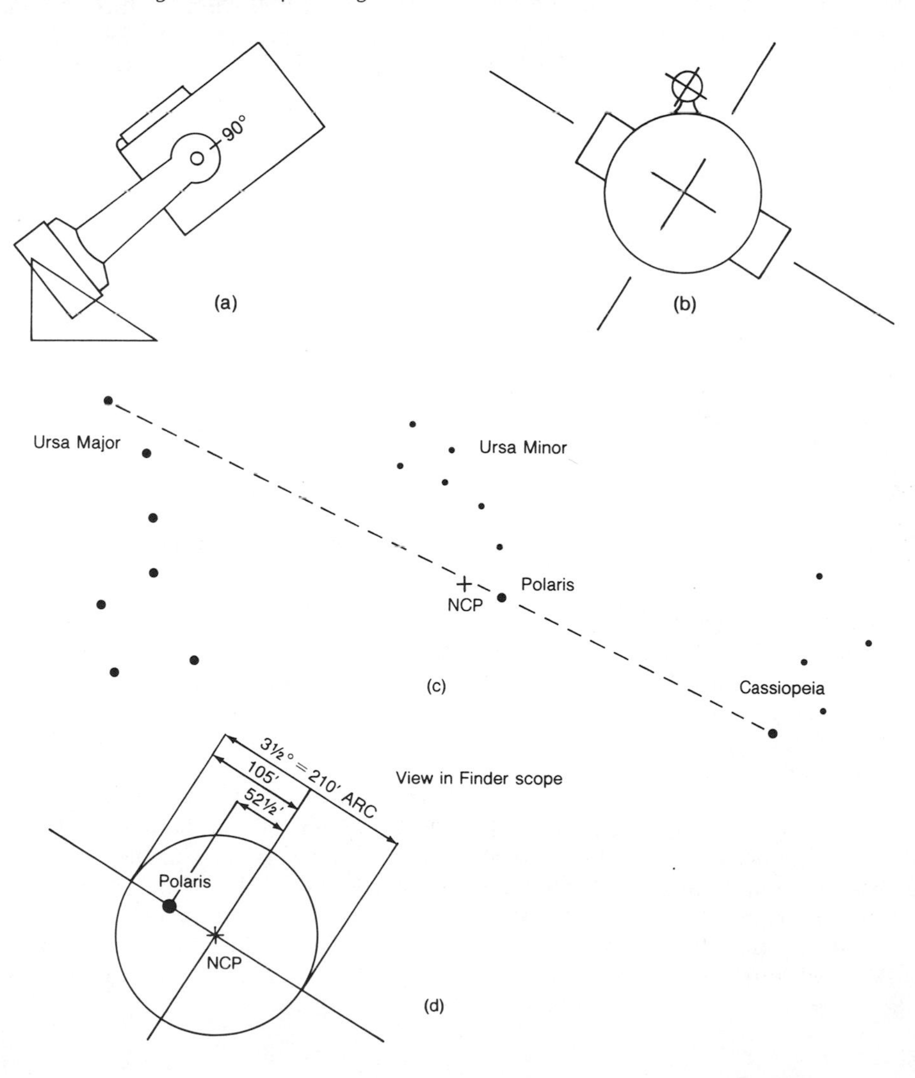

finders can show about 5°. The field of view of any optical system or finderscope can be found by timing the transit of a star from one edge to the other across the field of view. The star's apparent motion is 15° per hour, or 15′ arc per minute (1° every 4′). A handy reference table is in Appendix IV, which gives fields of view for required transit times.

Aligning the German Equatorial Mounting

The German equatorial mounting is a little more difficult to align to true north, primarily because the polar axis of the mounting and the optical axis of the telescope are not coincident. Theoretically, they *should* be coincident at infinity (the distance of the stars), yet in practice they are not.

A method similar to that described for the alignment of the fork mounting can be used for the German equatorial mounting. To find the proper position in the sky of the finderscope, the portable Newtonian or refractor telescope must first be aligned accurately the first time to true celestial north, using one of the methods discussed in the section on permanent installations (see page 31). Once alignment is achieved, the finderscope field can be noted along with the positions of the stars seen in it.

Once the telescope is properly aligned on the north celestial pole, find the theoretical line that intersects the end star of the Big Dipper, Polaris, and Epsilon Cassiopeiae. Next, orient the telescope so that the declination shaft appears to be perpendicular to this line when the telescope is pointing north and when the declination setting circle reads 90°. This procedure allows the telescope always to be positioned the same for subsequent alignments for any given time of night of any year.

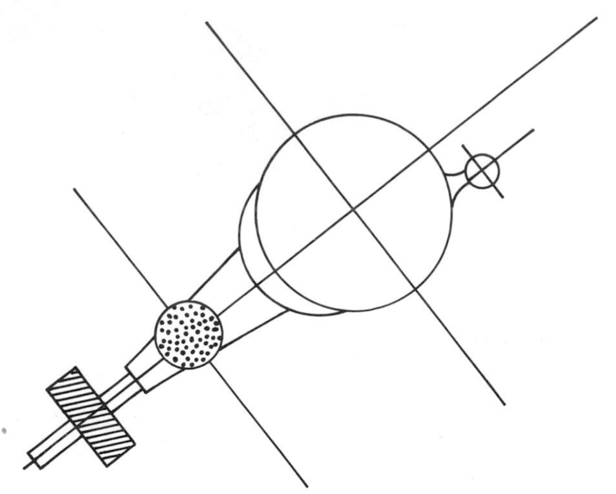

FIGURE 2-6. Axis position of German equatorial mounting for celestial north alignment. The procedure for using this mounting is the same as that given on page 29 for Figure 2-5.

Once precise alignment has been achieved, using one of the permanent installation procedures, the observer looks through the finderscope and notes the position of Polaris relative to other stars in the field of view. The location of Polaris should be sketched in relation to the cross hairs of the finderscope and other stars and also in relation to the overall field of view. Keep this drawing with you when setting your telescope up for a night's observing.

Once learned, any of the following methods of field alignment can be done quickly and precisely. Method 1: First, set the telescope up pointing north, with the declination circle reading 90°. Find the line between Polaris and the end star of the Dipper handle. Rotate the telescope 90° around the right ascension axis until the declination shaft is perpendicular to this imaginary line in the sky, as it was in the initial field determination. Then lock both axes tightly so that there can be no further accidental movement of the telescope. Look into the finderscope and adjust the tripod for altitude and azimuth, moving the tripod until the field of view properly matches that drawn in the sketch when the instrument was precisely aligned. Some shimming of the tripod legs or elevation may be necessary, so it helps to carry along some wooden plank cutouts for that purpose. With a little practice in using this method, you can achieve alignment efficiently in five minutes.

Method 2: German equatorial mountings can also be aligned fairly quickly by using known star coordinates. To do this, make an initial rough alignment on Polaris. Next, choose a bright star nearly overhead and note the coordinates of the star. Move the telescope in declination to read the declination of the star, and lock the declination axis. If the star is not centered in the field of a low-power view in the telescope (not the finder), adjust the elevation and azimuth of the mounting until the star appears in the field of view. Notice that the right ascension and declination axes of the telescope are not moved to acquire the star; only the altitude and azimuth of the equatorial mounting are moved.

Method 3: A more precise method is similarly performed. The difference is that after you set and lock the declination for the coordinates of the star, you then also set the right ascension setting circle for the proper star coordinates. Quickly rotate the telescope about the right ascension axis to the opposite side (east or west) of the pier, or tripod. At this point, reset the setting circles to the proper coordinates of the star for that position and lock both axes in place. Adjust the mounting at this point until the star is in the center of the field of view. Again, reset the right ascension circle after this adjustment to read the star's proper coordinates, and rotate the telescope to the opposite side of the pier, or tripod. Repeat the procedure, this time using a higher magnification. Several repetitions progressively increase the exactness of polar alignment.

ALIGNMENT OF PERMANENTLY MOUNTED INSTRUMENTS

Observatory, or permanently mounted, telescopes should be as accurately aligned as possible, even if the procedure requires a week or more. Accurate alignment makes possible the most accurate use of the telescope setting circles and can virtually eliminate any declination drift during long-exposure photography or precise research measurements.

The method described here can be used for *any* equatorial mounting—German, fork, or whatever. It requires a cross hair eyepiece (preferably illuminated) and a very high magnification.

After making an initial quick alignment on the North Celestial Pole, point the telescope to a fairly bright star that is both near the meridian passing overhead and very near the celestial equator. Some stars that fall along the celestial equator are the following:

Star	*Right Ascension*	*Declination*	
Alpha Aquarii	22h 03m	-00° 34′	Sadalmelek
Alpha Aquilae	19h 48m	+00° 44′	(Altair)
Beta Ophiuchii	17h 41m	+04° 35′	(Cheleb)
Alpha Virginis	13h 23m	-10° 54′	(Spica)
Alpha Leonis	10h 06m	+12° 13′	(Regulus)
Alpha Hydrae	09h 25m	-08° 26′	(Alphard)
Alpha Canis Minoris	07h 37m	+05° 21′	(Procyon)
Alpha Orionis	05h 53m	+07° 24′	(Betelgeuse)
Belt Stars of Orion			
Delta	05h 29m	-00° 20′	(Mintaka)
Epsilon	05h 34m	-01° 14′	(Alnilam)
Zeta	05h 38m	-01° 58′	(Alnitak)
Beta Orionis	05h 12m	-08° 15′	(Rigel)

Under high magnification, some drift of these stars will be noted in declination if the telescope is not properly aligned to celestial north. All drift in right ascension should be ignored. If the star drifts *south,* then the polar axis is pointing *too far east.* If the star drifts *north,* the polar axis is pointing *too far west.* Do not judge drift by apparent eyepiece movement; instead, make sure that the direction of drift is *direction in the sky.* Then adjust the polar axis in azimuth, and monitor the star again, adjusting as necessary until all drift is eliminated. The adjustment can be so refined that absolutely no drift will be noticable in a 20- to 30-minute interval.

After adjustment for drift has been made, center in the high-power field of view a bright star close to the east horizon, at an elevation of 15° to 20° and also near the celestial equator. Using the crosshair eyepiece again, monitor for declination drift as before, ignoring any drift in right ascension. This time, if the star drifts *south,* the polar axis is pointing *too low,* and if it drifts *north,* the polar axis is pointing *too high.* Then adjust the polar axis in elevation until no drifting is seen. Repeat the procedure as often as necessary.

This method accurately aligns the telescope mounting, regardless of type, to the north celestial pole, usually in only several hours. The higher the magnification used for the study, the faster any apparent motion in declination will be noticed, thereby reducing the time required for complete adjustment. The crosshairs should be set to run north-south and east-west by monitoring a star as the telescope is shifted in those directions. If the mounting is properly aligned, the star will appear to move along one line described by the crosshairs. Remember that if the drift increases when an adjustment is made, rather than decreases as it is supposed to do, the adjustment was too great or in the wrong direction.

COLLIMATING THE OPTICS OF A NEWTONIAN REFLECTOR

A well-collimated Newtonian reflector can equal or surpass the performance of telescopes of other types of equal aperture, yet few persons are familiar with the exact procedure of collimating optics.

Four steps are involved in collimating the optics of a Newtonian, each of which must be done in the following order:

1. Centering the secondary mirror in the middle of the telescope tube (Figure 2-7).
2. Centering the secondary mirror directly beneath the focuser and hence, the eyepiece (Figures 2-8 and 2-9).
3. Aligning the secondary mirror so that it is facing squarely toward the back of the telescope tube where the primary is located (Figures 2-9 and 2-10).
4. Making adjustments to the mirror cell of the primary (main mirror) so that the primary is aligned squarely to the secondary mirror.

Step Four is a bit bothersome and can be done with greater efficiency by two persons, as will be described. Once the entire process is learned, it requires at most about half an hour. Once properly aligned, a Newtonian will remain in adjustment for a long time unless it is bumped or dropped. Once Steps One, Two, and Three have been done initially, future collimating can be achieved using only one step (Step 4), which requires about 15 minutes when done by one person.

Step One

The support that holds the Newtonian secondary mirror must be centered in the telescope tube (see Figure 2-7). The distances marked by a, b, c, and d in the illustration must all be equal. This is achieved simply by tightening or loosening the spider support screws (or nuts) to move the support in the desired direction. A small ruler should

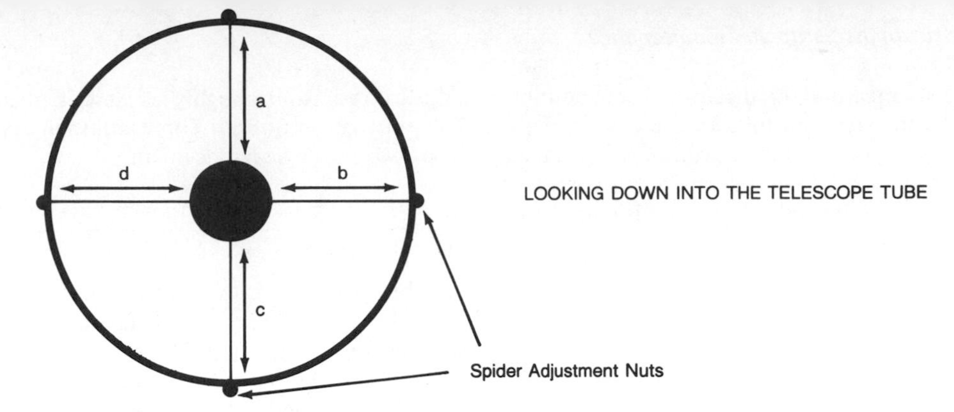

FIGURE 2-7. Step 1 in collimating. The distances marked by a, b, c, and d must all be equal; this centers the secondary holder inside the telescope tube. Spider adjustment nuts are used to make this adjustment. The illustration shows what is seen when one looks down into the telescope tube.

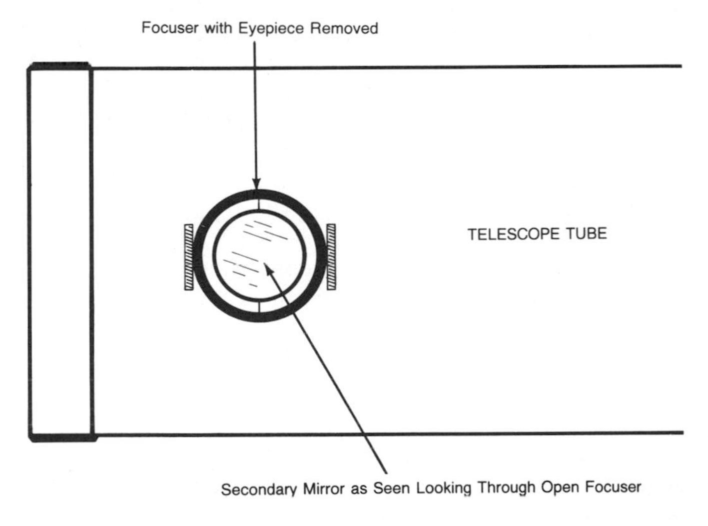

FIGURE 2-8. Step 2 consists of centering the secondary mirror beneath the focuser.

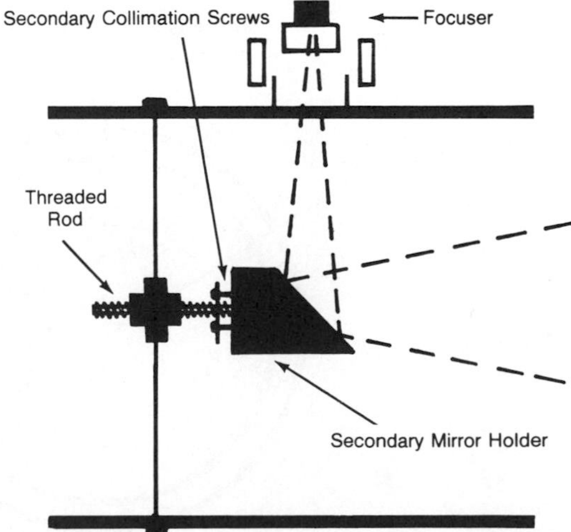

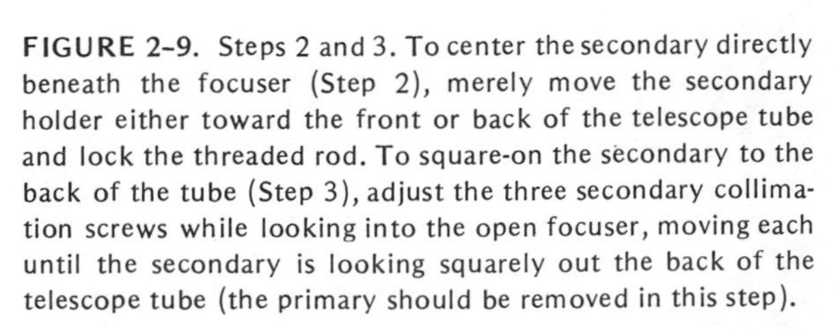

FIGURE 2-9. Steps 2 and 3. To center the secondary directly beneath the focuser (Step 2), merely move the secondary holder either toward the front or back of the telescope tube and lock the threaded rod. To square-on the secondary to the back of the tube (Step 3), adjust the three secondary collimation screws while looking into the open focuser, moving each until the secondary is looking squarely out the back of the telescope tube (the primary should be removed in this step).

be used to measure the distance from the inner walls of the telescope tube to the *center* of the secondary support. Once all distances [as measured down each of the four spider support veins (a, b, c, and d)] are equal, Step One is completed.

Step Two

After Step One is complete, the secondary mirror must be centered directly beneath the eyepiece focuser, as shown in Figures 2-8 and 2-9. To check for centering, merely back away from the focuser (Figure 2-8) until the inside wall of the focuser appears right on the edge of the secondary mirror. If the two are *concentric,* as shown in Figure 2-8, the secondary is properly placed. However, if the secondary mirror appears too far to the front or the back of the telescope tube, a bit of adjustment is necessary. Most commercial secondary supports come equipped with a long threaded rod (see Figure 2-9), which allows the support to be moved up or down the tube as necessary. Lock nuts on each side of the spider allow the rod to be locked in place once the proper position is achieved.

During Step Two, the secondary mirror edge should appear *circular* rather than elliptical to facilitate better squaring on. If it appears slightly elliptical, adjust it roughly as described in Step Two, but go on to Step Three and then return to Step Two for critical placement.

Step Three

It is best to remove the *primary* mirror and cell from the bottom end of the telescope tube for best results during Step Three (Figures 2-9 and 2-10). Once the main mirror has been removed, you are concerned with two aspects: (1) the outline of the telescope tube at the very bottom edge, and (2) the outline of the secondary with respect to the bottom edge. In the secondary mirror you will see the open end of the telescope tube as well as the bottom edge (Figure 2-10). Step Three requires that the image of the bottom edge of the tube be centered concentrically within the edge of the secondary mirror.

Because Steps One and Two have already been done, all that is required to complete the adjustment of Step Three is to turn the small adjusting screws behind the secondary holder (see Figure 2-9) by very small amounts. Turn one screw a slight amount while looking into the focuser. If the adjustment appears to move the secondary the wrong way, readjust using the other two screws until it appears to move in the

FIGURE 2-10. Step 3. To center the secondary where it is looking squarely down the telescope tube, the primary mirror is removed. The secondary collimation screws are each turned (as shown in Figure 2-9) until the outer edge of the bottom of the telescope tube appears concentric within the edge of the secondary mirror. This can be greatly facilitated by the use of the collimation eyepiece shown in Figure 2-14. If a collimation eyepiece is not available, you must locate your eye squarely at the center of the open focuser.

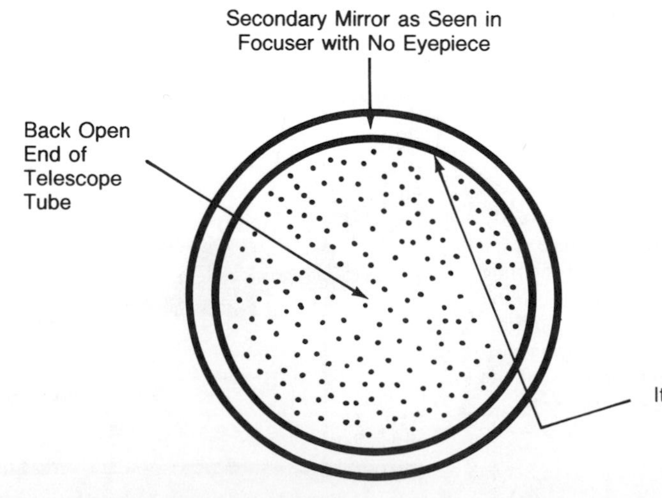

proper direction. Repeat, turning all three screws, until (1) the secondary mirror edge appears circular, not elliptical, and (2) the image of the bottom edge of the telescope tube is concentric within the secondary.

Step Four

Step Four can be very simple if a dot is placed in the *exact* center of the primary mirror (as shown in Figure 2-11), using pink fingernail polish and a small brush. The dot should be about 1/8 inch in size; it will not affect the performance of your scope in any way.

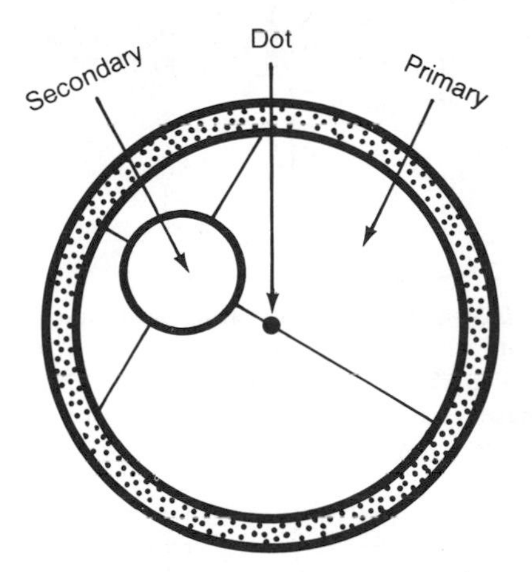

FIGURE 2-11. With the main mirror inserted, it is necessary to move the mirror cell adjustment screws until the dot is located exactly in the center of the reflection of the secondary.

To place the dot exactly at the mirror's center, use a template of posterboard. Use a compass to trace the exact size of the telescope mirror onto the posterboard and carefully cut (using a sharp knife or razor blade) out the outline. Check with the mirror to make sure it is exactly the size of the mirror. The tiny hole left by the compass point (pin) at the center of the template designates the *exact center*. Enlarge this hole with a nail or ice pick. Lay the template very carefully onto the front surface of the mirror and place the dot of polish through the center hole. Let the polish dry for about ten minutes.

After replacing the mirror in the tube, proceed to turn the three adjustment screws on the back of the primary mirror cell until the image of the secondary mirror seen in the primary is centered as closely as possible, as shown in Figure 2-12. At this point, you should see the dot ex-

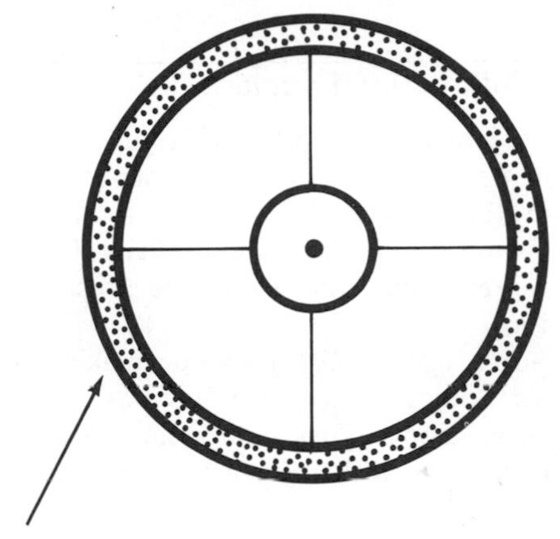

FIGURE 2-12. When you look into the open focuser, this is the appearance of a well-collimated optical system.

actly in the center of the image of the secondary when you look into the focuser. If a collimating eyepiece is not available, the centering can be more critically checked by backing away from the focuser, as shown in Figure 2-13.

FIGURE 2-13. To check for perfect collimation (if you are not using a collimation eyepiece), simply back away until the concentric rings merge.

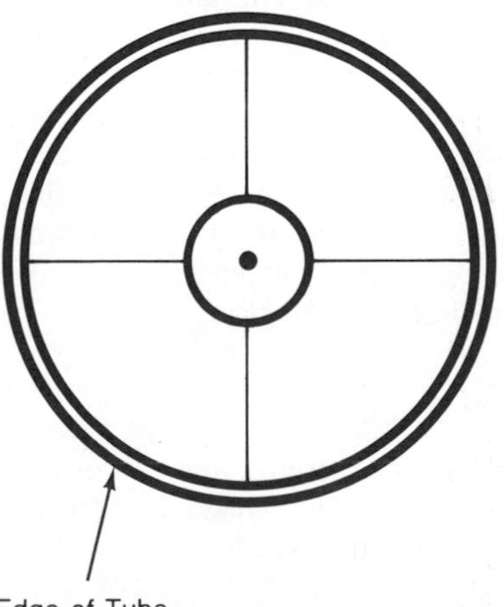

To speed up the process, it is best to use two persons, one looking into the focuser and the other making the adjustments at the bottom of the tube. The telescope tube should be pointing straight up during this process.

The Collimating Eyepiece

The collimating eyepiece (Figure 2-14) is a simple device, yet it is one that can critically aid in the adjustment of a Newtonian. Insert the eyepiece and use it throughout Steps Three and Four. Very fine adjustments are possible using this tool.

Some commercial telescope makers supply focuser "plugs" that can be used as collimating eyepieces. These plugs have a small hole in the center through which the observer collimates the telescope. The purpose of the small hole is to prevent the person from moving his head to one side, thereby changing his perspective during collimating. If a commercially made plug is not available, one can be acquired in two ways, as follows:

1. Make, or have made, a metal plug that fits as snugly as an eyepiece into the focuser. It is best to have a metalworking shop make one for you because the center hole *must* be in the exact center of the plug for accuracy. The hole should be about 1/16″ in diameter, drilled in a metal cap not thicker than 1/4″. A barrel (probably of 1 1/4″) is then attached at the base of this cap (Figure 2-14), which allows insertion into the focuser.
2. Find an old, or inexpensive, 4 mm eyepiece. Remove the lenses and use this "empty" eyepiece for collimation. The exit pupil of most 4 mm eyepieces is so small that it is sufficient to serve as a focuser plug.

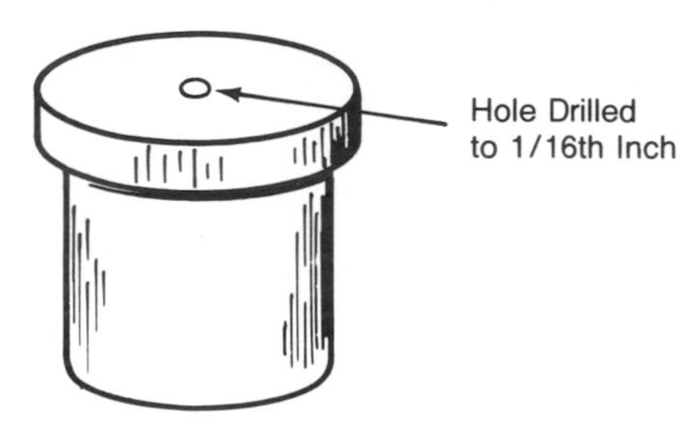

FIGURE 2-14. A collimating eyepiece made from a plug of metal with a 1/16-inch hole drilled squarely in the middle. If you cannot make such a device, a 4-mm eyepiece with lenses removed will do.

CLEANING TELESCOPE OPTICS AND EYEPIECES

The performance of any telescope depends a great deal on the transparency and/or the reflectivity of its optics. The amount of light that either passes through glass (lens or corrector) or reflects from a surface (mirror) can be seriously reduced if the optics are not as free as possible from dust, lint, spots, and film. Precautions can be taken to prevent excessive cleaning. Never touch any optical surface with the hands, and always keep surfaces capped as tightly as possible to ensure best maintenance.

Brushing Optics

By keeping your optics free of loose particles (dust, lint, etc.), you increase the performance of a telescope in two ways:

1. Dust particles block or divert light that normally would pass (or be reflected) as intended through the optics. Lenses and mirrors are corrected to extreme accuracy to bend light in a specified path. If dust blocks part of the light path, then the near-perfect correction of the optics is not realized.
2. In most climates, the night air is frequently filled with moisture. Because moisture collects readily around dust particles, dew and frost form quickly on optics covered with dust, but very slowly on optics free from dust particles.

To keep your lenses or mirrors free of dust, a camel hair brush is necessary. One can be obtained either from an art supply store (about 1/2 inch is a good size) or at a photo supplier. Those obtained from photographic supply stores usually come equipped with a soft rubber bulb which, when squeezed, blows a steady stream of air across the optics, thus helping to loosen particles from the surface.

Brushing optics should be done only when dust can be seen on the surface. (Examining optics at night with a bright flashlight is a good way to show dust accumulation.) Brushing should be done with delicate care. *Do not* bear down on the brush to remove stubborn particles, and *do not* brush in a circular motion. It is best to begin on one side of the optics and brush lightly to the other side in a straight path, much as one would mow grass with a lawnmower, allowing a little overlap on each successive path. Let the brush drag on the surface as you move in parallel to the optical surface. *Use no pressure when brushing.*

Optics that are examined and brushed when necessary will retain their precise optical surfaces and will remain clean for long periods of time. Optics must be brushed prior to cleaning.

Cleaning Optical Surfaces

Always brush optical surfaces as described in the preceding section *prior to cleaning* to prevent hairline scratches caused by rubbing dust particles across the polished glass.

Refractors and Catadioptic Telescopes

All modern refractors and several catadioptic telescopes have optics coated with magnesium fluoride to allow greater transmission of light. Extreme care should be taken when cleaning coated optics because this coating is very thin and will sleek or even chip away when cared for improperly.

Use only photographic lens cleaning solution, as supplied by Kodak, or the formula below. In addition, use *no cloth,* no matter how soft it appears to be. Kleenex brand white tissue should be used because it contains no silicone abrasives. Do not use a substitute. Follow the procedure for cleaning lenses and corrector plates without deviation.

A recommended solution for cleaning coated optics is the following, mixed together:

3 quarts distilled water, as supplied in most drug stores

1 quart pure-grade isopropyl alcohol

2 drops concentrated Ivory Liquid dishwashing detergent or equivalent

For lenses less than 2 ½" in diameter, use the following procedure:

1. Brush surface as described in preceding paragraphs.
2. Remove three sheets of lens tissue.
3. Squeeze one drop of cleaning solution onto center of lens (with lens in a flat position).
4. Using tissue beneath two fingers, gently distribute fluid uniformly around the optical surface. *Do not rub.*
5. Very quickly use second dry tissue under two fingers to wipe the lens gently and uniformly to remove excess fluid.
6. When lens appears fairly dry, use third tissue to polish the surface gently to remove streaks left after using second dry tissue.
7. Wipe edges of lens near cell to remove any excess fluid or spots.

For lenses and correctors three inches and larger, use this procedure:

1. Preferably remove lens from telescope (except Celestron, Questar, and similar scopes) and place flat on a table in its cell.
2. Squeeze two or three drops of cleaning solution on surface and follow steps 4 through 7 for lenses of less than 2½" in diameter.
3. In addition, use a fourth tissue to give a final polish to the surface.

Cleaning Mirrors

You will need the following supplies for cleaning a telescope mirror: (1) Ivory Liquid dishwashing detergent, (2) two towels, (3) high-quality cotton balls, (4) a large sink with running water, and (5) about 1 quart of distilled water. Use the following procedure:

1. Remove mirror from mirror cell and place on counter.
2. Brush surface as described previously.
3. Remove any rings or jewelry from hands and wrist.
4. Place folded towel in bottom of sink, and place mirror on towel.
5. Direct cold water from faucet onto mirror for two to three minutes.
6. Insert drain stopper in sink and add about one teaspoon Ivory Liquid.
7. Fill sink about half full with warm water. Let mirror soak four minutes.
8. Using a cotton ball, gently swab surface, starting from one edge and moving toward the other in straight paths until the entire surface is swabbed. YOU MUST KEEP THE MIRROR SURFACE *UNDER* THE WATER AT ALL TIMES.
9. Repeat step (7), but use a fresh cotton ball and use strokes perpendicular to previous paths.
10. Before removing drain stopper, turn on cold water and let it pour onto mirror surface. Remove stopper from drains for one minute.
11. Give a final rinse with distilled water while holding mirror on edge in sink.
12. Remove mirror from sink and prop it up at a sharp angle to allow water to run off. Mirror should be placed on folded towel on edge to prevent slipping. Let dry for about one hour.

Cleaning Eyepieces

Never take your eyepieces apart. This is very important. As long as they remain in the holders, there is no reason for dirt or film to accumulate between the tiny lenses. Many fine eyepieces have been ruined by persons trying to take out the elements. The only cleaning of eyepieces that should be done is of the *eye lens,* that lens to which your eye is exposed when you are observing. Several eyepieces have large *field lenses,* located inside the base of the eyepiece toward the bottom. Occasionally these lenses collect dust and lint and should be brushed gently with the camel hair brush.

Because most eyepieces are now coated with magnesium fluoride, it is necessary to clean them with Kodak (or similar) Lens Cleaning Fluid and lens tissue. The procedure to use is as follows:

1. Place one drop cleaning fluid on *tissue* (not eyepiece) and fold tissue so that it can be accessible to all areas of the tiny lens.
2. Rub gently, using a circular motion, making sure the extreme edge of the lens is cleaned.
3. Use a second, dry piece of tissue and rub gently with similar circular motions until the eyepiece appears dry when held up to a bright light.
4. If eyepiece still appears to have smudges, repeat step (3).

Eyepieces should always be stored in a moisture- and dust-proof box of some type. Dirty eyepieces are the cause of many nights of worthless observing. When you have your equipment outdoors, keep eyepieces in their box and keep the box closed tight.

3

OBSERVATIONS OF METEORS

Much practical astronomy can be learned and much valuable information concerning our solar system can be determined through the observation of meteors and meteor showers. Many principles of data reduction of meteor observations require the application of physics (meteor velocity), trigonometry (determinations of orbit and height), and chemistry (temperatures and composition). Many excellent texts are listed in this chapter's bibliography concerning such determinations. The observation of meteor showers, however, yields yet another face of astronomy to the amateur: the opportunity for an observer to spend an enjoyable evening under relaxed conditions viewing one of the most spectacular phenomena of the skies, while collecting valuable data of importance to the professional astronomer.

PREPARING FOR METEOR OBSERVATION

Meteor observations should be made under conditions as dark as possible, away from city lights, and preferably when the moon's light is absent from the sky. As for equipment, the observer needs only a comfortable position (such as that provided by a lawn chair, blanket, or sleeping bag), a set of star charts that can be written on, a flashlight covered with a red filter, and some method of keeping accurate time (preferably a shortwave receiver to pick up WWV time signals or a watch set to an accuracy of 1 sec by WWV). Because the light of most meteors is faint, it is essential that the observations be made when the moon is near new phase, or when it is absent altogether from the sky. Perhaps the most valuable equipment any meteor observer can have is the friends who share in the excitement of the meteor hunt and provide the good times and companionship necessary to make it through the late hours of night. In addition, the buddy system works quite well in meteor observing.

The sightings of meteors increases toward the morning hours because it is after midnight that an observer's location is aimed in the direction of the earth's path about the sun (see Figure 3-1). The earth is moving in space through its solar orbit with a velocity of 18.5 miles per second; during the evening hours only meteors with a velocity *greater* than that velocity can overtake the earth in its path and be seen by the observer. Those meteors that make head-on collisions with the earth's atmosphere are missed by the observer on the evening side of the earth. However, during the morning hours (i.e., after midnight) the observer is facing *into* the stream

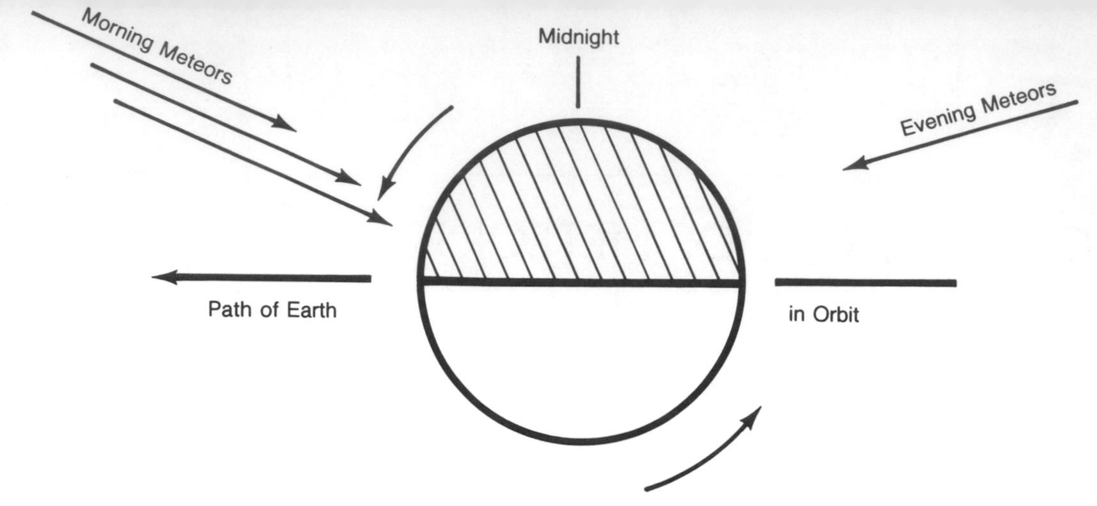

FIGURE 3-1. Earth in space during morning and evening.

of meteors, and the orientation results in the sighting of more meteors. And those that are seen seem to be brighter because of the opposing directions of travel of the earth and the meteoroid.

All meteors are apparently planetary in origin in that they are remnants of comets that are members of our planetary system. The streams of meteors orbit about in huge clouds much like comets, and many meteor showers are found orbiting within the orbits of known and recurring comets. The historic idea that some meteor groups were traveling from interstellar, rather than interplanetary, space has now been disproved through improvements in orbital calculations.

Most meteors are thought to be the debris of comets, primordial visitors from the "stuff" that caused the cosmic beginning of our solar system. Every year the earth passes through certain meteoroid material, thus causing what are known as *annual showers*. Table 3-1 represents most of the prominent annual meteor showers. In addition to these annual showers, many minor groups of meteors can provide spectacular displays one time and one time only. Likewise, there are indications of some very small meteor groups that either grow or diminish in number, and new showers (all very minor) are discovered quite often, usually by amateur observers.

SPORADIC METEORS

On any given dark, moonless night a patient observer might see as many as 12 or 15 meteors that cannot be traced to any particular known meteor shower predicted for that date. These renegades are the *sporadic meteors.* You should log them on separate charts. If you can determine that several are coming from the same point of origin; it is likely that you have found a new shower.

SOURCES OF METEOR ILLUMINATION

Meteors belong to a very select group of objects, the brightness of which is *intrinsic* (i.e., it is a physical property of the meteor itself). Objects such as planets, which shine by reflected light and not by light they themselves produce, are said to be *extrinsically* illuminated.

TABLE 3–1. Important meteor showers (reccurring) and associated comets.

Shower	*Radiant*	*Average Date of Maximum*	*Speed (kps)*	*Maximum Height*	*Comet*	*Number Per Hour*	*Duration (days)*
Quadrantid[a]	15h 28m +50°	Jan 3	41.5	—	—	40+	0.6
Virginid	12h 24m 0°	Mar 26	35	—	—	5	10.0
α Virginid	12h 24m 0°	Apr 9	35	—	—	?	20.0
Lyrid[a]	18h 4m 34°	Apr 21	47	—	1861–I	12	2.3
Aquarid[a]	22h 30m - 2°	May 4	67	—	Halley	20	18.0
Ophiucid	17h 50m −28°	Jun 15	28	105 km	—	?	25.0
Draconid[a]	15h 12m +58°	Jun 28	14	—	Pons–Winnecka	50	?
Sagittariid	20h 16m −35°	Jul 6		—	—	?	25.0
Capricornid	20h 20m −10°	Jul 22	25	—	—	?	25.0
Aquarid[a]	22h 30m 0°	Jul 29	42	—	—	20	20.0
Perseid[a]	2h 04m +58°	Aug 11	60	110 km	1861–II	50	5.0
Cygnid	19h 20m 55°	Aug 18	24	—	—	5	15.0
Giacobinid	17h 28m +54°	Oct 9	21	—	Giacobini–Zinner	20,000[c]	?
Orionid[a]	6h 20m +15°	Oct 20	67	—	Halley	25	8.0
S. Taurid-Arietid	3h 32m +14°	Nov 5	—	—	Encke	15	30.0
Bielid	1h 40m +44°	Nov 14	20	—	Biela–1836	10,000[d]	0.2
Leonid[b]	10h 8m +22°	Nov 16	71	155 km	1886	10,000[e]	4.0
Geminid[a]	7h 28m +32°	Dec 13	35	—	—	50	6.0
Ursid[a]	14h 28m +78°	Dec 22	33	—	Tuttle	15	2.2

[a] Indicates yearly showers of high numbers, very dependable each year.
[b] The Leonid shower reaches a maximum every 33 years, the last one being in 1966, when 10,000 meteors per hour were seen. The next maximum will be in 1999. Expect only 25 per hour during other years.
[c] In 1933.
[d] In 1885.
[e] In 1967.

Meteors' light is caused by meteoroids' intense friction with the atmosphere. The meteoroid is plunging into the air at anything up to 80 kps, causing great compression of air in front of it. Also, air is flowing around it. The collision of the meteoroid with the air causes the air in front to become extremely compressed and therefore to press back against the meteoroid, making it lose speed. Hence, the meteoroid's kinetic energy is transferred to the air. Also, by action and reaction, the air flowing around the meteoroid acquires some of the kinetic energy.

But so much kinetic energy is made available so rapidly in such a narrow, elongated volume of air that the air cannot radiate the energy away as fast as it is supplied. So the air becomes extremely hot. Once it is hot enough, the meteoroid's surface begins to melt and boil away. At this temperature, the volatilizing material is visibly glowing. Further, the heat is so great—especially in the bow shock wave—that the atoms start to become ionized, that is, lose their electrons. Now when the electrons recombine with the ionized atoms they must radiate away the energy that freed them from the atoms. Much of the energy that is being realized is visible light.

Almost all meteors recorded by visual observers are tiny particles rarely exceeding the size of a grain of sand. Such small particles are almost always totally vaporized in the upper atmosphere of earth at an altitude of 40 km (25 miles) and do not reach the ground. Very large meteoroids of 20 cm (8 inches) or larger can possibly survive the fiery trip through the atmosphere and strike the ground as meteorites. However, such large particles are quite rare and generally would have to enter in daylight hours or early evening hours to remain intact, because the friction from velocity would be substantially less during those times than in morning hours. For a meteor to be seen during the evening or daytime, its velocity would be such that it would have to overtake the earth's speed of 30 kps (18.5 mps) around the sun. The combined vector velocities of the earth and the meteor that must catch up would be near zero because most meteors rarely travel faster than 32 kps (20 mps) in space. Such a meteor has an effective entrance velocity into the earth's atmosphere of only 11.2 kps (7 mps).

OBSERVATION OF METEORS BY THE AMATEUR

Plotting the Apparent Path Of the Meteor

The first observation of any meteor can be recorded by marking the apparent path of the meteor as seen from earth on an appropriate map of the stars. A person familiar with the sky can become quite proficient at such plotting once the star patterns are learned to the limit of the naked eye. So you should first familiarize yourself with the night sky as it will appear during the night of a predicted meteor shower.

Using a pencil, trace the path of the meteor from the point at which it first appeared and stop the tracing at the point where the meteor seemed to disappear. Do this simply by interpolating the path relative to stars in the naked-eye background that appear on the star chart and through which the meteor appeared to travel. A straightedge or ruler of some sort aids greatly in making easily reduced tracings. However, make the tracing as quickly as possible if you are observing alone so that you can turn your attention back toward the sky for the next meteor. Record all other data for that particular meteor above that trace line so that there will be no mistake as to which meteor the data refer.

At this point, the value of having an observing companion becomes obvious. One observer watches the sky while the other records all pertinent information as voiced by the sky observer. The added advantage of this system is that the person observing the sky maintains his or her night vision and the area of concentration is not interrupted. Each hour, each member trades off. The one previously watching the sky takes the pencil and begins recording information while the other gets a chance to watch the sky.

Take note of the time (in Universal Time—UT) to the nearest *second* at which you first see

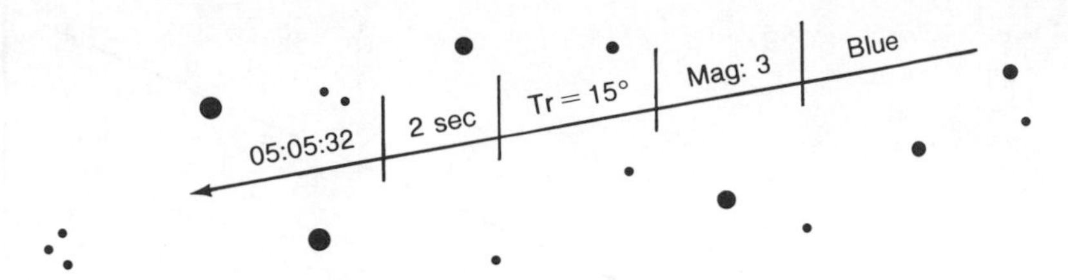

FIGURE 3-2. An example of a trace, with pertinent data.

the meteor. For this, use either an accurate watch set to Universal Time or a shortwave radio tuned to WWV radio transmission at a frequency of 5, 10, or 15 MHz. The use of WWV broadcasts is perhaps more efficient than the shortwave radio since the observer need not continue to look at a clock or watch during observation. Thus, each observation can be more rapidly recorded during a busy meteor shower. A meteor seen at 32 sec after 5:05 UT would be written first across the trace line as 05:05:32 (see Figure 3-2).

The Duration of the Meteor

The duration of a meteor is the interval between the time the meteor is first seen to the time at which it is last seen. It is noted on the observing form as the total number of seconds. Noting the duration of the meteor enables one to extrapolate the actual speed of the object as it enters and passes through the earth's atmosphere. You can compute the speed once you have determined the altitude and angle of entry, which generally requires observations from many stations, each equipped similarly and each with a well-determined position. The duration will rarely be longer than 3 sec. Many of the fireballs have durations of 2 to 3 sec, yet the interval will seem much longer to the inexperienced observer. You can best determine the true duration interval by noting the number of ticks on the WWV broadcast while watching the meteor, or by counting in one-second beats. A little practice allows for quite accurate determinations of duration. A duration time of 2 sec would be listed on the path trace line following the time of the meteor as

dur = 2s

The Length of the Meteor Train

The train of light that follows the brighter meteors is usually of consistent length, not changing as the meteor passes through the sky. By determining the length of this train in degrees as seen from one's observing station and by knowing the approximate altitude of the meteor (see Figure 3-7), it is simple to determine the actual length of the meteor train in space, using the formula

$$L = \frac{A^\circ \times D}{57.3^\circ}$$

where

L = linear size, or actual length in space of the train,

A° = maximum angular length as determined by observer,

D = known altitude of meteor as it enters the atmosphere

For example, suppose you observe a meteor train that has a total maximum length of about 25° (about the distance from one end of the Big Dipper to the other end), and you have determined that meteors from this shower enter the atmosphere at an average altitude of approximately 150 km. The actual length of the train is

$$L = \frac{25^\circ \times 150 \text{ km}}{57.3^\circ}$$

$$L = 39 \text{ km}$$

Angular diameters can quickly be determined by remembering the angular distance of a few known stars, such as those given in the following list.

Disk of full moon = 0.5°

FIGURE 3-3. Geometric circumstances for Observer A's sighting of a meteor.

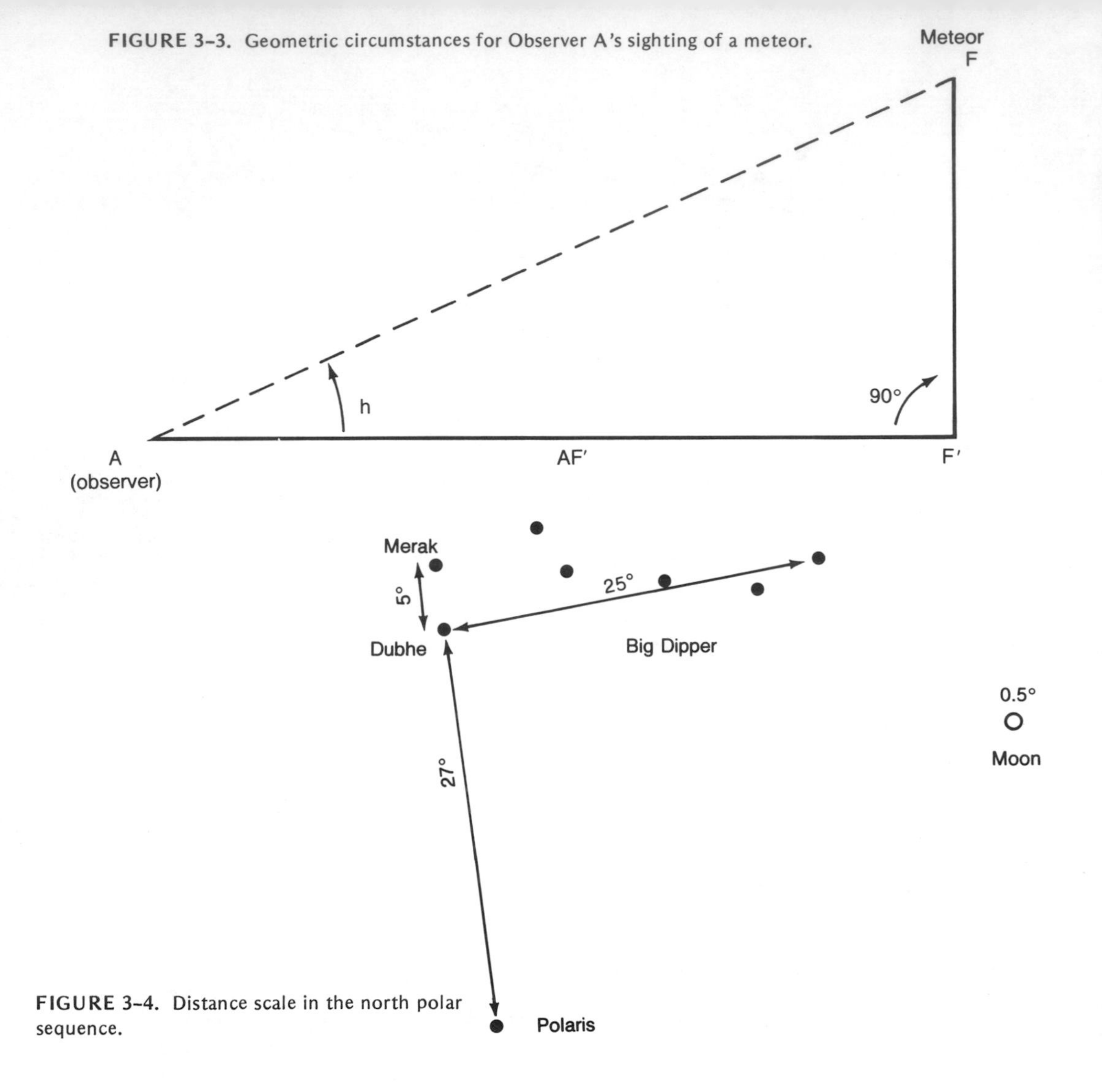

FIGURE 3-4. Distance scale in the north polar sequence.

Distance separating pointer stars (Dubhe and Merak in UMa)	= 5.5°
Distance from one end of Big Dipper to other	= 25°
Distance from Dubhe to Polaris	= 27°

Additional angular distances for stars in the north polar sequence are given in the following chapter on comets. Angular lengths can quickly be visualized by studying Figure 3-4.

Determining the Brightness Of a Meteor

You can easily determine the brightness, or magnitude, of a meteor by comparing the meteor's brightness with that of stars visible to the naked eye. However, first you must determine the magnitudes of several reference stars for each evening from a reliable source and mark them on your star chart. Estimates of fireballs and bolides might be a bit more difficult if there are no bright planets of known magnitude in the sky at the time of the meteor shower with which to compare them. In such cases it is necessary to estimate the brightness from memory, by remembering the brightness of planets, although such estimates can make for a considerable margin of error.

Record the estimate of brightness on the trace line following the other data thus far recorded. A meteor of magnitude 3 would be written

mag = 3

The Color of the Meteor

Reporting of coloration in the light you see in a meteor is a valuable start toward a rough determination of the chemical constituents of the meteoric material, as well as a good indicator of quick changes in the temperature of the fiery material. Base your color estimates on the head, or preceding end, of the meteor. To indicate the color, simply write whatever color best describes what you have seen, after you have written the estimate of the magnitude on the trace line (see Figure 3-2).

Unusual Behavior of the Meteor

As you observe, notice any unusual changes or sightings in the meteor, such as sound (hissing or sudden, loud reports following bright flashes), sputtering, color changes, or breakup. Some meteors will appear to break apart from heat of friction, which results in a spectacular sequence of events that certainly should be recorded.

At this point, the information that you should record on the path trace would be complete. An example of a meteor trace path on a star map with all pertinent information is shown in Figure 3-2. This sample indicates the direction (arrow), the path, the beginning and end of the meteor sighting, the time (05:05:32), the duration (2 sec), the length of the train (15°), the magnitude (3), and the color (blue).

Determining Hourly Counts For a Major Meteor Shower

One of the most beneficial functions of amateur meteor observing is the determination of hourly counts, which can reveal the peak time of the meteor shower, and thus the time at which the earth passed through the most dense portion of the meteoric cloud in space. Hourly counts are determined by *one observer*; observations from several observers should not be pooled for such counts. The *hourly count* is the number of meteors visible to one person concentrating on any given area of sky. It is acceptable to have a companion aid in timekeeping and recording of data, but the same observer must watch the sky throughout the night. The timekeeper and the observer must not switch roles.

Meteor counting can be done in one of two ways, one of which is most beneficial but sometimes impractical during very active showers. The first, and simpler method allows for the observer merely to note every meteor seen, either by making a tick mark for every meteor that passes or by using voice descriptions on a tape recorder. Every hour, on the hour, a new count begins. If the shower is very active, however, it is advantageous to begin a new count each 15 minutes to obtain greater accuracy in determining the time of the peak of the shower.

The second method is similar to the first, with counts beginning anew each hour. However, rather than merely counting meteors, the observer is equipped with six or seven sky charts, one for each hour of the night of observation. Like the previous method, a new chart is begun each hour on the hour unless the shower is quite active, when the observer begins a new chart every 15 minutes. Only the tracing of the meteor is recorded on the charts, along with a number representing the magnitude of the meteor. No other information need be recorded. This method has the added capability of determining the radiant, as well as the average magnitude of the meteors for any given interval during the shower. The data obtained from either method should be reported as total number of meteors seen during

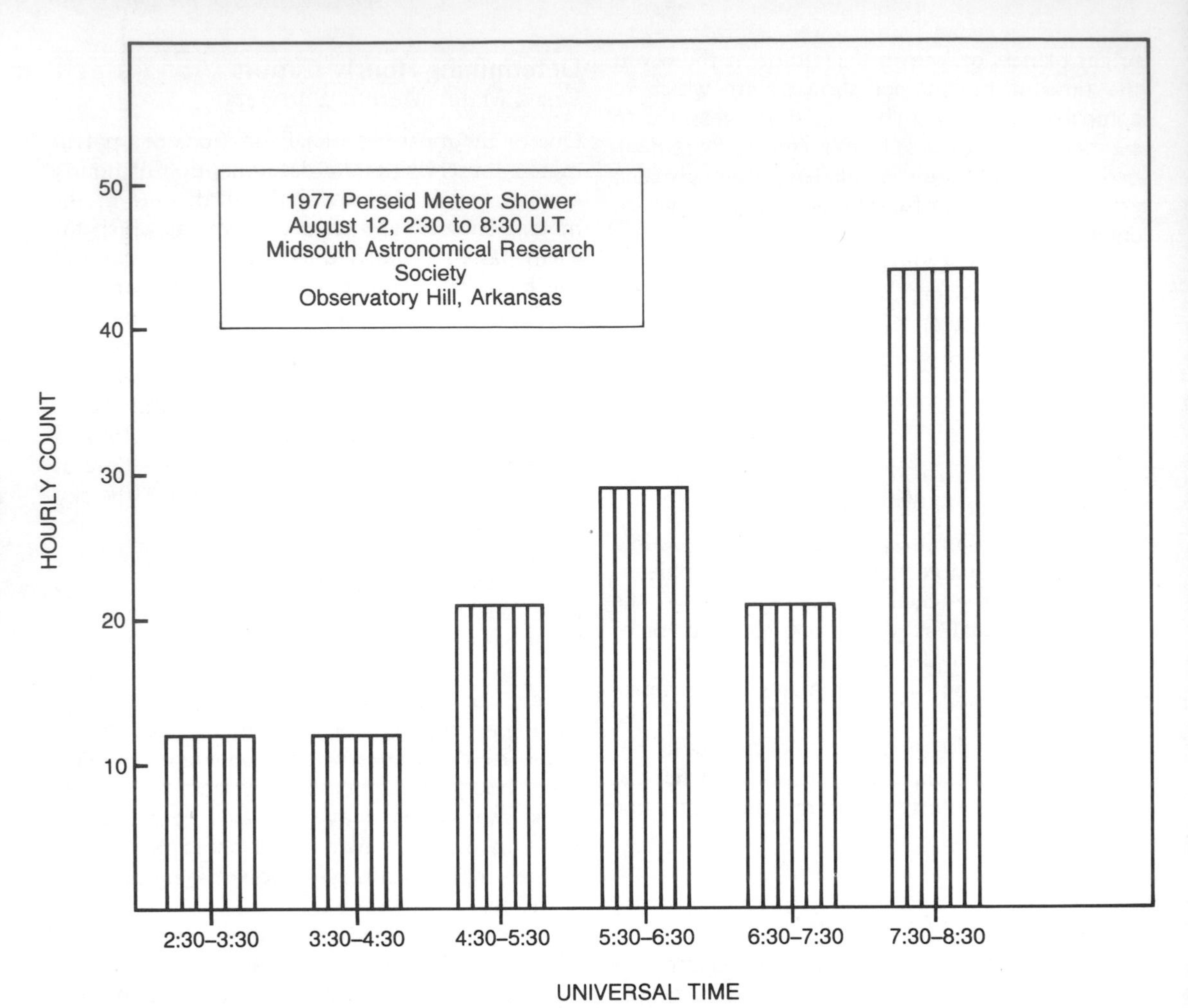

FIGURE 3-5. Graphic representations of hourly meteor counts. (Chart by the author)

each interval of time. From such methods graphs can be compiled showing the increase in meteor strength as the earth approaches the cloud and the diminishing number of meteors as the earth moves away again from the cloud. Figure 3-5 is a graph of hourly meteor counts.

THE REDUCTION OF DATA COLLECTED

Once all the data described in the foregoing paragraphs are collected, you can make several determinations. Send all reduced data to the American Meteor Society, Department of Physics and Astronomy/SUNY, Geneseo, NY 14454. A standard observing form for meteor observing appears as Appendix VI. Many determinations the amateur can make are (1) the point of the radiant and its motion is space over an interval of time, (2) the determination of altitude (somewhat difficult), and (3) the determination of the time (to within 0.25 hour) of the maximum activity of the meteor shower, perhaps the most needed information that can be derived from this study.

Determination Of the Meteor Shower Radiant

The radiant of any meteor shower can be traced from all directions, converging to the shower's point of origin. This point, the radiant, is the intersection of the earth's atmosphere and the cloud of the meteoric material in space. Figure 3-6 shows a simple and effective method for determining the radiant of any meteor shower. After all observations have been made for a given night, the paths of all meteors can be traced back to their points of origin. The point of intersection of most paths demonstrates the radiant for that night, assuming that all traces have been accurately rendered on the star chart. The intersection point is referred to by that point's right ascension and declination in the sky. Remember that many meteors are seen each night that are not part of any known shower, much less the one under observation. Several such meteors can be seen as traces in Figure 3-6. It is a good practice *not* to trace meteors not coming from the radiant point of the predicted

FIGURE 3-6. Determining the origin points of a meteor shower.

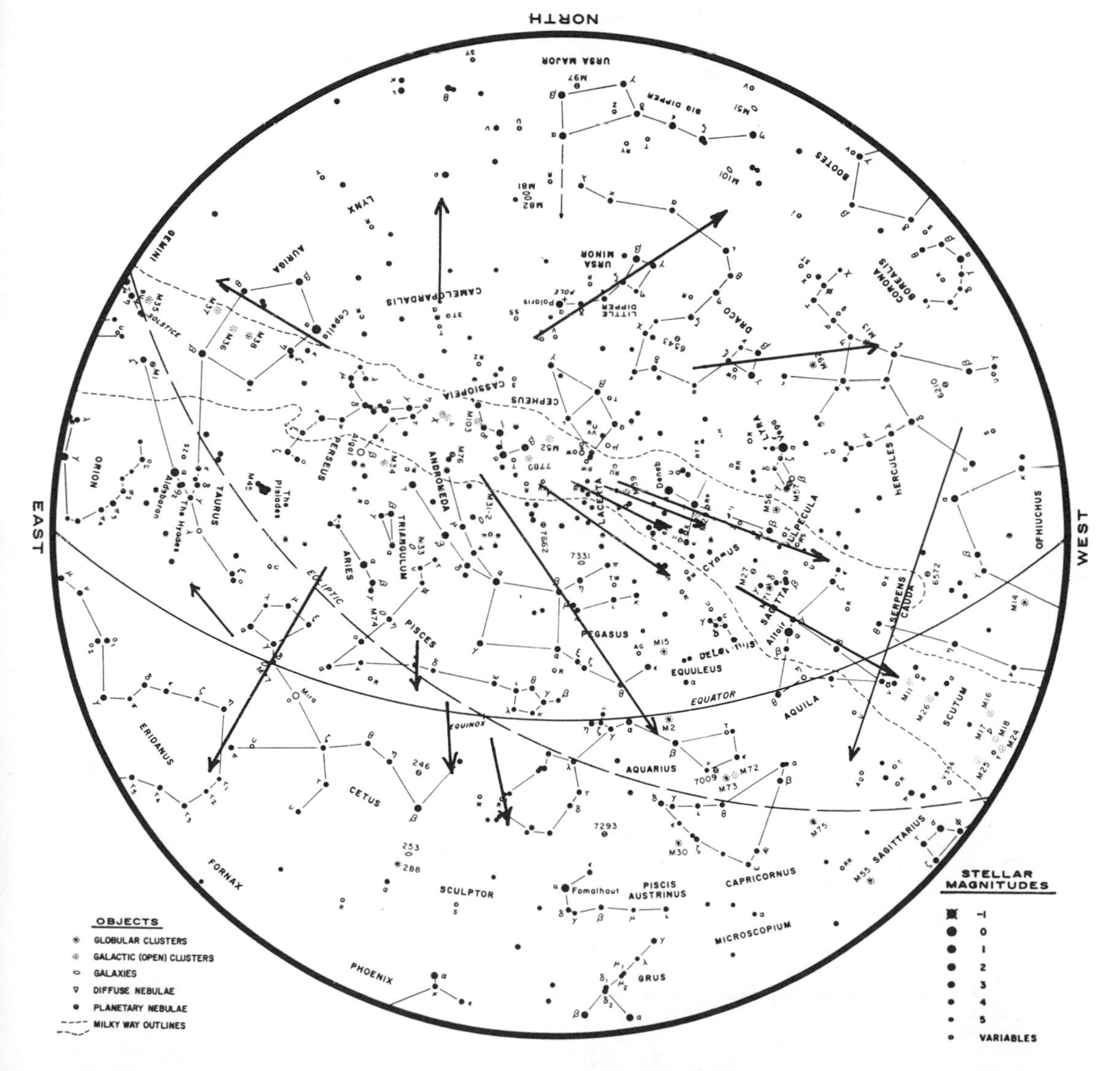

shower on the same chart used for the predicted meteor shower. A better method is to have two charts—one for the shower, and one for the sporadic meteors. The reasoning for plotting any sporadics is twofold: First, such meteors might be part of a minor shower that is peaking on the same night as a major shower, and information about these minor showers is vital. Second, sporadic meteors might lead to a radiant of a yet-determined meteor shower. Such new showers are generally short-lived, but nonetheless they are discovered often.

An interesting project is to record a shower for its entire duration (three days for the Lyrid shower, for example), plotting the meteors' radiant each night. If the meteors are seen over this period of time, and the radiant can be accurately reduced for each night, then the *motion* of the radiant in space can be determined simply by extrapolating the difference in the position from night to night. Remembering that the radiant is expressed as a point of right ascension and declination, you simply record the changing position from each night. This can provide a direction and an amount of motion of the meteor cloud in relation to the earth.

The motion and direction of the meteor shower radiant are of great importance to solar system astronomers, and coupled with the density of the cloud (simply the hourly counts of the numbers of meteors seen) for each night, enable astronomers to compute the true motion in space of this cloud of meteoric material as well as its size and density. Such information substantiates or refutes theoretical arguments as to the origin of the meteor clouds and their orbits in relation to comets.

Determination of the Altitude Of a Meteor

It is not necessary to have more than the most basic knowledge of trigonometry for two observers at two different locations to undertake a project to determine the approximate height of a meteor at the time of sighting. The two observers (*A* and *B* in Figure 3-7) must be situated 10 to

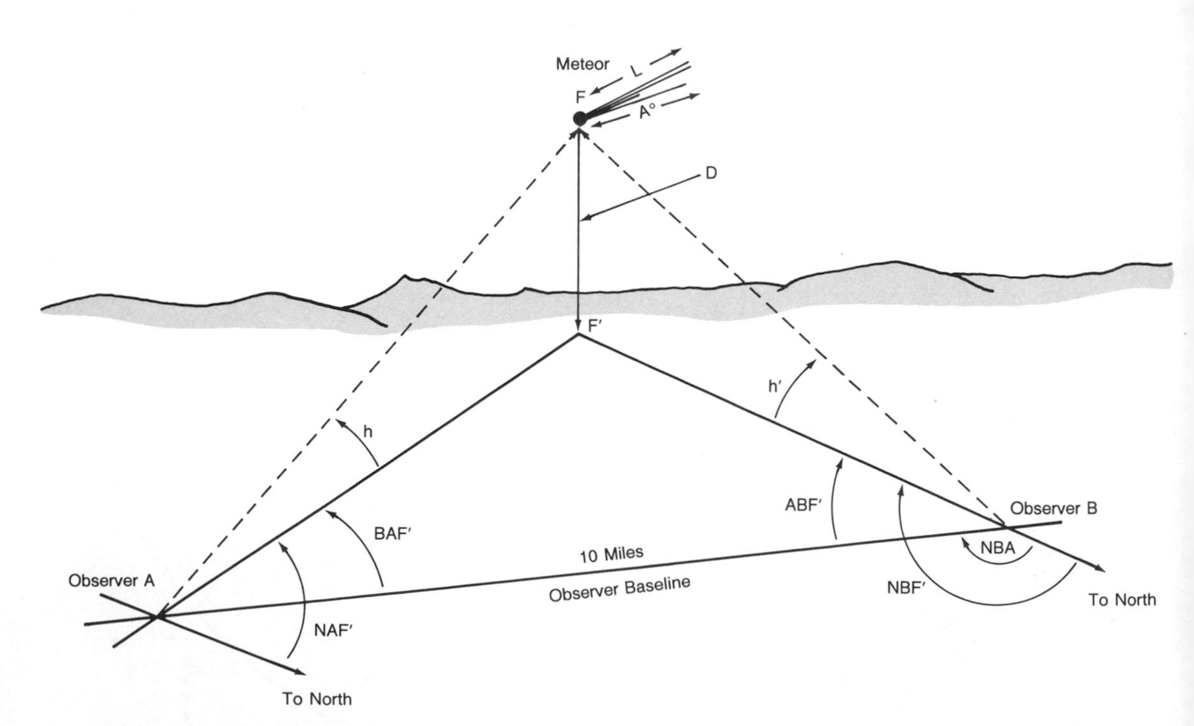

FIGURE 3-7. Triangulation of a meteor seen by two observers. Triangulation is used to determine height.

30 miles apart, with their positions well determined beforehand from a geographic survey map showing their locations. The base line between them should be well established and measured for distance, going by the map. It is helpful if the base line is in an exact north-south or east-west orientation to aid in the rapid reduction of data. If this is not practical, the directions of each observer from the other should be accurately determined, because all measurements are made of the meteor sighting relative to true north.

Each observer notes the angle from true north at which a certain meteor is sighted, as measured from north to east. North is 0°, due east is 90°, south is 180°, and west is 270°. In Figure 3-7, observer *A* notes the meteor in the direction AF; observer *B* notes the same meteor in direction BF. Because the base line (the distance between the two observers, and the direction of one observer relative to the other) has been determined beforehand and is accurately known, a simple trigonometric calculation allows for the determination of the height of the meteor. For this determination, it is necessary for each observer first to determine the direction on the ground (F′) directly beneath where each one sees the meteor. This measurement is an angle measured from *north*. Therefore, the angle for observer *A* is NAF′, and for observer *B* it is NBF′. Once these angles are determined, the observers have the following data:

BASE LINE (AB) = distance (i.e., 10 miles)

Angle A = NAF′-NAB = BAF′

Angle B = NBF′-NBA = ABF′

Since the two angles and an included side are now known, you can determine the distance from each observer to point F′ (line drawn from meteor down to the point on the horizon below which the meteor appeared). This is, however, not our final answer.

Assume that angle AF′F is a right angle and measures 90°. The angle F′AF (h) was determined as the altitude above the nearest horizon, and line AF′ was determined earlier through the above calculation. Again, using two angles and an included side of this triangle, one can determine the length of line F′F, or the height of the meteor above the ground, as noted in Figure 3-7.

RECURRING METEOR SHOWERS FOR EACH MONTH OF THE YEAR

Annual Meteor Showers for January

Not many meteor showers occur during January and February, but those nights on which they do occur in January are among the clearest of the year, providing a peek into deep space farther than ordinary vision can take us. An observer of meteor showers in winter undergoes a cold unlike that experienced by the hunter or sportsman. When stalking game or actively engaging in exercise, one can more easily bear the cold. Astronomers, however, lying quietly in one position under the cold winter skies, experience a cold like no other. Feet begin to numb, noses feel brittle, and speech becomes an uncomfortable hardship. Is it worth it? Some of us think so, or otherwise the night sky of winter would go virtually unnoticed. The sudden appearance of a spectacular fireball lighting the sky so brightly that shadows are cast onto the ground has a remarkable way of thawing the feet, warming the nose, and relaxing the frozen face.

January 4, Quadrantids This meteor shower radiates from the constellation of Boötes. It is a very short-interval shower, lasting less than one day. For observers in the United States, the radiant for this meteor shower does not rise until midnight, but the earth's turning brings it well overhead around sunup. The number of meteors seen each year varies considerably, from as many as 250 to less than a dozen. You should always expect around 60 meteors per hour.

There are some indications that this meteor shower is intensifying, and thus it is one well worth the amateur's attention. The meteors travel at a medium speed of about 41 kps and are usually very faint, with a distinctive bluish color.

To observe the Quadrantids it is best to be in a reclining position with feet toward the northeast and eyes concentrated overhead. A companion can help to distract your attention from the winter cold.

Position of radiant: Right ascension 15h 28m; declination +50°.

January 16, Delta Cancrids Rising about the time the sun sets and located directly overhead at midnight, this meteor radiant is one of the easiest to find. It is located just west of the large naked-eye star cluster, Praesepe, in Cancer. It is a minor shower, providing perhaps only four meteors per hour and moving very swiftly across the sky.

Position of radiant: Right ascension 08h 24m; declination +20°.

January 18, Coma–Berenicids Like the Delta Cancrid meteors, these meteors come from a radiant very easy to locate, and are also near a large naked-eye star cluster. The radiant, located in the constellation of Coma Berenices, rises in mid-January about 10:00 PM and is directly overhead at predawn. These are some of the fastest meteors known, entering the earth's atmosphere at about 65 kps; the disappointing part of this shower is the number: only one per hour.

Position of radiant: Right ascension 12h 30m; declination +19°.

Annual Meteor Showers for February

February is not much better for the observer of active meteors than is January. Only one weak shower occurs during February. But you should always be on the lookout for meteors that at first seem sporadic but later can be associated with other meteors seen the same night. You may see as many as 12 sporadic meteors each hour on a dark winter night, many of which must be associated with some diffuse cloud of meteoric material. From these sparse streams the new meteor showers are discovered each year.

February 26, Delta Leonids This meteor shower intersects the earth's orbit beginning as early as February 5 and leaves as late as March 19. There is a definite peak in late February, but it provides no more than five meteors at 24 kps. When you are observing, notice that the meteor radiant is located midway along the Lion's back between the stars Zosma and Algeiba and that it is overhead around 11:00 PM. Most meteors will be seen after that time if moonlight does not interfere.

Position of radiant: Right ascension 10h 36m; declination +19°.

Annual Meteor Showers for March

Although March brings somewhat warmer pre-spring temperatures and an increased urge to go out and explore the heavens, the month is unfortunately almost devoid of prominent meteor showers. But going out and observing gives practice in plotting meteors.

March 16, Corona–Australids This is a meteor shower of short duration, lasting only from March 14 to March 18. It is best observed in far southern latitudes. From the southern United States, the meteors' radiant is only about 7° above the southern horizon at about 4:45 AM on March 16. Nonetheless, the meteors sometimes travel northward from the radiant, with as many as five to seven meteors per hour being seen from this latitude.

Position of radiant: Right ascension 16h 20m; declination –48°.

March 22, Camelopardalids A small shower with a big name, the Camelopardalids have no definite peak, with only about one per hour being seen. The meteors of this shower are some of the most unusual of all; not only is the rate a bit unusual with about one per hour, but those seen travel more slowly than any other meteors known. They enter the earth's atmosphere at only about 7 mps, at a height of about 80 km (50 miles).

Position of radiant: Right ascension 07h 50m; declination +68°.

March 22, March Geminids This is a shower to which amateur skywatchers can make a valuable contribution. It was discovered in 1973 by several persons in Hungary, when as many as 43 per hour were noted, and it was confirmed in 1975. Just why the shower was not noted until 1973 remains somewhat a mystery; it is possible that the earth had never passed through the particle cloud until that time. These meteors, like the Camelopardalids, are quite slow and come from almost directly overhead at dark. By midnight, when the meteors generally are more frequent, the radiant will be low in the western sky. Observations by all persons are urged to help determine whether the March Geminids are, indeed, an annual meteor shower. If they *are* an annual shower, this could indicate the presence of yet another undiscovered comet.

Position of radiant: Right ascension 06h 22m; declination +23°.

Annual Meteor Showers for April

April brings showers of two types. This month is a favored month for meteor watching, particularly if the moon is either nearly new or has set early enough for observers to enjoy dark skies during the meteor's peak periods. Meteors come from at least five major constellations this month; among them are Leo, Draco, Virgo, Serpens, Lyra, and Boötes. You may see a meteor or two from any of the eight April showers in less than 15 minutes.

April 4, Kappa Serpentids This shower lasts from about April 1 to April 7, with no particular peak during that time. The radiant rises about 8:00 PM local time, just south of due east, and is highest at about 2:00 AM. If the moon is near new phase during this shower, several meteors per hour should be seen—the later you wait to begin observing, the more meteors you will see.

Position of radiant: Right ascension 15h 20m; declination +18°.

April 7, Delta Draconids Another shower with no definite peak, these meteors can be seen coming north of overhead from March 28 to April 17. The radiant is in the sky all night; therefore, meteors might be seen from sundown until sunrise. This shower was discovered in 1971, with several very slow meteors leaving conspicuous trails. You should set up around 10:00 PM local time and face northeast. Most of the meteors will be seen near midnight when the radiant will be directly overhead, so you should concentrate your vision near the zenith.

Position of radiant: Right ascension 18h 45m; declination +68°.

April 10, Virginids This is only one of three meteor showers coming from the constellation Virgo during April. This radiant will be just south of overhead from April 1 to April 15 about midnight, and as many as 20 meteors per hour might be seen in dark skies.

Position of radiant: Right ascension 12h 24m; declination 00°.

April 15, "The April Fireballs" From April 15 to April 30, several prominent fireballs will be seen. These April Fireballs have no known radiant, thus they have no constellation designation. The meteors are very, very bright and long lasting, and seem to originate somewhere in the southeast sky. Look for these fireballs late in the night, preferably after the moon has set. Several of these meteors have been known to reach the ground as meteorites.

Position of radiant: Between right ascension 20h and 24h, declination near ecliptic, but erratic.

April 17, Sigma Leonids This shower is actually now in Virgo, having moved out of Leo many years ago. The radiant is south of overhead about 9:30 PM local time.

Position of radiant: Right ascension 13h 00m; declination -05°.

April 22, The Lyrids This famous meteor shower is one of the oldest on record, having been logged by the Chinese astronomers in 687 B.C. However, the number of particles in this swarm has greatly diminished in modern times until the Lyrid shower now provides quite a poor show. These very bright, long-lasting meteors are remnants of a very famous comet, Comet Thatcher, visible last in 1861. The radiant, just on the border of Lyra and Hercules, rises about 7:30 PM and is just north of overhead about 4:30 AM local time. It is located near the bright star Vega. Because the radiant is so close to Vega, you should have no trouble zeroing in on this shower's point of origin. Perhaps the best viewing position for this shower is simply straight overhead, beginning about 11:00 PM local time.

Position of radiant: Right ascension 18h 04m; declination +34°.

April 25, Mu Virginids Another shower from Virgo, the Mu Virginids is overhead about 1:00 AM and should be visible most of the month. It originates from eastern Virgo. Expect to see as many as seven of these medium-speed meteors per hour, unless moonlight interferes.

Position of radiant: Right ascension 14h 44m; declination -05°.

April 23, Grigg-Skjellerups This meteor shower gets its strange-sounding name from a comet of the same name, which produced the debris resulting in the meteors we see. It has the distinction of being a "localized" meteor shower, visible in some parts of the world but not in others. In 1977 many meteors were seen in New Zealand and Australia, but none in the United States. You should really be on the lookout for these meteors, south of overhead in the early evening.

Position of radiant: Right ascension 07h 48m; declination -45°.

April 28, Alpha Boötids These meteors appear from very near the bright star Arcturus in Boötes, which is nearly overhead about 1:00 AM. It is a long shower, with no definite peak, lasting from April 14 to May 13. The meteors are very slow and leave very fine trails.

Position of radiant: Right ascension 14h 30m; declination +19°.

Annual Meteor Showers for May

May is a great month to observe meteors. The skies are beginning to clear more often, the weather is warm, and this month the moon, if in the sky during late May, will not interfere with sightings of meteors because most of the best showers occur early in the month.

May 1, Phi Boötids The Phi Boötid meteors last from April 16 to May 12 and from midnorthern latitudes the radiant is in the sky all night. This shower has moved in recent years out of Boötes into northwest Hercules. The best time to observe is about 2:00 AM when the radiant is overhead and as many as 6 per hour might be seen.

Position of radiant: Right ascension 16h 00m; declination +51°.

May 3, Alpha Scorpiids This radiant moves noticeably eastward night to night. It begins in early April on the Libra-Scorpius border, but by May 9 it will have moved into Ophiuchus. It rises in the southeast sky about 9:00 PM and will be overhead at 1:00 AM.

Position of radiant: Right ascension 15h 40m; declination -21°.

May 4, Eta Aquarids These meteors are probably the best meteor shower of spring. The meteors can be seen coming from the region of Aquarius from April 21 to May 12, but the peak can be either on May 3 or May 4, so you should watch each night. It is possible to see some faint meteors with binoculars about May 3; meteors of average brightness can best be seen on May 6 to the naked eye. Fireballs are expected from May 9 to May 11. The radiant of the Eta Aquarid meteors is located in the "water jar" in Aquarius and moves a little northeast each day. The meteors are on the horizon about 2:00 AM the morning of May 3 and are only about 50° from the east horizon by 7:30 AM. Expect at least 21 meteors per hour, most of which will be very bright yellow with glowing trails. Interestingly, this shower is debris from the famous Halley's Comet and was recorded by the Chinese as early as 401 A.D.

Position of radiant: Right ascension 22h 30m; declination -02°.

Annual Meteor Showers for June

The warm night of a June evening are filled with "shooting stars." No less than 13 meteor showers occur in June each year.

June 3, Tau Herculids This month-long shower begins in late May and continues until mid-June. The radiant is almost overhead at about 10:00 PM, making for ideal observations from northern latitudes. You should wait until after the moon has set, if it is in the sky, to see more than 15 meteors per hour, most quite faint.

Position of radiant: Right ascension 15h 12m; declination +40°.

June 4, Alpha Circinids This shower was discovered by Australian amateur astronomers in June, 1977, when 15 fast meteors were noted per hour. From northern latitudes this radiant is very low in the south about midnight, so only a few brighter members may be seen.

Position of radiant: Right ascension 14h 38m; declination -65°.

June 5, Scorpiids This is an interesting meteor shower because there are two streams of meteors rather than just one. Both radiants are nearly overhead at midnight. At about 3:00 AM, you may see more than 20 meteors per hour from both radiants. This can be a very pleasing meteor shower to the eye; not only are the numbers high, but the star field behind the radiant is one of the most spectacular of the night sky. Position yourself in a sitting position facing south. It is best to begin observations around 10:00 PM and continue until 3:00 AM, at which time the constellation and the radiant will be near setting in the southwestern sky.

Position of radiant: Right ascension 16h 40m; declination -17°.

June 7, Arietids Another month-long shower, the Arietids peak on this date when as many as 60 per hour have been picked up by radar. Less than this number can be spotted visually, but observers concentrating on the eastern sky about 3:00 AM should see in excess of 30 per hour. These meteors are slow, leaving fine trains, and many times they split into bright bolides.

Position of radiant: Right ascension 02h 56m; declination +23°.

June 7, Zeta Perseids On the same night as the Arietids, another shower, the Zeta Perseids, peaks. Although the maximum for this shower occurs after the sun rises, radar indicates that as many as 40 per hour enter our atmosphere, leaving a possible 15 per hour visible in predawn skies to observers.

Position of radiant: Right ascension 04h 08m; declination +23°.

June 8, Librid Meteors This is a minor meteor shower with perhaps five per hour normally visible. This is an important shower to astronomers because the comet from which the meteors originate is not known and is yet to be discovered.

Position of radiant: Right ascension 15h 09m; declination -28°.

June 11, Sagittariids This is a two-week-long meteor shower, beginning in early June. It was discovered in 1958 by Air Force radar. The radiant rises in the extreme southeast sky about 11:00 PM, after which as many as a dozen meteors per hour can be seen. Since the shower members are quite faint, any moonlight will prevent observation of all but the brightest members of this shower.

Position of radiant: Right ascension 20h 16m; declination -35°.

June 13, Theta Ophiuchids These meteors come from the borders of the constellations Ophiuchus, Sagittarius, and Scorpius, rising about 9:00 PM. They are just south of overhead at midnight. You may see only two per hour from this shower, but expect them to be quite bright and spectacular.

Position of radiant: Right ascension 17h 50m; declination -28°.

June 16, June Lyrid Meteors This is a companion stream to the prominent May Lyrid showers. Coming from near bright Vega, the radiant of these meteors is almost directly overhead at 1:00 AM. Occasionally you may see a bright meteor, but you will see most of the June Lyrids as faint, blue streaks through the sky. Any moonlight on this night will prevent your seeing these faint meteors; otherwise, if you observe at about 11:00 PM, you should see more than 15 per hour. This is just one more of many meteor showers that have been discovered by amateur astronomers, having been found in 1966 and seen every year since then.

June 20, Ophiuchids The radiant of this shower can be seen all night from northern latitudes, rising at sunset, highest (just south of overhead) at 11:25 PM, and setting around sunup. The number of meteors that can be seen per hour varies from 8 to 20, but the shower is capable of producing many more meteors than the predicted average.

Position of radiant: Right ascension 17h 20m; declination -20°.

June 26, Corvids This shower is of very short duration, lasting only five days. Because it was last seen in 1937 astronomers theorize that it is a product of a yet-undiscovered comet. If the meteors happen to come again, you should expect at least 10 per hour. The meteors originate near the little trapezoid of Corvus, the Crow.

Position of radiant: Right ascension 12h 48m; declination -19°.

June 29, Beta Taurids This shower is a daytime stream, and ham radio operators, or persons with long-distance shortwave radios, might be able to hear the deflections of radio waves reflected by the meteors' ionized paths in the upper atmosphere. Every year, radio observers have recorded at least 30 per hour.

Position of radiant: Right ascension 05h 44m; declination +19°.

June 30, June Draconids These meteors were known in the past as the Pons-Winnecke meteors, from the comet that left the meteor debris in its wake. In 1916 over 100 meteors per hour were seen from this spectacular meteor shower, but the number has waned somewhat in recent years. The shower is irregular; you can expect anywhere from 10 to 100 per hour—the exact number is not known. The radiant rises in the northwest sky at sunset, and is directly overhead about 9:00 PM. Although the Draconid shower is in the sky all night, you can see it best when it is directly overhead. You will see the greatest number of meteors if you position a reclining lawn chair with your feet to the north and concentrate your vision slightly north of overhead.

Position of radiant: Right ascension 15h 12m; declination +49°.

Annual Meteor Showers for July

A good way to cool down from the hot summer days of July is to leisurely watch the spectacle of the "shooting star." Although not many meteor showers occur in July, there are two major somewhat unpredictable showers that deserve careful scrutiny.

July 16, Omicron Draconids Peaking each year in mid-July, this is an almost month-long meteor shower with meteors seen as early as July 6 and as late as July 24. Coming from the head of Draco, these slow-moving meteors (24 kps) are a result of the breakup of comet Metcalf of 1919. This shower was not found until 1971, yet photographic records show that the shower has been visible since at least 1952. During recent years no members of this stream have been seen, so it could be that we have seen the last of this cloud in its nonreturning orbit.

Position of radiant: Right ascension 18h 04m; declination +59°.

July 28, Delta Aquarids This is a shower of very long duration that starts about July 15 and ends in early September. The peak of the shower lasts as long as a week. The radiant for these meteors rises in the east at about 8:00 PM local time and is near overhead at 2:00 AM. Recent studies show that the total amount of material in this great swarm of meteors is far greater than in any other known shower. It is only because the meteoric material is spread out over a large area that the total number of meteors seen seems small. It appears that this group of meteors as well as the Capricornid meteors, also of late July, originate from the same debris cloud, although they are distinctly separated. You will see most of the meteors after midnight, the number increasing rapidly until 3:00 AM. Face south and expect to see an average of about 27 meteors per hour. Because they intersect the earth's orbit sideways, they appear to be moving somewhat more slowly than most meteors, usually appearing very bright and sometimes leaving fine yellow trains in their wakes.

Position of radiant: Right ascension 22h 35m; declination −10°.

July 30, Capricornids The duration of this meteor shower is similar to the Delta Aquarids, and the radiant is very difficult to differentiate from that stream. However, the Capricornid meteors are considerably slower as they enter the earth's atmosphere than the Aquarids. It appears that these meteors are remnants of debris left by Comet Honda-Mrkos-Pajdusakova. The radiant for this shower is just south of overhead at midnight. It is in the sky all night long, so you should position yourself facing south. These bright yellow meteors often have accompanying fireballs, and anywhere from 10 to 35 meteors per hour can be expected.

Position of radiant: Right ascension 20h 20m; declination −10°.

Annual Meteor Showers for August

Sultry August, known in our part of the world for its blistering hot days and relievingly cool nights, is one of the most pleasant months of the year for the stargazer. The air is usually calm and the nights pleasantly warm. After sunset and the last light of dusk disappear, the dark night sky becomes rewardingly rich in bright stars. The bright band of light stretching from northeast to south, known as our Milky Way, appears as a diffuse summer cloud suspended from the stars. Binoculars and small telescopes show the Milky Way not as a cloud but rather as thousands and thousands of seemingly tiny stars, each a part of the vast spiraling system of our galaxy. Just before dawn's light, the mighty constellation of Orion, the Hunter emerges far in the southeast. It is one of the most majestic of all constellations and provides the summer skygazer with a reminder of the glories of the night sky of winter.

To top off the fascination of August, almost every clear night the sky is streaked with bright meteors. Perhaps the finest meteor show of modern times occurs each August, emanating from the constellation of Perseus, which is low in our northeastern sky at dark. For the casual observer, the Perseid meteors are the easiest to track back to their apparent point of origin in the sky.

In many countries, the Perseid meteors are referred to as the Saint Laurence Tears, in reference to his martyrdom on August 10, 258 A.D. A few Perseids can be spotted as early as July 25, and the darker the night sky is, the more that will be seen on each night. An interesting project for observers is to monitor the sky on each available night and simply make a count of the number of meteors that can be seen each hour. Slowly, as the month progresses, this count should climb until around mid-August, when the maximum number of Perseid meteors is expected. By the night of maximum, you may have a few hours of very dark sky, without the moon, during which the spectacular display of Perseids can be enjoyed. Even if the moon makes its debut on the night of maximum, you should continue to monitor the sky because the number of meteors always goes up dramatically after midnight when the dark side of the earth plunges head-on into the meteor swarm during the second week of August. Based on my past observations, you should be able to spot about 25 meteors on the date of maximum from about 10:00 PM until 11:00 PM and about 45 from 11:00 PM until midnight (with a dramatic increase at about 11:30). After midnight the true nature of the spectacular meteor shower can be seen, the fainter ones whipping across the sky like falling stars. Because the rate of meteors after midnight is usually well in excess of 100, you can probably expect at least 45 to 50 bright meteors per hour in spite of moonlight, since this shower contains more bright meteors than dim ones.

By 6:00 AM local time, the radiant of the Perseid shower is directly overhead from northern latitudes during mid-month, at which time you can expect the maximum number of meteors. These bright yellow Perseids frequently leave trails of smoke, and many times they explode midway, which results in several bright meteors flying away from the central body.

The Perseids are apparently debris left over from the deterioration of a comet first seen in 1862. This comet, known as Swift–Tuttle, has an average period of 120 years, meaning that it should return (if there is anything but debris left of it) in 1982. Interestingly, the comet and the Perseid meteors never pass close to any planet in our solar system except the earth. The result is that the orbit is stable, and the objects in that orbit are predictable. Early this century, E.J. Öpik estimated that the total mass of the material in the Perseid meteor stream might be around 1 billion tons, but later he revised that estimate upward to 10 billion tons, and even again up to 200 billion tons! His estimates were based on the number and the frequency of the meteors that observers had recorded. Even though it appears that his estimate is fairly accurate, however, the comet itself could not have been that massive. The modern theory is that the Perseid meteors do not come from that single comet but from several comets that sweep through the same orbit, each deteriorating a little each time it passes the sun.

We do not know how long the debris that accounts for the Perseids has been in space. The earliest recorded observations of this shower were made by the Chinese in 36 A.D. when the shower peaked, not in August as it does today, but on July 17. From 714 A.D. until the present year, the Perseids have been recorded every year. It does appear, however, that the large swarm of particles is decreasing in number because each year the total number observed drops off somewhat.

Position of radiant: Right ascension 3h 04m; declination +58°.

August 20, Kappa Cygnid Meteors In addition to the dependable Perseid meteors, two other fine meteor showers are visible in August. The first, known as the Kappa Cygnids, is in the sky all night during this month and is almost directly overhead at midnight. Discovered in 1042 by Chinese astronomers, the Kappa Cygnids have provided up to a dozen meteors per hour every year since then. Even though that number may not seem outstanding, the shower is really a fine spectacle, providing many fireballs in brilliant white-yellow flashes. Because the meteors travel quite slowly, they frequently leave interesting trains of smoke that persist for several minutes.

Position of radiant: Right ascension 19h 20m; declination +55°.

August 31, Andromedid Meteors Close in the sky to the Cygnid meteors are the third shower of this month, the Andromedid meteors. One of the longest meteor showers of the year, the Andromedids last from August 31 to November 29. They move slowly through the sky, often being fiery red in color. This is probably the most unpredictable of all meteor showers. In Japan in 1798, over 400 meteors per hour were noted, and the "stars flew like snow." In 1885 in most locations there were reports of nearly 13,000 per hour. Since the shower has such a long duration, it is almost impossible to predict when the actual peak will occur. From data I have located, it seems that the most likely times of peak activity might be either August 31, October 3, or perhaps November 14. The latter date seems to be the best possibility. The Andromedid meteors are fragments of a very famous comet, Biela's Comet, which split into two large comets after passing close to the sun and earth early in 1845. Even though there may be many millions of fragments left in orbit, it is possible that the great gravity of Jupiter has perturbed those fragments into a different orbit in recent years, robbing present-day skywatchers of this splendid shower. Look for the Andromedid meteors almost directly overhead at about 3:00 AM on August 31 and into early September. Since all three August meteor showers are located in the same region of sky, it might be difficult to differentiate among them.

Position of radiant: Right ascension 01h 40m; declination +44°.

Annual Meteor Showers for September

The advent of crisper skies and cooler temperatures tempts many skywatchers outdoors. When the days of September are unbearably hot, the nights seem remarkably cool, adding to the pleasure one attains when searching the early fall skies. Most of the annual, or recurring, meteor showers that occur in September are minor streams, with sparse concentrations of meteors. However, some are interesting and notable, and most are a bit unpredictable.

September 1, Aurigids Although this shower has been seen only once—one night in 1935—it might be worth watching for, as it could be a massive recurring group. In that year, up to 35 meteors per hour were seen coming from the constellation of Auriga in the northeast sky, all of which were very fast and quite bright. Each year the shower would peak at about 3:00 AM near September 1. Observations of this curious shower are badly needed. You should try to recover this stream.

Position of radiant: Right ascension 05h 38m; declination +42°.

September 6, Lyncids An insignificant meteor shower, the Lyncids were once a major sky show. It was recorded by the Chinese in 1037 and 1063 as "rains of stars," and it was similarly logged in Korea in 1560. However, the swarm of debris comprising this shower apparently passed close to one of the major planets in recent times, thus preventing the major portion of its meteoric material from encountering the earth.

Position of radiant: Right ascension 06h 40m; declination +58°.

September 7, Epsilon Perseids A usually dependable meteor shower of up to a dozen meteors per year, this group can be obscured from view by the light of the moon. Look for these meteors low in the northeast at about 10:00 PM.

Position of radiant: Right ascension 04h 08m; declination +37°.

September 21, Kappa Aquarids This meteor shower is almost overhead at 11:00 PM local time, and the absence of any moonlight will aid in detecting these very faint meteors. Very few if any bright meteors have been noted from this shower. Probably their slow treks through the atmosphere prevent them from becoming bright objects.

Position of radiant: Right ascension 22h 35m; declination -05°.

September 23, Alpha Aurigids Probably the best meteor shower in September, this group rises in the northeast about 8:00 PM and is nearly directly overhead about 5:00 AM. Usually the Alpha Aurigids are quite fast and leave very nice trails in their wake.

Position of radiant: Right ascension 04h 56m; declination +42°.

Annual Meteor Showers for October

October was originally the eighth month of our year, March being the first. Throughout the world there are references to this month as being the clearest and "weather free-est" month of the year. Several fine meteor showers occur each year during October, and you can take advantage of the crisp, clear skies to view faint meteors that might be missed during other months. October can be quite chilly, giving hints of the spectacular winter skies approaching.

October 7, Piscids This is a quite long meteor shower with members seen as early as September 25 and others as late as November 2. The radiant is near the Pisces–Aries border. Take caution not to confuse these slow meteors (29 kps) with the Andromedids of November. Because the Andromedid radiant moves rapidly through the sky in October, during this time it will be located in the constellation of Pisces. It is thought that the Piscids are debris left in the wake of the famous comet Encke. On nights when the moon is not present, you can expect to see perhaps 15 per hour from this shower, which is overhead about 10:00 PM.

Position of radiant: Right ascension 01h 40m; declination +14°.

October 9, Draconids Unlike the Piscid meteors in October, the Draconid shower is of short duration, lasting only from October 7 through October 10. In any year when the earth passes directly through the center of this meteoric cloud, the numbers seen can be astonishingly high. The main group of debris particles is still located directly behind the parent comet (Comet Giacobini–Zinner), so the years in which the comet passes close to the earth's orbit are also good years for sighting many Draconid meteors. In 1933, for example, 30,000 meteors per hour were seen when the earth encountered the center of the meteor swarm. In 1946 when the comet had passed the earth's orbit only 15 days prior, 1000 per hour were seen. In any given year, when the comet is even reasonably close to the earth, as many as 200 to 500 per hour can be expected.

The Draconid radiant is up any time, day or night, and the shower can occur during the daylight hours. You should go out as soon as possible after evening twilight ends and begin observing in a reclining position, feet pointed north, and eyes concentrated directly overhead. The actual hour the greatest number of meteors is to be seen is uncertain, so the longer and more diligently you observe, the better are your chances of seeing a great number of meteors.

Position of radiant: Right ascension 17h 28m; declination +54°.

October 19, Epsilon Geminids This is a week-long shower beginning on October 14 each year, with no definite peak. The radiant is located near the twin Castor, in Gemini. The radiant rises about 10:00 PM local time and is near overhead at the first break of dawn. You can expect to see only two to three meteors per hour, those entering the atmosphere very fast (70 kps).

Position of radiant: Right ascension 06h 50m; declination +27°.

October 20, Orionids One of the best meteor showers to occur each year, the Orionids begin on October 2 and last until November 7. The radiant is positioned in the prominent constellation of Orion, near his club, rising about 9:30 PM local time. Conveniently, the radiant will be just south of directly overhead when most meteors are expected—5:00 PM in the morning. These also are very fast meteors, entering the earth's atmosphere at 67 kps. Like many other meteor showers the Orionids have as their origin the famous Halley's comet. You can expect to see about 30 meteors per hour in dark skies. However, since most of the meteors are faint and are moving fast, only observers dedicatedly watching the skies will see so great a number. Orionid meteors present many varied colors; about one out of five leaves a smoky train that can persist for as long as 2 sec after the meteor vanishes.

Position of radiant: Right ascension 06h 20m; declination +15°.

Annual Meteor Showers for November

When the pleasant constellations of summer set in the western sky and the mighty square of Pegasus looms overhead, you are presented with a preview of winter during early November when the great constellations Taurus, Orion, and Gemini make their debut. Even though November usually has poor viewing weather, openings in the clouds can allow skywatchers to see one or more of the prominent meteor showers that peak during this month.

November 5, Taurids This shower actually begins about September 5, but peaks two months later. You should see about 10 meteors per hour each year, with numbers increasing after midnight. The Taurids are part of a debris cloud left behind the famous comet, Encke. The meteors are very slow, and most are bright fireballs leaving fine trains. Look for the meteors from a point near the Pleiades star cluster.

Position of radiant: Right ascension 3h 32m; declination +14° (south). Right ascension 04h 16m; declination +22° (north).

November 9, Cepheids The radiant for this shower is about overhead at 8:00 PM, and you can see meteors from November 7 through November 11. Observations of this meteor shower are badly needed because it was discovered only in 1969 when nearly 50 meteors were seen within 15 minutes. On the date of maximum, you may expect to see at least eight meteors per hour, and possibly more.

Position of radiant: Right ascension 23h 30m; declination +63°.

November 12, Pegasids The radiant for these meteors is nearly directly overhead at 8:00 PM on November 12, but meteors can be seen from the Pegasids as early as late October and through late November. This group of particles was at one time a very active shower, being derived from an old comet, Blanplain, in 1819.

Position of radiant: Right ascension 22h 54m; declination +19°.

November 14, Andromedids This long-lasting annual meteor shower actually begins on August 31 and does not cease until late November. These are spectacular meteors; many of them are red and leave reddish trains. The meteors come from the now-lost Comet Biela, which split up into two objects in 1845, causing the majority of what we now witness as Andromedid meteors. Shortly after the breakup, the meteors put on a spectacular show in 1885, with as many as 13,000 per hour having been seen. However, shortly after that the shower passed uncomfortably close to the massive planet Jupiter and was perturbed in such a way that only scant numbers of meteors are presently seen. Some of the fragments of this shower have been verified as reaching the ground as meteorites.

Position of radiant: Right ascension 01h 40m; declination +44°.

November 17, Leonids This shower has provided the most spectacular meteor display on record. However, the Leonid shower peaks to its maximum only every 33 years, the next occurring in 1999. On November 16, 1966—the date of the last primary maximum—observers were dumbfounded by the great number of Leonids, and efforts to determine accurate hourly rates were difficult. Reliable estimates gave more than 150,000 per hour visible to the naked eye on an average, and at some times as many as 500,000 per hour—140 a second—were seen. The great planet Uranus perturbed the Comet Temple-Tuttle in such a way as to provide us with this spectacle. The meteors have been seen in great numbers for many centuries. In 902 A.D. the Arabic nations signified that year as "The Year of the Stars" in honor of the Leonid spectacle. In 1799 a great astronomer and explorer, Humboldt, wrote that the meteors fell "like snowflakes" from the sky of the mountains of Venezuela.

The Leonids emanate from a point near

Gamma Leonis in the sickle of Leo. The meteors enter the earth's atmosphere faster than all other meteor showers, at about 71 kps. Because of this swift passing through the earth's air, most of these meteors are burned up before ever reaching the ground. Notice that the Leonid radiant does not rise until after midnight and at that time the earth is meeting the meteors head-on; the greatest number of meteors can be expected in the predawn hours. The years between the 33-year intervals provide only about 20 meteors per hour. However, there are apparently large "knots" of material other than the main swarm still traveling in the comet's orbit that the earth occasionally encounters. Consequently, *any* year can be a good year for the Leonids. On November 18, 1999, the great cloud will return.

Position of radiant: Right ascension 10h 08m; declination +22°.

Annual Meteor Showers for December

". . . now lies the earth
  all Danae to the stars,
And all thy heart
  lies open unto me.

"Now slides the silent meteor on,
  and leaves
A shining furrow,
  as thy thoughts in me."

(Tennyson, song from *The Princess*)

December 10, Monocerotids This is a long-lasting shower that begins about November 27 and lasts up to December 17. The radiant for these meteors is located right on the border of Monoceros and Gemini. The meteors rise low in the southeast sky at dark and are overhead and to the south at about 1:00 AM. Possibly a dozen meteors from this shower can be seen if the moon is absent from the sky.

Position of radiant: Right ascension 06h 50m; declination +10°.

December 10, Chi Orionids Interestingly, in many years this two-part meteor shower peaks on the same night as the Monocerotids. Also, both showers come from the same general direction of sky and are possibly related in origin. The Chi Orionids actually come from two points in the constellation of Orion, one close to the horns of Taurus, the Bull, and the other a little farther into Orion.

Position of radiant (average): Right ascension 5h 38m; declination +21°.

December 11, Sigma Hydrids From the head of Hydra, the water serpent, these fast meteors rise in the southeast about 11:00 PM and are overhead at sunrise. These meteors are quite swift, and many can be seen just before morning twilight. The meteors are usually a bit faint because they enter the atmosphere so quickly, but you can possibly see a dozen per hour.

Position of radiant: Right ascension 08h 32m; declination +02°.

December 14, Geminids This is a superb annual meteor shower. The radiant, normally providing at least 60 meteors per hour, rises about dark in the east and is overhead at midnight. Most of the meteors are quite bright, but only about 3 % leave a trail in the sky. The meteors are whitish. It is believed that the Geminid meteors are derived, not from a known comet, as most meteor showers are, but from *'Icarus,* a very peculiar asteroid that swings close to the earth during some of its passes.

Position of radiant: Right ascension 07h 28m; declination +32°.

On the same night as the Geminid meteors, a minor shower known as the Leo-Minorids (from Leo Minor, or Small Lion) peaks at about 5:00 AM on December 14. This shower, consisting of not more than 10 meteors per hour, was discovered by casual stargazers in 1971. The radiant rises about 8:00 PM local time. Any observations of meteors are badly needed.

Position of radiant: Right ascension 10h 24m; declination +35°.

December 16, Piscids In 1973 these faint meteors were first seen emanating from Pisces, the Fish. None of the members left trains, and only about 8 per hour were seen. Since this is a new shower, observations of it are very important and welcome from all observers. Look in Pisces, just west of overhead at dark for these faint meteors.

Position of radiant: Right ascension 01h 42m; declination +09°.

December 20, Delta Arietids This is a good shower for persons who like to observe in the early evening. The radiant, from which as many as 12 meteors per hour be seen, is directly overhead at 10:00 PM in the tiny constellation of Aries, the Ram.

Position of radiant: Right ascension 03h 35m; declination +25°.

December 22, Ursids One of the few meteor showers that never sets, the Ursid meteors originate from within the Little Dipper, Ursa Minor. The meteors are highest in the sky just before dawn. Derived from Comet Tuttle, these meteors move at medium speed, giving rise many times to spectacular trails in their wake. Observers may expect up to 17 to 20 meteors per hour, most fairly bright, coming from almost due north.

Position of radiant: Right ascension 14h 28m; declination +78°.

BIBLIOGRAPHY

Abell, George, *Realm of the Universe*. New York: Holt, Rinehart & Winston, 1976. Paperback.

Farrington, O.C., *Catalogue of the Meteorites of North America to January 1, 1909*. Memoirs of the N.A.S., Washington, DC: Government Printing Office, 1915.

Hawkins, G.S., *Meteors, Comets, and Meteorites*. New York: McGraw-Hill, 1964.

Harvard College Observatory, Annals of, "Meteor Heights from Photographs," Vol. 87. Cambridge, MA, n.d.

Middlehurst, B.M., and G.P. Kuiper, *The Moon, Meteorites, and Comets*. Chicago: University of Chicago Press, 1963.

Moore, Patrick, *Astronomy Facts and Feats*, pp. 128-131. London: Guinness Superlatives, Ltd., 1979.

Muirden, James, *The Amateur Astronomer's Handbook*, pp. 211-224. New York: Thomas Y. Crowell, 1974.

Olivier, Charles P., *Meteors*. Baltimore: Williams and Wilkin, 1925.

Richardson, Robert, *Getting Aquainted With Comets*. New York: McGraw-Hill, 1967.

Roth, Gunter D., *Astronomy: A Handbook*. Cambridge, MA: Sky Publishing Corporation, 1975. Paperback.

Watson, Fletcher G., *Between the Planets*. Cambridge, MA: Harvard University Press, 1956.

Whipple, Fred L., *The Collected Contributions of Fred L. Whipple*, Vol I. Cambridge, MA: Smithsonian Astrophysical Observatory, 1972.

4

COMETS: A GUIDE TO OBSERVATION, PHOTOGRAPHY, AND DISCOVERY

The study of comets is an area in which the amateur can make quality, scientific observations. The professional astronomer limited to a schedule is generally unable to turn the giant telescopes of earth toward comets—those objects that have always intrigued and frightened people on earth. It is somewhat ironic that the very objects in our solar system about which we know the least are also the most neglected. Amateur astronomers, using their modest equipment, have a field of study apart from those of the professionals to which they can devote effort that is sure eventually to result in important scientific data.

If the study of comets is so well suited to the equipment and knowledge of the amateur astronomer, why are not more observations made or reported than those occasional few in astronomical journals? This question could best be answered by examining the literature on the study of comets by visual and photographic methods if there were any. However, anyone who has searched through the archives of libraries, observatories, universities, or astronomy journals knows that there is little sound information to help the amateur astronomer achieve success in the study of comets. And, on the other hand, there is a great deal of information available that explains the physical nature of comets and the many current theories about them.

The purpose of this chapter, therefore, is to aid you as an amateur astronomer to develop the proper techniques for locating, observing, photographing, and recording comets. In addition, the discussion will help you to conduct searches for new comets by describing the proper search techniques as well as the best areas of the sky in which to search. The extrapolation of your observing records into a usable form that is acceptable for evaluation by professional observatories is discussed. If you wish to be a serious observer of comets, it is important to learn the proper way to submit your data. However, no proper study can be done without the proper optical equipment, nor can the results be recorded accurately without the aid of a good set of star charts. We begin, therefore, by describing effective equipment and materials for the detailed observation of comets.

THE TELESCOPE

As in all areas of study of astronomy, no one telescope can serve all purposes of cometary observation. There are times when you need extremely low power and a wide field of view to study the structure of the comet tail or perhaps to aid in locating a faint comet, and there are times when you need high power and excellent resolution for studies of changes within the head and nucleus of the comet. There is no telescope capable of both functions, no matter what claims are made by the suppliers of commercially made instruments.

If your budget is small, there is no simple solution to finding the *one* telescope for the overall study of comets. Much is printed about the proper selection of an astronomical instrument, and it is also discussed in other chapters of this book. The discussion in this chapter concentrates on the choice of the proper instrument for the study of comets.

The Refractor

The refracting telescope is somewhat limited and generally not well suited for the study of comets. The long focal length prohibits photography of faint comets, and the field of view is entirely too restrictive. In addition, instruments with longer focal lengths provide very faint images in proportion to the overall focal ratio of the objective lens or mirror. A refractor with an aperture great enough to provide for the study of comets that are dimmer than magnitude 10 is usually beyond the means of most amateurs. A 4-inch refractor, costing more than $1600, will just barely show a comet of magnitude 8 or brighter well enough for detailed study. The price more than triples for every added inch of aperture if you buy the scope from commercial sources.

If you have a rich field (f/6 or faster) refractor, however, there are many distinct advantages inherent in the refractor design. My personal comet telescope is a 6-inch f/5 refractor that cost me only $550 to construct in 1980. The refractor allows for very dark background sky against the comet, which makes for added contrast, and it has a field of view better than 2.75° with a 32-mm Erfle eyepiece. The limiting magnitude of such an instrument is about 11.3 for comets, and resolution near the nucleus of the comet is a bit better than the resolution I get with my 10-inch Newtonian reflector. The 6-inch refractor is quite portable. It does not need an equatorial mounting because the field of view is sufficiently wide so that the comet being observed does not rapidly drift out of view. Thus, constant shifting of the instrument to keep up with the comet is unnecessary.

The Reflector

If you must choose a single telescope to serve all the functions of a comet telescope and still be useful for other areas of astronomical study, you should choose a Newtonian reflector with a focal ratio of f/6 or faster. If it is equipped with an accurately figured mirror and a clock drive, the Newtonian reflector can be used for high-magnification work near the nuclear center of a comet as well as for moderate wide-field views of the inner portion of the tail and head. Because of the ease of construction and low cost of such a telescope, it can be made much larger than a refractor telescope for the same money. For comet studies (as well as other fields of interest to the amateur astronomer) both a large aperture and a wide field of view are desirable for study of fine linear plumes and changes within the nucleus. However, increasing the aperture and shortening the overall focal length of a telescope require a correspondingly darker sky. In most suburban areas, little is gained in comet study if you use an instrument in excess of 32 cm (12 inches).

Many commercial firms supply either complete telescopes or finished mirrors of exceptional quality in f/5 and faster focal ratios. A 10- or 12-inch Newtonian reflector of about f/5 is the most useful telescope for comet observation during pre- and post-perihelion. During perihelion, the great aperture and resolving power of such a telescope allows you to scrutinize the nucleus for possible splitting or eruptions. For wide-field studies of the head and tail regions, however, as well as for magnitude estimates, I strongly recommend telescopes of a smaller aperture for reasons that I discuss later.

Binoculars

Even a good pair of binoculars can be used for the scientific study of comets, provided you are well informed as to their proper use. The usefulness of binoculars is somewhat limited to estimates of magnitude and estimates of coma size and tail lengths, but little more. Large binoculars, say, 4 to 5 inches (100 to 125 mm), are quite useful for comet searching and study, but their cost is nearly prohibitive for most amateurs. In dark skies, a standard pair of 7 X 50 binoculars or a pair of 11 X 80 binoculars can provide spectacular views of bright comets, and can discern enough physical character of the comet to be useful as a research instrument.

In summary, the sophisticated instrumentation that is required in many other amateur studies is not needed for comet studies. Comet research and discovery can be carried out to some degree of *any* instrument, but certainly not by any *one* instrument, for all areas discussed in this chapter. No matter what instrument you may have, if you have the desire to study comets, you can do so, and you can contribute to the scientific knowledge of these ethereal bodies.

STAR CHARTS

The best astronomical telescope is of little help in the organized study of comets if you have no method whereby what you see in the telescope (or photograph) can be plotted directly on charts for accurate evaluation. A good star chart should always accompany you to your telescope or other instrument whether your study involves an existing comet or the search for a new discovery.

The usefulness of any good star chart is greatly increased if it is accompanied by a *star catalog* that lists the precise magnitudes and other pertinent data about the stars on the star chart. You can use these data for comparison with any comet you sight. As Figure 4-1 shows, there is some scarcity of atlases with such catalogs. Perhaps the Smithsonian Astrophysical Observatory (SAO) atlas and catalog are the best for the detailed study of fainter comets, but the atlas is now out of print, and the catalog is in increasingly short supply. Most modern star atlases and catalogs have been compiled by data and reference material taken from previously published older catalogs, and most of the errors that went uncorrected in the original work have been passed on to the newer atlas. Such is the case with the SAO atlas and catalog. Figure 4-1 shows the way three of the more popular star atlases represent the star cluster Pleiades. The dif-

FIGURE 4-1. Some characteristics of selected star charts.

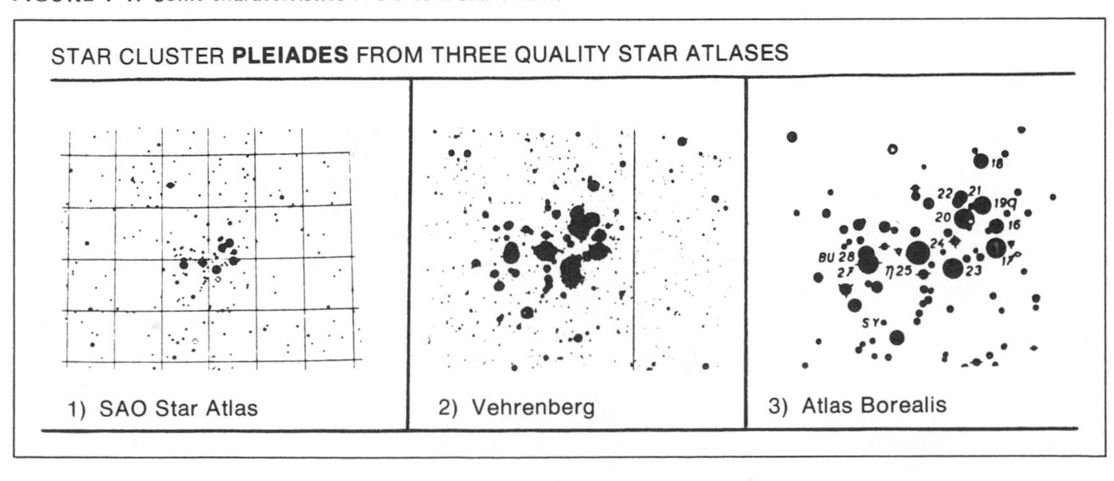

	Maps				Stars			Nonstellar			Catalog
Author/Title	Epoch	No.	(°/cm)	Size (cm)	Sym.	Mag.	No.	Sym.	No.	Const. Recog.	
Inglis, Popular Star Atlas	1950	8	3.9	19 × 32	2	5½	2,500	1	50	Fair	No
Brown, Mag 5 Star Atlas	1970	6	3.2	26 × 38	1	5	1,000	3	110	Good	No
Menzel, Field Guide	1950	54	3.3	9 × 11	1	–	Photo.	5	300	Fair	No
Norton, Star Atlas	1950	8	3.1	25 × 41	3	6⅓	8,400	1	600	Fair	No
Sweeney, Seasonal Charts	1950	8	4.3	25 × 18	3	5¾	3,000	2	170	Good	No
Howard, Telescope Handbook	1970	14	4.0	23 × 15	7	6½	9,000	5	1,150	Fair	No
Moore, Color Star Atlas	1950	8	3.5	25 × 31	3	6½	9,000	6	200	Fair	No
Vehrenberg, Handbook Const.	1950	57	2.4	25 × 18	3	6	6,000	6	1,050	Fair	No
Becvar, Field and Desk Eds.	1950	16	1.9	36 × 26	14	7¾	32,000	7	1,850	Poor	Yes
Becvar, Skalnate Pleso Deluxe	1950	16	1.3	53 × 38	14	7¾	32,000	7	1,850	Poor	Yes
SAO, Star Atlas (out of print)	1950	152	0.6	28 × 35	1	9½	260,000	4	1,300	Poor	Yes
Becvar, Atlases *Borealis*, *Eclipticalis*, and *Australis*	1950	79	0.2	35 × 50	7	9½	300,000+	–	–	Poor	No

FIGURE 4-1. (continued)

ferences in both scale and magnitude range in each are quite evident. Notice that each example is exact scale, relative to the others.

To estimate the magnitude of bright comets using the naked eye or binoculars, the most accurate and most widely used list of bright stars is in the *Arizona-Tonantzintla Catalogue*, which gives photoelectrically derived magnitudes that are the equivalent of visual interpretation. This reprint is available at a modest cost from:

Sky Publishing Corporation
50 Bay State Road
Cambridge, MA 02138

Very faint comets will occasionally pass through known fields of variable stars. On these occasions you can derive accurate estimates of magnitude by using the variable star charts supplied by the American Association of Variable Star Observers (see Chapter 12). You can extrapolate the paths of comets to find such occurrences so that you can obtain the proper chart beforehand.

Star atlases and their accompanying catalogs are useful for comet work for the following reasons:

- ☆ Determination of total and nuclear magnitude of the comet.
- ☆ Plotting the approximate position of the comet from night to night.
- ☆ Determination of the *position angle (PA)* of the comet tail if it is evident.
- ☆ Determination of the total length of the comet tail and any secondary components, and possibly the determination of the size of the comet head, or coma.
- ☆ Notation of the degree of movement of the comet each day.

If you find or suspect that you have found a comet, the star charts are essential for two additional reasons:

- ☆ Determination of the direction and amount of the drift or movement of the comet.
- ☆ Quick reference to make sure that the "comet" is not actually some known nebular or diffuse object.

For the person searching out new comets, it is necessary to have a chart that lists physical information (e.g., the size and magnitude) of diffuse objects to aid in immediately disqualifying such objects as being comets. In my opinion the two best atlases for comet searching are the Becvar *Atlas of the Heavens* and Vehrenberg's *Handbook of the Constellations.*

THE NATURE OF COMETS

Before beginning the study of comets, you must first become familiar with the nature of comets, as well as the astronomical jargon used in describing some of the phenomena of a comet.

The comet is an important piece of the great puzzle of the solar system. It is a sort of missing link in our understanding of what created the planetary system of which the earth is part. To the best of our understanding comets are debris left from that creation, and have not changed in the great span of time since that event. Only when the comet passes within the orbit of Jupiter does it develop a tail, in most cases. For amateur studies and for the basis of this book, we need only to define the comet in a physical and observational sense and leave the chemistry and physics of comets to the many excellent texts listed in the bibliography following this chapter.

Basically a comet has three components that must be studied: (1) the nucleus, or center condensation, (2) the head, or *coma,* and (3) the tail (see Figure 4-2). The nucleus is often, but not always, seen. It is the principal supply of the material from which the comet originates and is the orbiting body that encircles the sun. Emanating

FIGURE 4-2. Diagram of a comet.

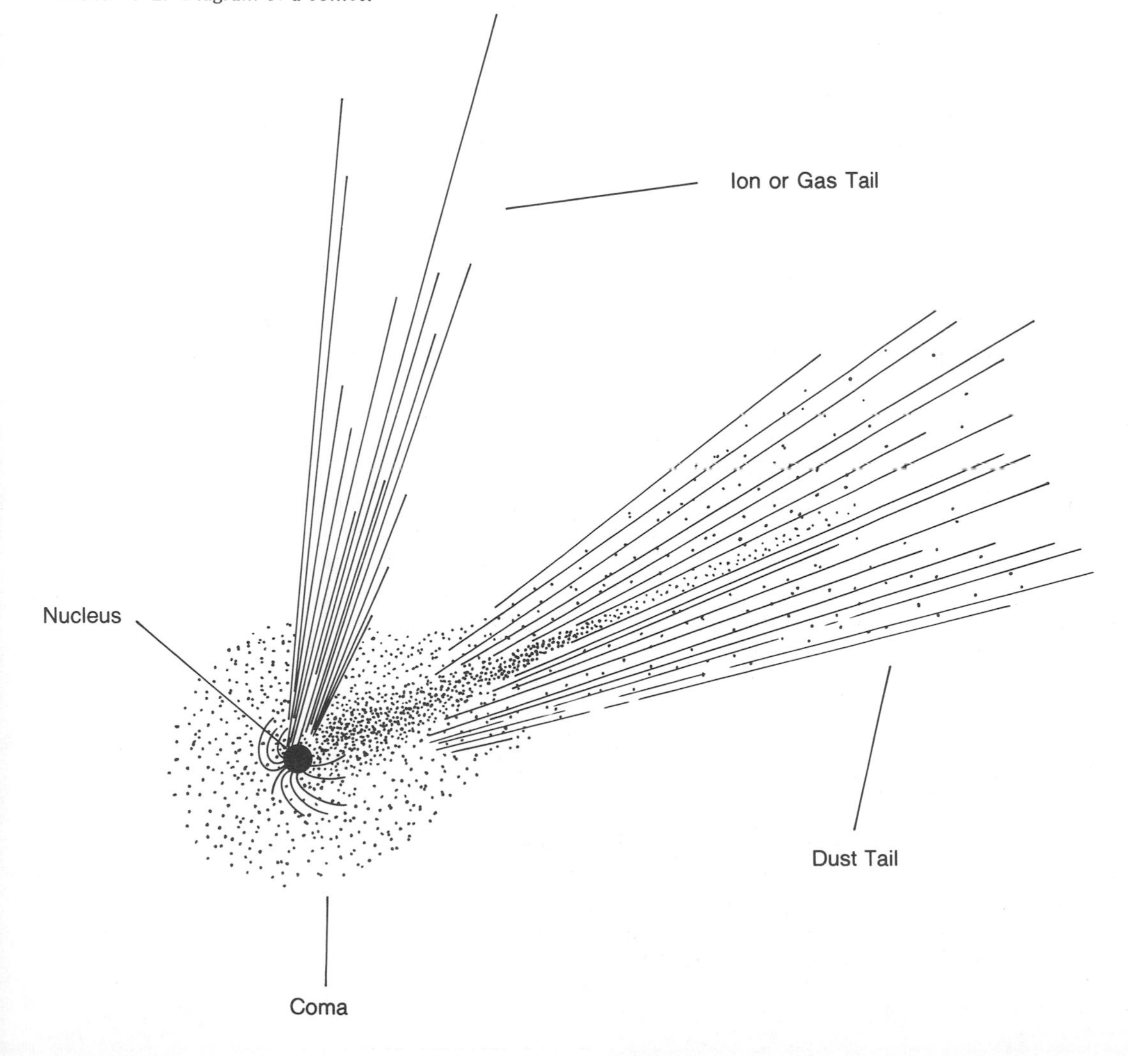

from the nucleus is the *coma*. The coma can become quite large, many times having a diameter equal to the distance between the earth and moon. Usually this diffuse cloud appears as a halo surrounding the nucleus of the comet, and it is this portion from which you should make estimates of the comet's magnitude. Many times the coma is seen clearly differentiated, with the inner portion being clearly divided from the periphery by a sharp delineation in brightness. Being pushed away from the nucleus and coma of the comet by the forces of the solar wind is the tail of the comet, that part which has been likened to swords and axes in our recent past by those persons fearing the bright comets. Tails of highly active comets can stretch as far as several hundred million kilometers from the coma, and many times they can be seen to separate into two or more components. The two main components of most comet tails are the *ion tail*, which consists of ionized atoms from the gases locked up in the previously frozen comet, and the *dust tail*, which consists of heavier dust particles. The ion tail is blown away from the comet, in a direction straight away from the sun. This is because the ions are so light they are easily accelerated outward by the solar wind and sunlight. The dust tail, sharing in the orbital motion of the comet, is made of particles 12 to 18 orders of magnitude more massive than the ions. Hence, the tail is not as readily blown out by the solar wind and sunlight, and tends to remain in the comet's orbital path, trailing behind the coma. Comet tails are usually referred to by one of three classifications:

1. Type I—a very straight and thin tail that might coexist with a wide diffuse fan-shaped tail.
2. Type II—a tail that is fan-shaped and quite wide, and usually curved.
3. Type III—less wide than Type II, but strongly curved.

NAMING A COMET

A comet goes by many names, the most important of which to its discoverer is his or her surname. Many bright comets that frequently return retain their popular names, such as Comet Encke and Comet Halley. (Halley did not discover the comet, but he predicted its orbit and return.) Frequently, two or more observers discover a comet simultaneously. The names of the first three observers to report their discovery properly are affixed to the new visitor (e.g., Comet Honda-Mrkos-Padjusakova). In addition to such names, the comet is assigned a letter of the alphabet (a, b, c, etc.) that coincides with its order of discovery in some particular year. Comet Kohoutek, 1973f, was the sixth comet discovered in 1973 and was discovered by Lubos Kohoutek. Eventually, even this letter designation is dropped for a longer lasting system that is related to all comets yet discovered. This final designation system specifies, with roman numerals, the year and the order in which each comet passes its closest point (perihelion) to the sun. Thus, Comet Kohoutek, 1973f, eventually is referred to as 1973 VII, which indicates that it was the seventh comet to pass perihelion in that year.

COMETS OF SHORT AND LONG PERIODS

A comet is further classified according to the time it requires to complete one orbit about the sun. Long-period comets require more than 200 years to return to the sun's vicinity, and they are the most common. Short-period comets, on the other hand, require 200 years or less for their circuit and are among the best known of comets. Table 4-1 lists only a few of the known short-period comets.

TABLE 4-1. Important short-period comets of more than one appearance.

Comet	*Maximum Distance From Sun*	*Period (in years)*
Encke	2.21 AU[a]	3.30
Grigg-Skjellerup	2.87	4.90
Honda-Mrkos	3.01	5.21
Temple II	3.02	5.27
Tuttle-Giacobini	3.11	5.48
Pons-Winnecka	3.35	6.12
Giacobini-Zinner	3.45	6.42
Perrine-Mrkos	3.47	6.47
Schwassmann-Wachmann II	3.49	6.53
D'Arrest	3.55	6.70
Brooks	3.56	6.72
Faye	3.80	7.41
Whipple	3.80	7.42
Arend	3.93	7.79
Comas Sola	4.20	8.59
Crommalin	9.19	27.87
Westphal	15.62	61.73
Olbers	16.92	69.57
Halley	17.95	76.03

[a]Astronomical unit.

AN OBSERVING PROGRAM FOR THE AMATEUR

Estimates of Magnitude

Perhaps the first, and surely the simplest, undertaking during your session of observing a comet is the determination of the magnitude of the comet. Such a determination consists of two estimates—one of the head, or coma, region of the comet, and the other of the nucleus if it is visible. Both estimates are important in extrapolating the activity of a comet because the magnitude is the best indicator of physical change that we presently have.

When estimating magnitudes, your primary consideration should be *consistency*. The method for obtaining magnitudes that may be easiest for one observer may not be for another. Whatever method you choose, use it exclusively and do not alternate it with other methods. The following method is perhaps the most reliable and widely used, regardless of optical instrumentation:

1. Position the comet in the center of the field of view and examine the brightness of the coma in focus.
2. Locate stars on the star chart that are in the appropriate magnitude range and that are in proximity to the comet. Ideally, the comparison stars should be within the same field of view as the comet.
3. Reposition the telescope or binoculars to center on the comparison star.
4. Turn the focuser until the star image, which will now be *out of focus,* appears to be the same size as does the comet's coma *in focus.*
5. Determine if that star is brighter, dimmer, or of the same brightness as the comet.
6. If no star of the same brightness as the comet is nearby, establish an arbitrary scale relative to nearby stars (at least one that is

brighter and one that is dimmer than the comet), and make a relative estimate.

7. Lightly indicate on the star chart the comparison star(s), by drawing a thin circle around the star(s). If you have used a relative scale, this fact should be indicated on your records.
8. Use a reliable catalog (such as those previously described) after all observations have been made, and from that determine the magnitudes of each star. You will thus find the brightness of the comet.

Many observers choose to examine *both* the comet and star *out of focus* simultaneously, making both appear the same size out of focus and determining the brightness of the comet while seeing both together defocused. This method is reliable because an extended object such as a comet will not appear to change significantly in brightness whether it is in or out of focus.

When making the comparisons, it is essential that you repeat the procedure several times to lessen the chance for error. If a catalog is available, you should include in each day's report of the comet the star number (GC number in the *Atlas of the Heavens* catalog, or SAO number in the SAO catalog, for example).

It is obvious by now that estimating magnitudes can be easier if you use a wider field of view in a telescope or binoculars. If the comet is of magnitude 6.5 or brighter, low-power binoculars are the instrument of choice for magnitude estimates for the following reasons:

☆ Binoculars allow a much wider field of view and brighter image, thus facilitating easy acquisition of suitable comparison stars within the needed range of magnitude.

☆ All magnitudes are converted by professionals (and amateurs) for use in published material to correspond with a standard instrument, with an aperture and brightness factor closely matching that of standard 7 X 50 binoculars. Thus, little or no conversion is necessary. In published material, such as the International Astronomical Union Circulars, the magnitude estimates from various observers always appears along with the instrument used to make that estimate so that conversions can be made by persons using those data.

Any good pair of binoculars is adequate for the magnitude studies of brighter comets, if they are 7 X 50 or larger. As mentioned earlier, the new 11 X 80 binoculars, which have become popular in recent years and are moderately priced, are excellent for this purpose.

Exceptionally bright comets must be estimated by the naked eye. You generally can determine them in focus by using bright planets as comparison objects, or—in some rare instances—by using the moon. The major problem with estimating bright comets is that planets of suitable brightness to use as comparison objects are not always visible in the same sky as the comet. Consequently, a bit of estimating by memory must suffice, which leaves considerable margin for error resulting from poor atmospheric conditions, poor memory, or simply from carelessness.

Magnitude of the Nucleus

Nucleus magnitudes are usually difficult to determine due to the faintness (or even the ill-defined boundary) of the tiny nucleus and to the dispersion of light from the coma. However, you should make estimates of nuclear magnitude at every opportunity if you can see a distinct bright center of the comet. If the nucleus is bright enough, make the comparison in much the same manner as you would make a variable star estimate, using appropriate comparison stars. Because the nucleus is usually fainter than the limiting magnitude of star catalogs, however, a precise magnitude (to 0.1 magnitude) is difficult to ascertain. Normally, you must determine the magnitude with both the nucleus and the comparison star(s) in focus. If you cannot find the precise magnitude of the comparison star, you can use a relative scale between two stars, noting the right ascension and declination of each star. An experienced observer of variable stars can, with reasonable accuracy, examine stars in a known variable star field on each given night and then turn the telescope to the comet, using a mental image of the comparison stars as they appeared in the telescope to estimate the magnitude of the nucleus.

If a comet has a multiple nucleus from splitting, it is quite important to observe the magnitude of each component *daily,* as well as the changing position of one component relative to the other. In such cases, the brightest component at the time of splitting is considered to be the primary particle, and all measurements are made from that component. As in the determinations of the position angle of a comet's tail as discussed later, measure each secondary component as an angle in degrees from north toward east from the main component. Also, each night determine the changing distances (expressed in seconds of arc) of each secondary particle from the primary component. Such information provides the astrophysicist data from which can be determined the approximate mass of the nucleus components, their relative velocities, and consequently the force of expulsion that led to the initial splitting.

You can roughly determine the distance between nucleus particles by knowing the approximate field of view of the eyepiece in use and extrapolating the percentage of that field between components, or by using a reticle eyepiece with a relative scale (depending on the telescope in use) from which separations can be more accurately determined. The most efficient method of determining these changing distances, if a filar micrometer is not available, is to use the drift method, as described on page 72. Using this simple method, you can make quite accurate determinations by allowing the first component to drift out of the field of view and timing the interval from that instant until the second component also drifts out. Make at least three timings of each interval and determine a mean value. The conversion of this timing is discussed in the following section dealing with estimating the diameter of the coma.

Nucleus Condensation

Many times a comet will not exhibit a clear nucleus but rather a gradual or "stepped" condensation toward the center of the head. In such cases, it is important to record such a condensation each night on an established scale, which is widely accepted, as indicated in Table 4-2.

TABLE 4-2. A scale for the degree of coma condensation.

Value	*Description*
0	Coma diffuse; shows no indication of condensation, with brightness uniform.
1	Coma diffuse, possibly some degree of brightening toward the center half.
2	A definite brightening in the center of the coma as compared to the outer area.
3	Coma well differentiated with two levels of brightness.
4	Definite condensation, though diffuse, in the center of coma; condensation is greater than a quarter of the coma's diameter.
5	More condensed than above, appearing as a bright spot in the center.
6	Condensation very evident; coma uniform in brightness except for center; nucleus round.
7	Condensation appears as a fuzzy star that cannot be focused sharply.
8	Nucleus appears much as a bright star might appear in a scope on a night of unsteady air; definitely star-like, but not distinct.
9	Stellar in appearance.

Photoelectric Determinations Of Comet Magnitudes

The photoelectric photometer is an electronic device for measuring the intensity of light. Normally, the intensity is measured relative to a source that has a constant magnitude value, usually a star. As well as determining quite accurately (usually to 0.01 magnitude) the brightness of a celestial object, the photoelectric photometer can also determine the brightness in various wavelengths of light by using special *interference filters* that allow only small portions of the spectrum to pass on to the detector. Such an instrument sounds ideal as a research tool for astronomers, and many of these instruments are now in the hands of amateur astronomers, being used primarily for variable star research. Although the functions of this instrument are ideal for the study of comets, the photoelectric photometer cannot be adequately used for magnitude estimates and is useful only as a relative determiner of *color index.*

The color index is a numerical rating that compares the intensity of an object (in this case, the comet or its nucleus) in one wavelength of the spectrum to its intensity at another wavelength. Wavelengths of light frequently used by both amateurs and professionals for such studies are as follows:

Ultraviolet	(U)	320 to 420 nm
Blue	(B)	400 to 500 nm
Visual	(V)	500 to 650 nm[a]
Red	(R)	560 to 750 nm
Infrared	(I)	750 to 950 nm

[a]Most closely matches the visual magnitude as seen with the human eye (V).

The most common color indexes are U-V, B-V, V-R, and R-I. In any case, the index can be either negative or positive. Color indexing from standard wavelengths is discussed in Chapter 13.

Determining the magnitude of either the coma or the nucleus of a comet by using the photoelectric photometer presents many difficulties. Accurate determinations of the nucleus can be made using a suitable comparison star and the V filter, but the estimate will not be without some scatter from the cloud of the coma, which surrounds the immediate vicinity of the nucleus. However, the estimate is better than no estimate at all. When determining the total, or coma, magnitude, set the photoelectric detector *off center,* away from the brightness of any central condensation, and check several different points across the coma. Use a mean of all measurements as the value that you report.

If photoelectric equipment is available, and if it can be used on a consistent basis, by all means take such measurements. Nonetheless, the photoelectric photometer has by no means yet replaced the human eye in efficiency for the determinations of cometary magnitudes, although the human eye is not nearly so accurate in its determinations as the sophisticated photometer.

Plotting the Comet on the Star Chart

Before you undertake any further serious work in your study of comets, you must determine where the comet will be, or where it is at the time of observation, by plotting it on the appropriate star chart. If you wish to undertake long-range studies on a particular comet, it is essential that you plot the predicted path of that comet on the chart prior to observing. This is important for several reasons.

First, you must be aware of any changes in the comet's predicted path, although such changes are rare and most deviations that do occur are beyond the range of amateur detection. Second, prior plotting allows you to determine in advance suitable comparison stars that will be in or near the field of view, from which magnitude determinations can be made. Valuable time can be lost at the telescope if you have to search around the sky for suitable comparison stars. Third, you can readily examine the path of the comet relative to the background stars for close passes, or even occultations, that might occur. A direct occultation of a star by the comet would be an extremely rare event, and one deserving high priority, because timing the nucleus as it passes in front of a star would provide direct data for determining the size and mass of that comet's nucleus.

To make plots of a comet's path you must have access to the *ephemerides* (i.e., predictions of positions and physical data) of the comet.

Such data usually appear in astronomical publications and journals, but the appropriate issue may not reach you until it is too late for early observations during the comet's early apparition. Rarely, a faint comet will be discovered photographically long before it becomes visible to amateur instruments, thus allowing for sufficient time for full ephemerides and predictions in many journals, but this is not often the case.

If you are a dedicated observer of comets, you should subscribe to the International Astronomical Union (IAU) Circulars, which announce new discoveries, including comets, as they are made. In addition, many vital data concerning predicted magnitudes, positions, dates of heliocentric (solar) opposition and of geocentric (earth) closest approach are in the circulars. Usually in less than one week's time, any change or unpredicted occurrence in a comet will be available to you. A subscription to the circular can be ordered through:

IAU Circular Service
Smithsonian Astrophysical Observatory
60 Garden Street
Cambridge, MA 02138

Whatever source you use to obtain the position for a comet on any given day, you must note that all positions (usually one per day) are given for each date at 00 hours Universal Time (UT). Allowance must be made if the comet is moving swiftly and any time has elapsed since 00 hr UT. Such a case might be early morning observing of a comet near sunrise from the United States, at which time almost 12 hours have elapsed since the time of prediction. To obtain local time from Universal Time, refer to Appendix I.

To plot the comet accurately you must determine the physical center of the coma. If the comet has a definite condensation or nucleus, use it as the center by which you make subsequent measurements. Some observers use a graduated reticle inserted into their eyepiece to determine the relative distance of a comet center from the field of stars. Such a method of plotting position is probably sufficient for amateur endeavors, although determination of position using a bifilar micrometer is much more desirable and usually within the reach of amateurs. You can also use a photograph to determine the position of a comet if all essential data are recorded at the time the photograph is made, including the data, UT, and the type of instrument used to obtain the photograph. Once you measure the position (using a simple millimeter rule) from the photograph, it can easily be converted to scale and plotted on the star chart.

You should not spend too much time attempting to record the position of a comet. The professional astronomers with their astrographic and Schmidt cameras are able to obtain accuracy of position down to 1 micron with the benefit of special photographic plates and measuring devices from which the determinations are made. Nonetheless, plotting the comet's position is important. Record data at the telescope as quickly as possible. If the measurement or plot can be done reasonably well at the telescope using whatever available method, do it then. Later, if you wish to confirm photographically what you have seen, that is fine. If you have neither a camera nor a reticle eyepiece, you can still plot a comet's position with sufficient accuracy by using a low-power, wide-field eyepiece. Merely use certain series of stars that appear both on the map and in the telescope as stepping-stones from which to interpolate a position. The degree of accuracy that can be attained if you are proficient at this method is somewhat surprising. After you have plotted the comet, make a note of the following data in your notebook before proceeding to any other areas of study:

- ☆ The date and time, expressed in UT.
- ☆ The method used to make the estimate.
- ☆ The optical aid used, whether a telescope or binoculars.

You are now ready to proceed to other areas of cometary study made possible by your simple plotting, including the following:

- ☆ Rough estimates of any discrepancies in the predicted vs. actual position of the comet.
- ☆ Approximate measurements of position, extrapolating from the positions of field stars as given in star catalogs.

☆ Direct measurement of the tail and any secondary components.

☆ Direct measurement of the Position Angle (PA) of the tail and any secondary components.

It becomes obvious, when you examine the possibilities for study listed above, that the star chart is indeed an essential item for the study of comets. The procedures for these and other areas of cometary study are described in detail in the following pages of this chapter.

Estimating the Size of the Coma And Other Features

There are several methods by which you can determine the size of the coma and tail of a comet, including the accurate drift method, micrometer measure, direct measurement from a photograph, and by simply knowing the field of view of a telescope and eyepiece combination.

The simplest measurement, although somewhat less accurate than the first three mentioned in the preceding paragraph, is also the quickest. Select a low-power eyepiece with a clear field whose field of view you know, or see Appendix IV for the method of finding the field of view of any eyepiece. When you are observing, have a list of each eyepiece's magnification and field of view conveniently at hand. Another method by which the field can be quickly checked would be to locate some bright field in the eyepiece (e.g., the Pleiades cluster or the belt stars of Orion). With the aid of a star chart, draw the extent of the field that can be seen in the eyepiece on the chart. Then, by knowing the scale of the chart in millimeters per degree of sky, you can quickly determine the approximate field of view as drawn on that chart. For example, if the circle drawn is 2° of sky, then there will be 120′ of arc, so the total field would be 120′.

Once the field of the eyepiece has been determined by one of the methods given in Appendix IV, it is a simple matter to roughly estimate the size of the comet's coma in relation to that field, provided that the coma is fairly large in relation. Simply estimate the fraction of the total diameter of the eyepiece field that is the diameter of the coma. Suppose that the coma under observation covers a span equivalent to the distance from one edge of the field to about one-fourth the distance across the field. The diameter of the coma in minutes of arc would be only one-fourth of the total field, or 30′ of arc. Although this method seems almost too simple to be effective, you can use it efficiently when you do not have time (because of twilight, clouds, or the comet's setting) to use a more accurate method. Accuracies of around 5′ of arc are possible using this method when viewing quite large comets.

Use of a Graduated Reticle Eyepiece To Determine Size

A more reliable and accurate method for determining the distances across a comet's coma or tail is the reticle measurement. Many eyepieces can be fitted with a graduated reticle, the gradations of which signify fractions of the field of view. Such reticles work very well with eyepieces of Kellner and Ramsden design and can simply be attached inside the lower barrel of the eyepiece. The reticles are quite inexpensive and can be purchased from a number of large optical supply houses. You will have determined from the table in Appendix IV the field of view of the eyepiece in which the reticle is used. Each gradation of the reticle is simply a fractional part of that field and can be accurately used to determine the size of the coma or tail of the comet.

The Drift Method of Determining Size

The most widely used and perhaps best accepted method of determining the size of the coma of a comet is the *drift method.* The drift method requires two timings, and it can be done in the following sequence:

1. Center the coma in an eyepiece that has a field large enough to contain the coma easily, with room to spare.
2. Disengage the clock drive of the telescope (if one is being used) and allow the comet to drift toward one edge of the field as a result of the earth's rotation.

3. As the comet approaches the edge of the field, time the precise instant to the nearest second that the coma touches the edge of the visible field of view.
4. Allow the comet to continue to drift out of the field until the last instant the coma can be seen. At that instant also time to the nearest second when the coma is last seen.
5. Determine the total interval in seconds between the first and last timings and multiply by the factor given below. Notice that the factor used depends on the comet's declination in the sky.

Approximate Declination	*Factor*
0.0°-12°.2	0.25
12.3°-20°.4	0.24
20.5°-26°.2	0.23
26.3°-30°.9	0.22
31.0°-35°.1	0.21
35.2°-38°.9	0.20
39.0°-42°.4	0.19
42.5°-45°.7	0.18
45.8°-48°.8	0.17
48.9°-51°.8	0.16
51.9°-54°.7	0.15
54.8°-57°.4	0.14
57.5°-60°.1	0.13
60.2°-62°.7	0.12
62.8°-65°.2	0.11
65.3°-67°.7	0.10

6. The value obtained is the diameter expressed in minutes (′) of arc.
7. Repeat the sequence twice, and determine a mean value for the actual diameter.

Photographic Measurements To Determine Size

An often overlooked but quite efficient method of measuring coma diameter is direct measurement from a photographic print. There are two major drawbacks of this method:

1. The length of time required from the initial photograph, through film processing and final printing, to the final measurement does not provide quick results.
2. The possibility of drift during an exposure can result in the measured value's being somewhat larger than the coma really is. Likewise, scatter of light onto photographic emulsions tends to increase the size of an image.

To make a photographic measurement, use a simple millimeter rule to measure, in millimeters, the diameter of the coma as it appears on the final print. Two brighter stars that can be also identified on the star atlas are likewise measured, both on the photograph (to determine scale) and on the star chart (to determine the number of minutes of arc between the two). Once the scale of the photograph has been determined, a simple calculation provides the diameter of the coma:

$$\frac{y}{a'} = \frac{b'}{x}$$

or

$$x = \frac{a'(b')}{y}$$

In both equations, y is the measurement in millimeters of the distance between the two stars; a' is the known number of minutes of arc separating the two measured stars; x is the sought value for the number of minutes of arc of the coma; and b' is the measurement of the coma in millimeters, as measured on the photograph.

A photographic determination does have the distinct advantage of allowing the faint peripheral portions of the coma cloud to be recorded on film, whereas this might be totally beyond the perception of the visual observer.

Determining Size with a Filar Micrometer

The most efficient and readily acceptable means of determining the diameter of the comet's coma is the *filar micrometer.* Unfortunately for the amateur, the sources for such instruments are quite limited, resulting in many homemade units of poor efficiency. Units of high quality have the added advantage of a *position circle* for determining position angle. This is a valued measurement as well, necessary in determining the direction of the tail and secondary streamers, as will be discussed, and also for measurements to derive the angles of movement of any particles that split away from the comet's nucleus.

Measurements of coma diameter with a filar micrometer are fairly simple, yet the exact procedure for reducing those measurements depends greatly on the construction of the micrometer. Basically, the micrometer consists of a measuring screw, two fixed webs, and one movable web. Two webs are fixed perpendicular to each other. They are normally positioned to cross at the center of the optical path. The movable web is on a track attached to the measuring screw (see Figure 4-3).

The design of modern filar micrometers has been standardized so that commercial measuring screws are readily obtainable, and calculations from these measuring screws are quite simple.

FIGURE 4-3. (a) Photograph of a commercial filar micrometer. (Courtesy of Ron Darbinian Micrometers) (b) The view through a filar micrometer eyepiece.

(a)

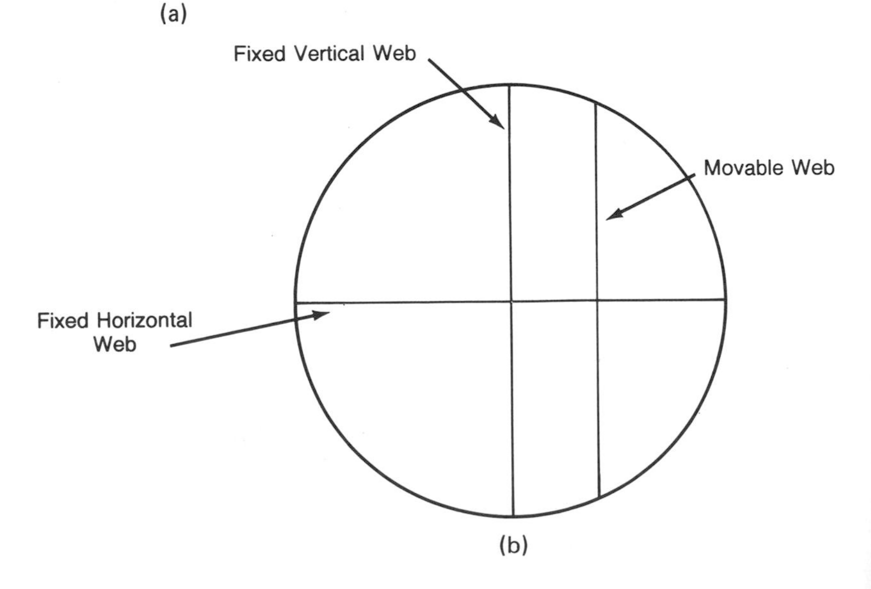

(b)

Normal measuring screws of high precision have threads that measure 0.5 mm each and a total travel of about 24 mm. The one web that is movable can be determined to have a correlation during one complete turn equal to a precise distance in degrees of sky. This value is known as the *screw value* and depends on the type of telescope you have.

$$\text{Screw value} = \frac{57^\circ.2958\ (\text{thread of measuring screw in mm})}{\text{telescope focal length in mm}}$$

Assuming that a common commercial measuring screw is used, then the thread of the measuring screw can be assumed to be 0.5 mm. If a telescope of 2000 mm focal length is used in the determinations, a screw value of 0.01432° is obtained. This value is constant as long as the same system is in use. All values obtained with the micrometer (the thread value, or that shown on the graduated scale of the measuring screw) are multiplied by that value to determine the diameter in seconds of arc. To convert to minutes of arc, merely divide the result by 60.

Measuring the Position Angle Of a Comet's Tail

The measurement of the changes in the position angle (PA) of a comet's tail is an important area of study that is best done visually or photographically with wide-field instruments. The angle of the comet's tail from the head depends on the sun-earth-comet positions in space, and these values continually change as the earth and the comet orbit the sun. Some changes or distortion in the tail and its position angle may arise from the comet's velocity around the sun as well as from solar activity, as evidenced in the solar wind.

Measurements of position angle are quite simple to determine if you are equipped with a good set of star charts. Plotting the center of a comet is essential, and should be done as described on pages 70–72. Once the comet has been positioned on the star chart, you can commence measurements to determine both the position angle and the total length of the tail. Position angle measurements are always determined from north (0°) toward east (90°), as shown in Figure 4-4. When most star charts are held upright, north is at the top, east to the left, south at the bottom, and west at the right.

Use the following procedures to determine the position angle:

1. As the tail passes away from the comet, find a star in line with the tail that is visible in the field of view and is also on the star chart.
2. Assuming that the position of the center of the comet's coma has already been marked on the chart, draw a faint line from that point in the direction of the star to indicate the direction of the tail. If no star lies in a direct path from the coma, use stars on either side of the tail and estimate the position.
3. With north representing 0°, east 90°, south 180° and west 270°, measure from north (using a protractor) to the line representing the comet's tail.
4. Record the angle in degrees. For example, in Figure 4-4 the main tail is about 48° from north. Therefore, its position angle is 48°.
5. After your eyes have become fully dark-adapted look carefully for secondary tail streamers, and measure any that you find. The antitail in Figure 4-4 is 182° from north (B), hence its position angle is 182°.

Recording the changes in the position angles of a comet tail over the duration of that comet's appearance is a valuable contribution that can be made even with a very small telescope or a pair of binoculars.

Measuring the Length of a Comet Tail

Generally, the length of a comet tail is expressed in degrees and decimals of degrees for a large comet, whereas minutes of arc are used when referring to comets with very short tails. The procedure for determining the total length of a comet tail is quite simple, as follows (see Figure 4-4):

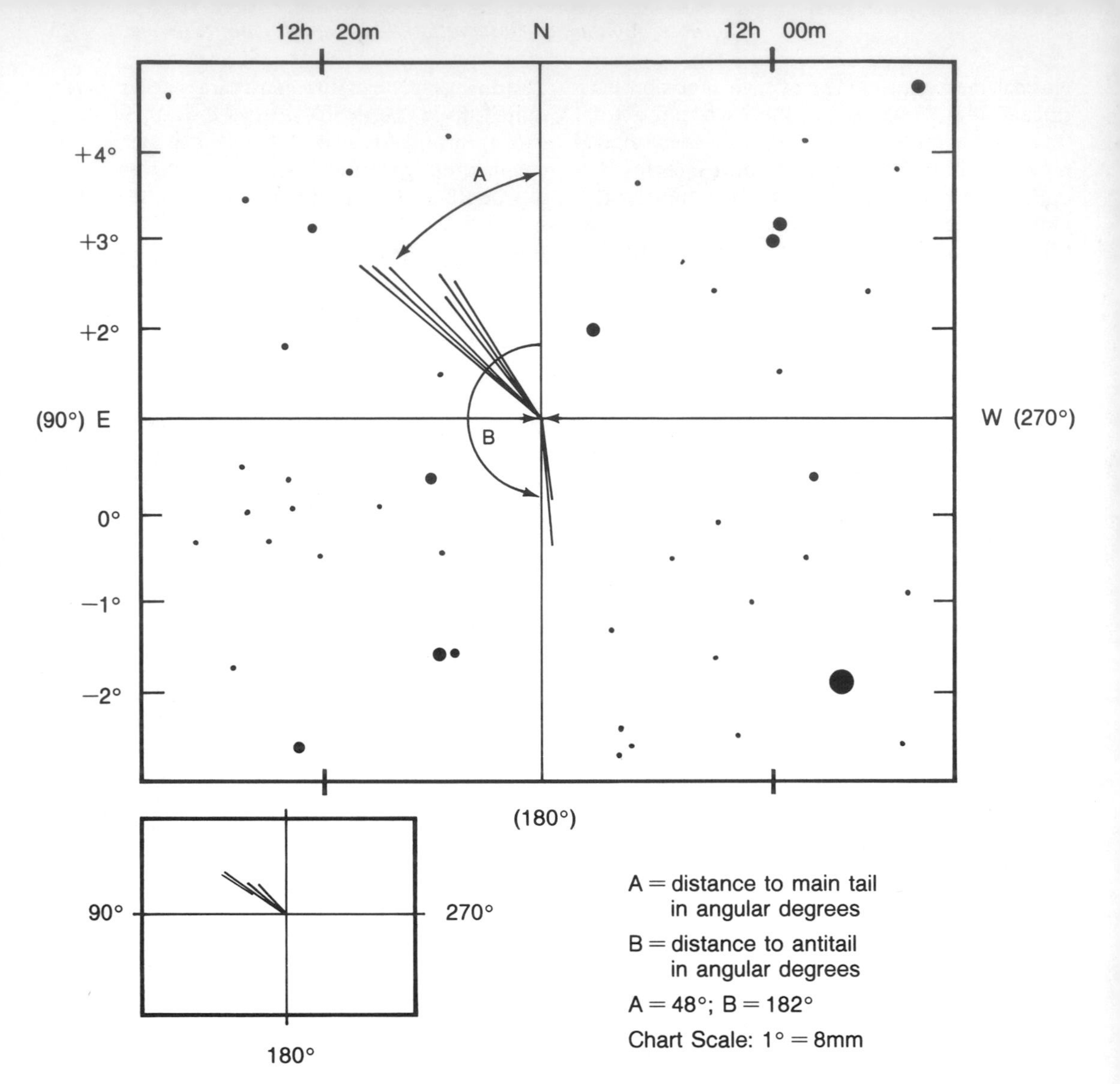

FIGURE 4-4. Plotting the tail of a comet.

1. Suppose that the comet depicted in Figure 4-4 is the one to be measured.
2. Sketch the tail of the comet just as it appears in the telescope (or binoculars), carefully noting the faintest extensions toward the end of the tail.
3. Using field stars for reference marks (as was done in measuring the position angle), you can accurately place the tail in relation to the comet's center on the chart by stepping off the distance from stars appearing on the chart and in the eyepiece.
4. Once you feel that the tail is rendered at the proper length, measure it on the chart, using a millimeter scale. Notice that the example chart in Figure 4-4 has a scale of 1° = 8mm.
5. The millimeter scale reveals that the comet tail is 33 mm in length on this chart. A simple proportional equation gives:

$$\frac{1^\circ}{8\text{ mm}} \times \frac{\ell \text{ (actual tail length in }^\circ)}{33\text{ mm}}$$

or,

$$33\text{ mm} \times 1^\circ = 8\text{ mm} \times \ell$$

where

$$\ell = 4^\circ.1$$

Remember that all star atlases have different scales; the exact scale of each must be determined starting your calculations. If secondary tails (i.e., antitails), emanate from the comet's head, measure their length too. In Figure 4-4 the antitail measures 18 mm. Thus its length is 2°.2.

Another quick method by which the length of longer comet tails may be measured is to know or to have readily accessible the known distances in degrees of several bright stars. As a starting point, remember that the full moon's disk measures 0°.5. The distances between stars given in Table 4-3 are useful.

Figure 4-5 shows the north circumpolar sequence and is helpful in visualizing the positions of the stars given in Table 4-3 in relation to one another. The advantage of this sequence of stars is that from the midnorthern latitudes to the North Pole, these stars are always above an unrestricted northern horizon, and thus can always be used for comparison.

Determination of Position

As previously noted, as an amateur, you need to make visually only rough determinations of position—and these only for the sake of plotting the comet and for record. The professional can determine position with a precision of better than 0.1″, so it is fruitless for the amateur to attempt to attain an accurate measurement. The position that you determine should, however, be included in your complete report, as well as any note of large discrepancies from the predicted position. The position of the comet in Figure 4-4 can be measured as: right ascension = 12h 10m Declination = +1°. For the amateur to achieve accuracy of position exceeding 2′ in right ascension and 5′ arc in declination is quite difficult, and need not be attempted.

Other General Information

Drawings of Comets

Try to make drawings on every date if time allows, but don't attempt these drawings at the sacrifice of other quality information. A drawing can be made inside a circle of about 3.5 inches (89 mm) diameter; the circle will represent the field of view. Drawing pencils of good quality aid in the rendering of the fine contrasts that exist in the coma and tail. If the nucleus of the comet splits into component parts, sequential drawings are of great importance because photographs do not reveal the fine activity near the center of the comet.

Occultations of Stars by Comets

If it appears that the track of a comet may result in its head or nucleus occulting (i.e., eclipsing) a star, give notice of this possibility to any observatory in your area. If you are reasonably certain that a star occultation will occur, set up a tape recorder and a shortwave

Table 4-3. Angular distances between stars.

Stars	*Distance*	*Stars*	*Distance*
Alpha Cas to Gamma Cas	4.7°	Beta Cas to Epsilon Cas	13.3°
Alpha UMa to Beta UMa	5.4°	Alpha UMa to Epsilon UMa	15.2°
Gamma Cas to Beta Cas	6.2°	Polaris to Gamma Cas	28.6°
Gamma Cas to Epsilon Cas	7.3°	Polaris to Alpha UMa	28.7°
Beta UMa to Delta UMa	10.1°	Polaris to Delta UMa	33.6°
Alpha UMa to Delta UMa	10.2°	Polaris to Beta UMa	34.1°

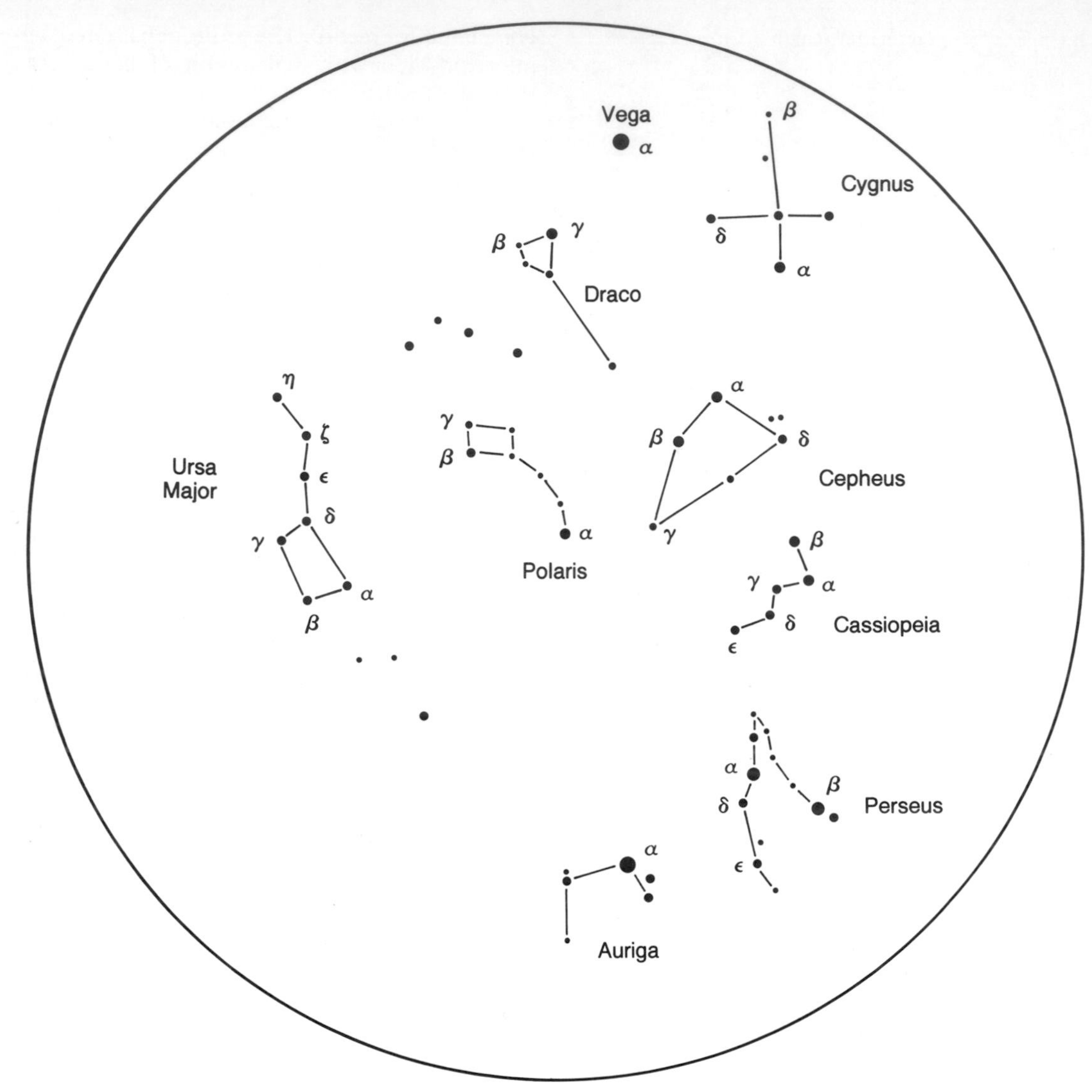

FIGURE 4-5. North circumpolar sequence.

radio that broadcasts WWV time signals at the observing site. Start the recorder to record the radio signals prior to the predicted time of the event. If any dimming or disappearance of the star occurs while it is passing behind the comet's coma or nucleus, indicate the exact moment by voice signal. Do the same if the star reappears or brightens. You can later extrapolate this information from the recording to accuracy of within one-tenth of a second.

Unusual Activity

Many comets exhibit unpredictable activity, such as bursts in brightness, explosions, sudden dimmings, splitting nuclei, strongly distorted tail structure, and so forth. In the event of such activity, send immediate notification to the IAU Circulars at the address previously given, and make subsequent observations of any unusual activity at every available moment.

Keeping Records

Meaningful observations are made by an observer who keeps a good set of records. Give top priority to your system of keeping records and the reporting of your observations of comets. For every new comet, it is a good practice to begin a new notebook devoted only to that one comet. Your astronomy library will soon be filled with your own observing reports.

When at the telescope, do not trust your memory to hold what your eye has seen. Record the information when you see it. Although every observer keeps records in somewhat of a personalized fashion, you should record the following data on every night of observation. These items are the summary of all the research thus far discussed:

- ☆ Name and address of observer.
- ☆ Optical equipment used.
- ☆ Seeing conditions on a scale of 1 to 5 (1 being the worst, 5 the best).
- ☆ Transparency on a scale of 1 to 6. These numbers correspond to the magnitude of the faintest star visible to the unaided eye. That is, if you can see only to the fourth magnitude, transparency is 4.
- ☆ Degree of nucleus condensation.
- ☆ Tail length in degrees or minutes.
- ☆ Position angle of tail and components.
- ☆ Nucleus magnitude (if any).
- ☆ Total (coma) magnitude of the comet.
- ☆ Size of the coma, in minutes of arc.
- ☆ General description, including unusual activity.
- ☆ A drawing of the appearance of the comet if time allows.

Appendix VII is an example of a standard reporting form as used by the Association of Lunar and Planetary Observers for comet observing. Such a form is ideal for adoption by any amateur astronomer. For those observations to be of any value to the scientific community, they must be seen by those who can use the information. Each week, or every five dates of observations, send complete data to the following:

Smithsonian Astrophysical Observatory
60 Garden Street
Cambridge, MA 02138

Comets Section
ALPO
Box 3AZ
Las Cruces, NM 88003

PHOTOGRAPHY OF COMETS

Other than a scientific contribution to research, perhaps the most rewarding experience of the amateur comet observer is the compilation of a series of photographs that show the approach and recession of a great comet. In many cases sophisticated equipment is not necessary, nor can it be used to any advantage in such a project. Although celestial photography is discussed in detail in Chapter 14, the photography of comets has unique aspects that should be discussed here to emphasize the primary area of comet photography suitable for amateur research: the wide-field cataloging of comet development using a standard camera format.

As in visual studies of comets, no photographic study is of any value if accurate records are not kept of the following data:

- ☆ Date and time (UT).
- ☆ Visual magnitude of the comet.
- ☆ Approximate position.
- ☆ Exposure length.
- ☆ Film type and ASA rating.
- ☆ Camera format and f/ratio of lens.

A comet is unique for every apparition, whether it is a short- or long-period visitor. Normally you do not have a second chance with comets. You must take full advantage of every opportunity to photograph a comet, or valuable data are lost forever. Therefore, if you wish to compile a scientifically useful catalog of the development of a comet's approach to the sun, you must be prepared to devote a great deal of time to this project. Not only is time required to actually take the photographs, but it is also required for the processing and printing, as well as the cataloging of each print. The effort is great, but so are the reward and satisfaction of having contributed to the information on such a little-known subject of space. Once the catalog is compiled (or the photographs are obtained), results should be forwarded to the addresses previously given on page 79.

Instrumentation for Comet Photography

The simplest photograph of a comet that I have seen is an excellent portrait of Comet Bennett (1969) made by a Polaroid Land Camera. The fast (4000 ASA) black-and-white film enabled the photographer to record about 10° of tail, not by a time exposure but by the shutter-controlled electric eye of the camera. Because the exposure was short (less than 1/8 sec), there was no reason for the photographer to mount the camera on a tripod. A snapshot of a bright comet was fully developed in only 3 min.

Such a compelling arrangement may be suitable for bright comets with long tails, but it is not suitable if you wish to compile the comet catalog described here and to record as many pre- and post-perihelion images as possible. Most amateurs own or have access to a 35 mm, single-lens reflex (SLR) camera. Such cameras are now moderately priced, within most amateurs' budgets and can be used in a variety of ways—both astronomically and terrestrially—that more than justify the expenditure.

Besides a suitable camera, little else is needed in the way of instrumentation for a cataloging project. It is advantageous, particularly for photographing the fainter comets, to have the camera mounted equatorially and for it to be motor driven. A telephoto lens is sometimes an added advantage. However, if these are not available, the camera can be mounted on a tripod, and exposures of 3 to 5 min with a normal lens (e.g., 50 mm f/1.8) can be made with little drift being evident for the duration of the exposure.

The Single-Lens Reflex Camera

The single lens reflex (SLR) camera has many advantages for celestial photography including those discussed in the following several paragraphs:

Through-the-lens focusing and framing. Such focusing and framing enable the photographer to frame and focus precisely on the object to be photographed. However, as discussed in the following pages, if the lens is in good adjustment, all comet and celestial photographs are taken with the focus set on infinity (∞). Thus, only one setting is needed for focusing once the setting has been checked for adjustment.

Capability of "bulb" or time exposures. Almost all makes and models of SLR cameras have the capability for time exposures. Because comet photography requires exposures that are usually longer than 1 min, you must make some provision in the camera system to hold the shutter open for a required length of time. With the "bulb" setting (marked "B" on the camera exposure knob) that is on most cameras, you must use a cable-release mechanism that can lock the exposure button of the camera down, thus keeping the shutter open as shown in Figure 4-6.

Figure 4-6b shows the setting dial typical of most cameras that has been set to B, or bulb, for a time exposure. Once this setting has been made, you need only push the cable button (B in Figure 4-6a), which subsequently depresses the camera button (C). While you hold button (B) down with your thumb, tighten the locking knob (A) to hold the cable down for the desired length of time. It is essential that you secure the locking knob or the cable will not hold the camera's shutter open. Release the locking knob gently at the end of the required exposure. When tightening or loosening the locking knob, take great care in handling the cable to prevent any vibration or shake that might ruin an otherwise good photograph.

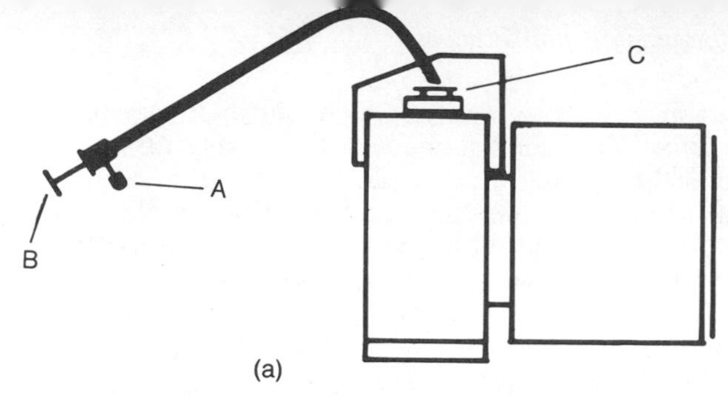

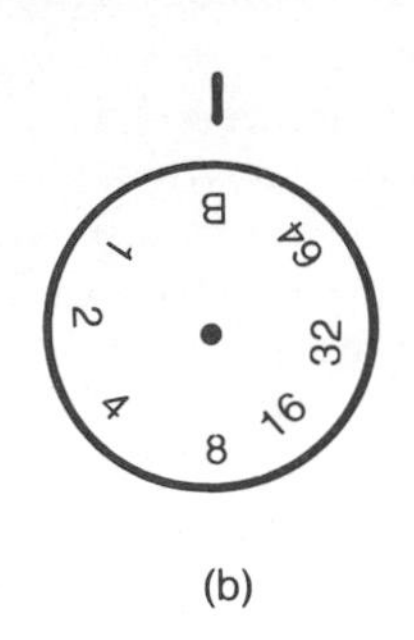

FIGURE 4-6. Camera with cable release (a) and bulb setting (b).

Aperture and focal length of lens. Most SLR cameras feature the capability of being interchangeable with a wide variety of lenses, thus increasing the utility of the photographic system for comets. Standard lenses on almost all SLR cameras have apertures of about 50 mm and focal lengths of about the same, distinct advantages of the SLR over the "instant" type of camera. Such lenses are able to reach faint stellar magnitude near the limit of 7 X 50 binoculars in good skies, and their resolution is quite good. The amateur astrophotographer who is shopping for a camera should certainly choose an SLR from which the lenses can be removed and interchanged.

Availability of films. The 35-mm film format is perhaps the best known and most widely used of all film sizes, a fact that has resulted in the commercial marketing of hundreds of specialized black-and-white and color emulsions, all of which are comparatively inexpensive and quite readily found in most stores. Emulsions ranging from exotic spectroscopic to infrared can be obtained in every major city.

Ease and cost of processing. Processing 35-mm film is an easy and exciting amateur endeavor. The film is the easiest to handle and one of the least expensive to process and print because the equipment and quantity of chemicals necessary are quite small when compared to those required for larger film formats.

Choosing Lenses for Comet Photography

Any lens can be put to some use in comet photography, but telephoto lenses are like telescopes—every type has distinct advantages and disadvantages. The notion that a "big one" is better than a "little one" is not necessarily true.

The f/ratio (i.e., f/1.8) of the camera lens represents the aperture in proportion to the length of focus of the lens. Larger lenses, such as telephoto types, usually have a greater f/ratio than smaller, standard lenses. The larger the f/ratio, the slower the lens will record the light of a comet. A lens of f/1.4 will record the comet much faster than will a lens set to f/3.5. Speed is a desirable factor to consider when you choose a comet lens. However, speed is only one criterion of comet photography; you also need to consider the *resolution* of the system and the effect that *sky glow* from city lights will have on the faster lens. A fast lens is affected more quickly by sky glow from artificial lighting or haze than a somewhat slower lens, allowing the brightness of the sky glow to be recorded much faster than the light of the comet.

Any lens used for comet photography should be used wide open, or on the fastest f/ratio setting. However, very fast lenses give poor star images when used at their fastest settings. Very expensive f/1.4 or f/1.2 lenses are the poorest choice for star photography. If such lenses are used, it is best to stop down by one f/stop to correct for some of this aberration.

When you compile a photographic history of a comet, it is essential to estimate the length of the tail for any given date and use a suitable lens that can adequately include that total length. Trial and error result in wasted time and film, so the following table is included to help match the comet length with the focal length of the lens best suited to provide the optimum image of the entire comet.

However, if you choose to photograph a *consistent history* of the development of a comet, it is best to stay with one lens that will offer the widest chronological span so that the scale of all photographs, and thus the scale and actual size of the comet, remain consistent. Which lens you choose not only depends on the length of the comet but also on the magnitude range and the size you wish to cover. The longer the focal length of the lens, the fainter the comet can be when recorded during guided time exposures, and the smaller the image that can be recorded. The following lists give the minimum magnitude of comets and the appropriate lens size sufficient to record detail suitable for a scientific catalog:

Comet Tail Length	*Proper Lens Size*
45°-70°	28 mm
28 -40	50
16 -24	85
13 -19	105
10 -15	135
9 -13	150
6 -10	200
5 - 8	250
4 - 6	300
3 - 5	400
2 - 4	500

Generally speaking, brighter comets are also the longest and will not suitably fit in the fields of telephoto lenses of very long focal length (i.e., 500 mm and greater). Because the comet is usually most active while at its brightest (i.e., during perihelion passage) and during the week prior to and following perihelion passage, a photographic patrol is valuable during that time—about two weeks. When photographing a bright comet in that period, a standard lens of 50-mm focal length that combines fairly fast recording power with a wide angle of view is the best choice.

Minimum Magnitude of Comet	*Lens Focal Length*
+2.5	28 mm
+4.0	52
+6.0	105
+8.0	300

However, I have consistently found that the best lenses for all comets I have seen are those in the 200-mm and 300-mm range of focal length (see Figure 4–7).

Checking a lens for infinity focusing. Before you use any lens for astronomical photography, it should be checked to see that it will focus to

FIGURE 4–7. Diagram of camera lenses and equivalent fields of view.

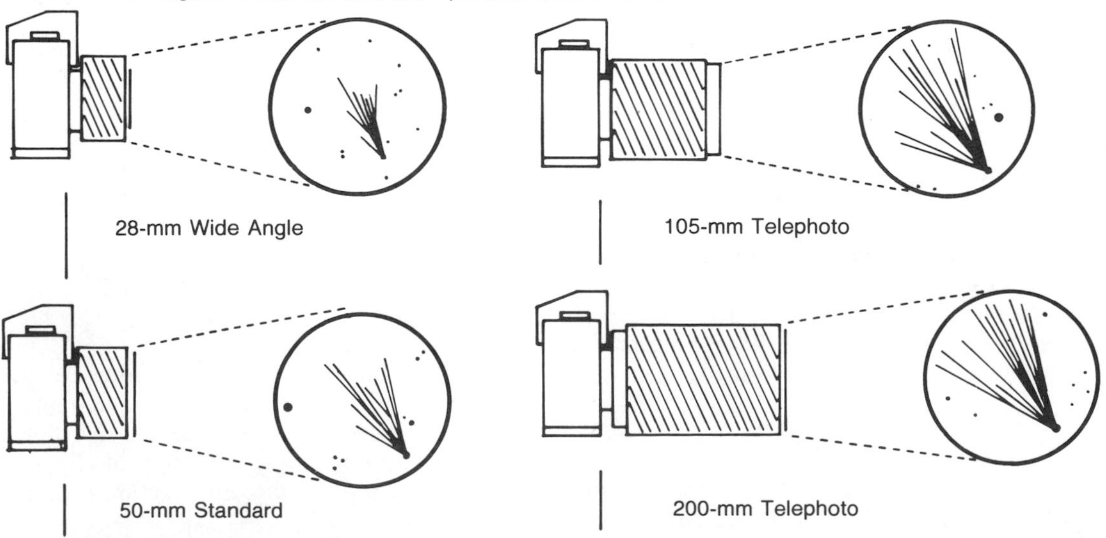

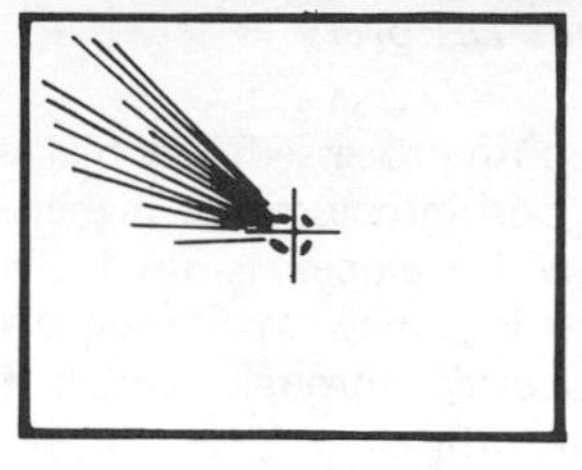

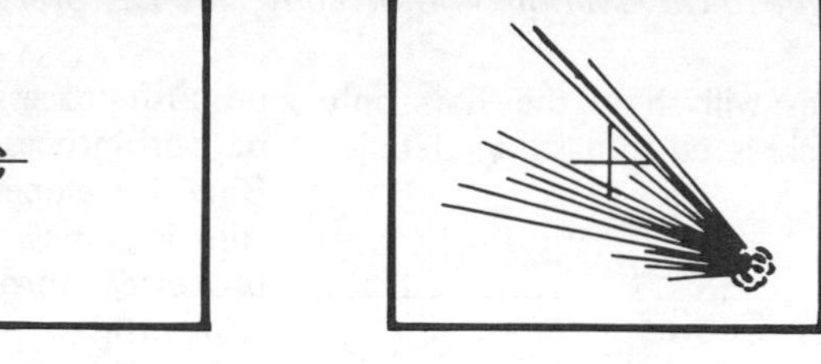

Incorrect Correct

FIGURE 4-8. Positioning a comet in the camera view finder.

infinity, the setting for celestial photography. You can check for infinity focusing by placing a camera on a tripod and viewing a bright field of stars, such as the Pleiades, with the lens setting placed at infinity. If the stars appear very sharp, the lens is focused at infinity. If the stars appear fuzzy, however, slowly back the focus away from the infinity point and see if the focus improves. (Remember that the focusing screens in most cameras cause star images to be a bit fuzzy anyway.) If it does not, then the lens suffers from improper factory adjustment for the stop at infinity and needs to be focused past the point where the lens stops. If this is the case, the lens can be reset only at the factory, and no adjustment should be attempted at home. A guided photograph of a star field, with the lens set to infinity, is the best method by which this check can be made. It is certainly worth the effort to make the check before the appearance of a bright comet.

Taking the Photograph of a Comet

Tripod-Mounted Camera

The use of a tripod-mounted camera is suitable for exposures using the standard 48-mm to 52-mm lenses with a focal ratio of about f/1.8. Comets brighter than magnitude 3.0 can be suitably recorded with such a system in five min or less, assuming that a fast black-and-white film is used. For any camera not equipped to guide on the comet, it is best to choose the fastest film possible, such as ASA 400 black-and-white or color film, in order for the image to be adequately recorded in the short exposure. Longer exposures merely allow the image to trail.

The only special attention that must be given to comet photography using a tripod is to avoid any vibration in the system. Make more than one photograph each night, recording all pertinent information for each exposure, to assure a back-up shot should one fail. Use a cable release.

Take additional care to position the comet properly in the viewfinder of the camera. Remember that when you use an SLR camera, what you see in the viewfinder is precisely what will record on film. Many comet photographers make the mistake of positioning the *head* of the comet in the viewfinder because the head is the center of the comet. As much of the physical structure of a comet as possible should be included in any cataloging project, so it is essential that the camera be positioned in such a way as not to chop off a section of the tail (see Figure 4-8).

Equatorially Mounted Camera

Basically, the same procedures that apply to a tripod-mounted camera apply to the equatorially mounted camera, except that there is the distinct advantage of being able to use telephoto lens with the equatorial mount, thus increasing the range of contrast, resolution, and magnitude.

For best results, take all comet photographs from some type of piggyback mounted system (as described in Chapter 14) that is motor-driven. Guiding these photographs—other than the operation of the clock drive—is not necessary provided you have aligned the equatorial mount well to your location and keep the exposures to 10 min or less. However, guide any lens greater than 105 mm (regardless of the length of the exposure) by a telescope or some other method of tracking. Guide such exposures by using the *comet* as your target. Do not make the field stars the target, as is done in other types of astrophotography, because the comet's drift against those stars as it rapidly circles the sun can be appreciable. If you guide on the comet, the resulting photographs will show the stars only as slight trails, which is customary in detailed studies.

Exposure and Film Choice

The length of exposure needed for a comet photograph depends on the focal length of the lens, as previously mentioned, and is subject to the following factors as well:

- ☆ The speed of the film (ASA).
- ☆ The brightness of the comet.
- ☆ Angular distance of the comet from the horizon.
- ☆ Moonlight.
- ☆ Sky glow, haze, artificial lighting.

Choice of Film

Choose only fast (ASA 400, or faster) film for comet photography; film that is somewhat red-sensitive is best. The film types given in Table 4–4 are recommended. All the films in Table 4–4 except Spectroscopic 103aE (red-sensitive) can be push-processed. The black-and-white films can be push-processed to higher speeds best if Ethol Blue Developer is used. The Ektachrome 400 film is somewhat red-sensitive and can be push-processed through Kodak to twice its normal ASA rating.

TABLE 4–4. Films for comet photography.

Black and White	*ASA*	*Color*	*ASA*
Spectroscopic 103aE	125	Ektachrome 400	400
Kodak Tri-X	400	Ekta. 400 Pushed	800
Tri-X Pushed	1600		
Kodak Plus-X	125		
Plus-X Pushed	600		

Comets record well in the red region of the spectrum, and consequently the red-sensitive films are recommended. Another way to enhance a comet photograph is to use a Wratten 25 (red) filter over the camera lens and a red-sensitive film such as the black-and-white Spectroscopic 103aE, or the color Ektachrome 400, which give the added advantage of blocking haze and extraneous lighting in city conditions. Most such interference is emitted in blue.

Angular Distance of the Comet Above the Horizon

Another factor that greatly affects the length of exposure of a comet photograph is the amount of absorption of the comet's light by the earth's atmosphere. Comets are always quite close to the horizon in night skies, so a great deal of *extinction* of the comet's light is always evident. The amount of extinction depends on the height above the horizon at which the comet is seen, the geographical location of the observer, and the condition of the air during the photograph. A photograph of a bright comet only 5° above the horizon might require four times the exposure that it would when at a height of 25°.

Table 4–5 gives approximate values for exposure lengths using various film and lens combinations in relation to the magnitude of a comet. This table assumes average conditions and an approximate altitude of the comet of 20°.

SEARCHING FOR NEW COMETS

Every amateur astronomer dreams of discovering a comet—not just for the worldwide recognition, but for the thrill of seeing for the first time what no person has ever before seen. Comet discovery usually does not just happen by luck or chance. The reward for such a discovery is the satisfaction that comes to those who search the sky long and arduously for that new visitor. The prize is the result of long hours spent during months or even years of cold morning scanning, devoted concentration, and many disappointing "comets" that turned out to be nebulae, optical reflections, or atmospheric glows.

Instrumentation for the Comet Seeker

If you wish to search seriously for new comets, a telescope with low power (about 4x for every

TABLE 4–5. Estimated film and lens combinations in relation to the magnitude of a comet.

Comet Magnitude	*Lens Type[a] (f/ratio)*	*Film ASA*	*Exposure Time (in min)*
0	1.8	125	2– 5
	1.8	400	1– 5
	1.8	1600	1
0	3.0	125	5–10
	3.0	400	2– 5
	3.0	1600	1– 3
+1	1.8	125	3– 7
	1.8	400	2– 5
	1.8	1600	2
+1	3.0	125	7–12
	3.0	400	3– 7
	3.0	1600	2– 5
+2	1.8	125	5–15
	1.8	400	3–10
	1.8	1600	3
+2	3.0	125	10–15
	3.0	400	5–10
	3.0	1600	5– 7
+3	1.8	125	10–20
	1.8	400	7–12
	1.8	1600	4
+3	3.0	125	15–20 (103aE only)
	3.0	400	7–12
	3.0	1600	5– 7
+4	1.8	125	15–20 (103aE only)
	1.8	400	10–15
	1.8	1600	5
+4	3.0	125	25 (103aE only)
	3.0	400	10–15
	3.0	1600	7–12
+5[a]			
+5	3.0	125	30 (103aE only)
	3.0	400	15
	3.0	1600	10
+6	3.0	125	35 (103aE only)
	3.0	400	18
	3.0	1600	12
+7	3.0	125	45 (103aE only)
	3.0	400	22
	3.0	1600	15

[a]The f/1.8 lens is assumed to be a standard lens of about 50 mm focal length. For comets of magnitude +4 and fainter, such a lens is not suitable because of the small image scale of the comet at those magnitudes.

inch of aperture of the objective) and a wide field of view should be used. Almost any telescope can be used for comet searching, but the search is laborious and tedious if a narrow field of view is used. If a standard astronomical telescope (f/8 to f/15) is used for searching, use a power as low as possible so that you can sweep a greater area of sky in a minimum amount of time. An equatorial mounting is not needed, and it is actually a hindrance in methodical sweeping. A simple but sturdy altazimuth mounting is preferred so that the field of sweeping follows parallel to the horizon and not in arcs, as does the equatorial mount.

At the telescope, it is necessary that you alternate between low power (4x per inch of aperture) for sweeping the sky, to moderate power (10x per inch of aperture) for examining any objects suspected of being comets. In addition, you should have ready access to high-power eyepieces so that you can attempt to find the true nature of a nebulous object that might actually be groups of stars. A very faint close double star can appear nebulous even to an experienced comet observer.

It is essential that the comet searcher operate in comfort. This does not mean that you should take an easy chair outside and relax so much that you become lazy or fall asleep. A comfortable chair can be designed to accommodate both winter and summer viewing by having warm winter padding on the seat that can be simply unsnapped or turned over for warmer weather. When you are using this chair, the telescope pedestal or tripod should be fixed in height so that your eye is comfortable behind the telescope when the telescope is pointed from 15° to 25° above either the east or west horizon. Because most searching for comets is done in these directions, begin there for your comfort while observing. Seek to achieve as little physical and mental fatigue as possible in your observing setup.

Star Charts and Miscellaneous Equipment

For comet searches, put together a kit so that all necessary accessory equipment and charts are contained in one place, including the following:

- ☆ Star charts.
- ☆ Observing notebook.
- ☆ Flashlight with red filter.
- ☆ Eyepieces giving
  4x, per inch aperture
  10x, per inch aperture
  25x + per inch aperture.
- ☆ An accurate watch or clock.
- ☆ Pencils.

Your observing accessory kit should accompany you to the telescope for every observing session. Suppose for a moment that you are observing and do not have an accessory kit with all the right gadgets. Suddenly you think you have found a comet. Quickly, you run to the house to get the star charts to confirm the nature of the object and its position; and you return just as the "comet" is drifting off the edge of the telescopic field of view. By recentering and checking with the star charts, you decide that the object is, indeed, a comet. You dash back to the house to find a pencil to mark the position of the comet on the star chart. Frantically, you search drawers and finally find a broken nub of a pencil, you have no sharpener handy, so you rush into the kitchen to find a knife to sharpen your pencil. In the process of groping around in the dark for a knife, you cut your finger, which is then hurriedly washed and bandaged. Quickly, you continue to sharpen the pencil and rush back outside to the telescope only to find that dawn began 10 minutes ago and that the comet is lost in morning twilight.

The star chart is used to confirm the nature of any suspicious objects; hence, it should show the positions and relative magnitudes of all clusters and nebulae that might seem comet-like when viewed through a small telescope. The Skalnate Pleso *Atlas of the Heavens*, by Becvar, is excellent for comet searching; it contains stars to magnitude 7 and almost all nebulae and clusters visible in amateur instruments. The deluxe edition of this atlas is a bit cumbersome to use at the telescope, but a smaller field edition is available that shows white stars on a black background. Another favorite atlas is Hans Vehrenberg's *Handbook of the Constellations*, which lists stars to binocular limit and clusters and nebulae to magnitude 12. The advantage of this atlas is that a table faces each chart to aid in singling out every constellation individually. It is my personal preference for such work.

In addition to quick confirmation of the nature of a cometary suspect, a good atlas provides a rough position of the object to be plotted, as well as the direction of any drift, both of which are essential when reporting a claim or attempting to relocate the object on the next clear date.

The Searching Procedure

Make searches for comets in the western sky about 45 minutes after sunset, and in the eastern sky three hours before sunrise. Most comets in the morning sky have been discovered visually, most of them being concentrated along the ecliptic. This fact might make you think you should have little trouble locating a new comet in just a few days. Because of the faintness and scarcity of comets, however, most are found after at least *100 hours* of searching. If you live in a temperate area, weather conditions might make it possible for you to observe on 55% to 60% of the mornings each year, a total of about 219 dates on which to search (including those in which the moon might be present). If you were to search *every clear morning of every year,* statistically you should discover 6.5 comets each year. However, any seasoned comet searcher will tell you it is just not so. T. Seki of Japan required *10 years* of dedicated searching before he found his first comet.

In the evening sky, when no moon is present, begin to scan just as twilight ends (about 45 minutes after sunset). Begin the search 45° north of the sunset point and scan slowly across the lowest point of the horizon at a rate no faster than 0.5° per sec. Continue sweeping in a straight (horizontal) path, moving north to south until the telescope points about 45° south of the sunset point. One scan from north to south requires about 3 min if you do not stop to examine objects more closely. If you complete a scan in less time, you are moving too fast. Now, raise the telescope up from the horizon about half of the field of view and begin sweeping back toward the north. Continue this overlapped sweeping until an altitude of about 45° above the western horizon is attained. If you have a 2° field of view, every search will require about 2.5 hours, if you do not stop for any reason.

The procedure for searching the early morning sky is similar, except that you begin above the eastern horizon, about three hours before sunrise, 45° *above* the horizon, and work *down* rather than up, sweeping from 45° south of the sunrise point to 45° north of it. To repeat: the chances of discovering a comet in the morning are about 75% greater than in the evening. Morning observing, of course, requires waking at about 3:00 AM every clear morning, which is incompat-

FIGURE 4-9. Searching for a comet on the western horizon.

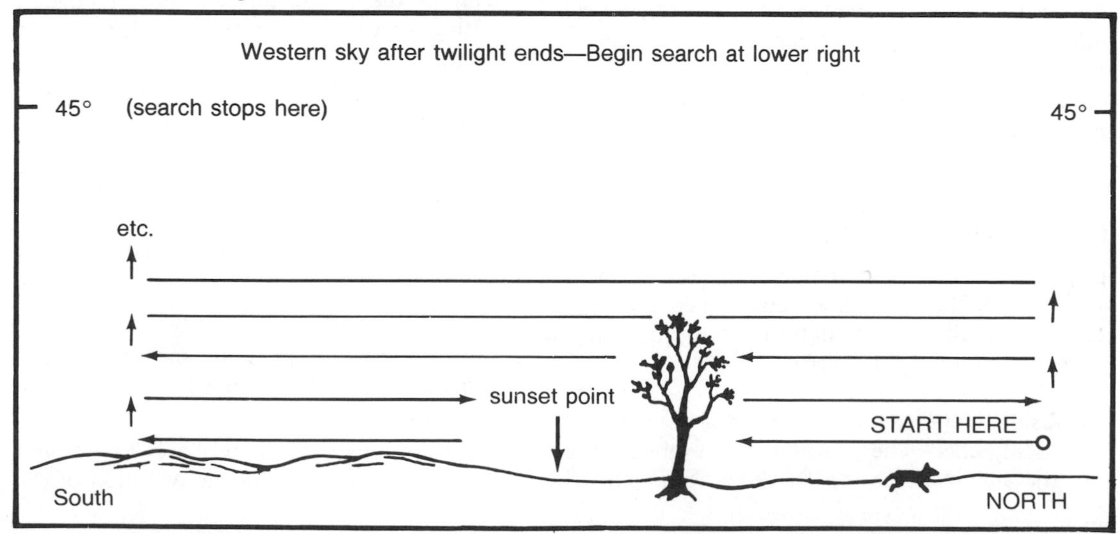

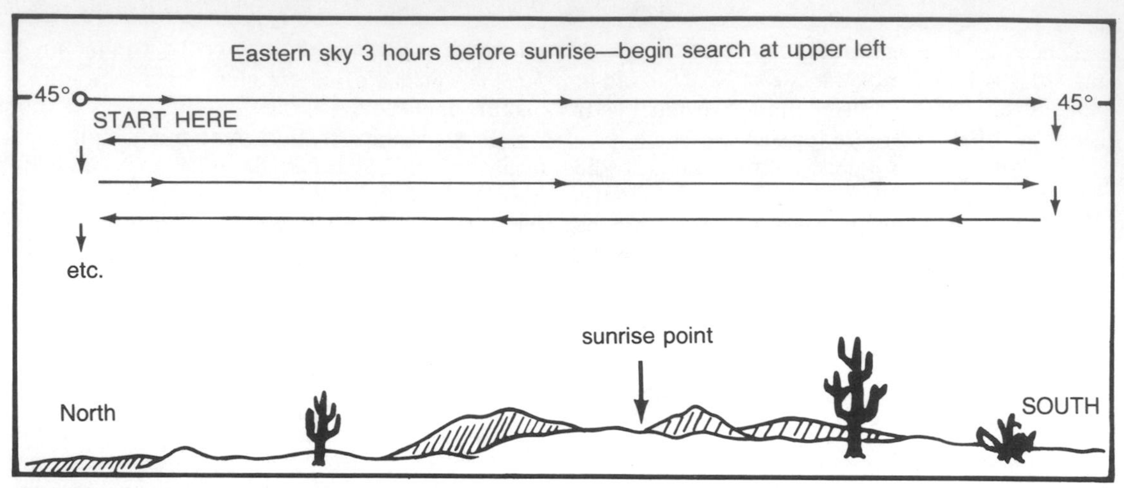

FIGURE 4-10. Searching for a comet on the eastern horizon.

ible with the nature of many humans. Search time for morning sweeps also requires about 2.5 hours. Don't rush it.

Another area of the sky in which you may search for comets is that *opposite the sun*. Concentrating first on the ecliptic, scan about 30° east and 30° west of a point that will be overhead around midnight (12h of right ascension from the sun). Determining this location beforehand on a star chart will provide you with a starting and stopping point (using reference stars) and convenient boundaries in which to confine the search. The opposition point for such comets will be on the eastern horizon at sunset, directly overhead near midnight, and setting at sunrise. Another area in which to search for these opposition comets is near the celestial pole around midnight. Persons with instruments of large aperture (i.e., 12 inches or larger) undertake regular searches of these additional areas, which are promising fields for the discovery of faint comets.

Confirming a Suspected Comet

If you sight an object that might possibly be a comet, take the following steps immediately to confirm or disprove its nature before notifying the world of your "discovery":

1. Using reference stars (if setting circles are not used), check the position of the object on the star chart to see if any known nebulae, galaxy, or cluster is near the position you have determined for this "comet".
2. If not, check with higher magnification to see if the object can be resolved into stars; many times tiny multiple star systems appear nebulous.
3. If no known object exists at the position of the object, you have found, and if high magnification reveals no stars, mark the position of the "comet" on the star chart. If an object is on the chart at that position, don't become discouraged and go inside. Continue your scheduled sweeping.
4. Once the position of the suspected comet is designated on the chart, sketch the star field seen in the telescope on a piece of paper, noting very carefully the position of the object relative to close faint stars.
5. Estimate the magnitude of the "comet" using field stars, and indicate the magnitude in the observing notebook. The approximate magnitude of stars can be extrapolated from the charts, or from your catalogs.
6. Note any tail, nucleus, or unusual nature of the object.
7. Make a note of the correct Universal Time.
8. Wait as long as possible and notice any apparent drift of the object relative to the stars drawn in Step 4.
9. If drift is apparent, you probably have found a comet. Note carefully the direction of the

drift, being very sure of directions in your telescope. Directions are inverted (i.e., they appear as they would in a mirror) in many telescopes.

10. If drift is *not* apparent, it does not necessarily mean that the object is not a comet, particularly if no object is known to exist in that position. It could be a very slow-moving distant comet. Making sure of its position, locate the "comet" again on the following night. If it has moved relative to the stars in that length of time, it most likely is a comet. If it has not moved relative to the stars, as noted the previous night in Step 4, it most likely is not. If motion is evident, go back to Step 9.
11. Check journals and IAU circulars to make sure that the comet is not a known periodic or recently discovered comet.
12. If it is neither, then you are ready to report the discovery of *your* comet.

Reporting the Discovery of a New Comet

If you are priviledged to find a new comet, by all means do not shout it out loud to the world too soon. Staking claim on a comet is like staking claim to land: once you find it there are always those who might take it away from you. Such unethical people are known as *comet rustlers* or *cometbaggers*. The only method by which a new comet should be claimed and reported is by telegram. After the telegram has been sent, you are then safe in telling anyone you wish. The telegram should be sent *in code* only in the method described here. Your completed telegram should resemble Figure 4-11. Notice carefully the sequence of numbers shown in Figure 4-11. The following list explains the code of Figure 4-11:

- ☆ bradfield comet bradfield—This simply means that someone with the surname Bradfield is reporting on Comet Bradfield.
- ☆ 19501—Designates that positions are based on 1950 epoch star positions.
- ☆ 60303—The number 6 stands for the year 1976; the next 03 stands for the month of the year (March); the last 03 represents the date of the month.
- ☆ 778—This is the time of the observations in Universal Time (see Appendix I), which has been converted into decimal form (See Appendix II). Consequently, this is actually 0.788 UT, or about 18:41 UT.
- ☆ 21301—This is the right ascension of the comet (21h 30.1m).
- ☆ 14702—This is the declination of the comet (-47° 02′); the number 1 preceding 4702 tells the receiver of the telegram that the reported declination is *south* of the celestial equator. If the comet had been discovered at +47° 02′, this number would have been 04702.
- ☆ 01094—This number represents a brief description of the comet at the time of its discovery. The 01 means that the magnitude was estimated; the next 09 indicates magnitude 9; the 4 describes the degree of condensation of the comet's nucleus. The degree of condensation is judged on a scale of 0 to 9, on which 0 would mean a perfectly uniform coma, with no condensation; 4 would mean condensation; and 9 would mean that the comet had a distinct stellar appearing nucleus.
- ☆ 94701—This number is a check sum, to make

FIGURE 4-11. Announcing the discovery of Comet Bradfield.

bradfield comet bradfield 19501 60303 778 /
21301 14702 01094 94701 37097

19501 60303 799 / 21306 14703 01094
96807 37103

sure that all the numbers thus far discussed have been correctly recorded by both the discoverer and the telegraph service. To determine the check number, simply add up all the preceding numbers in the telegram (19501, 60303, 778, 21301, 14702, and 01094), as shown below:

```
  19501
  60303
0.778
  21301
  14702
+ 01094
 194701
```

Be sure to align the numbers as shown above, because of the decimal point. Then list the *last five* digits (94701) of this sum. The person receiving the telegram will also add the numbers as above. If that person does not arrive at 94701, it will be assumed that one of the figures has been given incorrectly. The discoverer will be contacted to clarify the numbers.

☆ 37097—This is another necessary check sum. It is derived exactly as was the first check sum except that only the *last three* sets of numbers before the check sums (21301, 14702, and 01094) are used. The sum of these three is 37097.

Notice that in the telegram illustrated in Figure 4-11 Bradfield has included a second observation below the one just interpreted. The code is identical, and you can see that the numbers have changed very little. The purpose of the second set is to indicate the amount and direction of the motion of the comet. If possible, these facts should be included with any comet claim. Check the second set of numbers, and see if you can interpret the data included in that set of numbers as was done for the first set.

Once you have determined all the data needed for the telegram and have converted them to proper code, you are ready to telephone Western Union to relay the message of your comet discovery. Make sure the information is accurate and that the operator reads that information back to you. The toll-free number for international telegrams is 1-800-324-5100, which can be reached 24 hours a day, seven days a week. The telegram should be sent to the following address:

Central Bureau for Astronomical Telegrams
Smithsonian Astrophysical Observatory
Rapid Satellite Cambmass
60 Garden Street
Cambridge, MA 02138

Do not try to condense the telegram in any way to save money, and do not make any abbreviations. To do either would only make your data ambiguous. The discovery of a comet is a rare and rewarding experience; do not sacrifice it by pinching pennies.

The Central Bureau, upon receipt of the telegram, will publish the claim in an effort to confirm your discovery. Once confirmed, the comet will bear your surname as well as the chronological label that designates the *year* and the *order* in that year of the discovery of the comet. For example, the first comet discovered in 1981 is designated *1981a,* the sixth comet *1981f,* and so on.

Not every comet searcher has had the exhilarating experience of sending a telegram of discovery. The comets appear every year, seemingly at random and unannounced. Those rugged individuals who look methodically for comets will *someday* find one. The experience of realizing that discovery is surely the greatest in all of astronomy.

BIBLIOGRAPHY

Abell, George, *Realm of the Universe*. New York: Holt, Rinehart & Winston, 1976. Paperback.

Becvar, Antonin, *Skalnate Pleso Atlas of the Heavens*. Cambridge, MA: Sky Publishing Corporation, 1979. Paperback.

Beet, E.A., *Mathematical Astronomy for Ama-*

teurs, pp. 67–71, "The Orbit of a Comet." London: Norton, 1972.

Fredrick, L.W., and R.H Baken, *An Introduction to Astronomy* (8th ed.), pp. 176–184. New York: D. Van Nostrand, 1974.

Jones, Aubrey, *Mathematical Astronomy with a Pocket Calculator*, pp. 179–186. New York: John Wiley, 1978.

Moore, Patrick, *Comets*. New York: Scribner's, 1976.

——, *Astronomy Facts and Feats*, pp. 118–125. London: Guinness Superlatives, 1979.

Morris, Charles S., "The Reduction of Comet Magnitude Estimates and Comet Observing," *Journal of the Association of Lunar and Planetary Observers*, Vol. 24, Nos. 7 and 8 (1973), 150–156.

Muirden, James, *Astronomy With Binoculars*. London: Faber and Faber, 1979.

——, *The Amateur Astronomers' Handbook*, pp. 193–210. New York: Thomas Y. Crowell, 1974.

Middlehurst, Barbara M., and G.P. Kuiper, *The Moon, Meteorites and Comets*, Vol. IV, "The Solar System." Chicago: University of Chicago Press, 1963.

Marsden, Brian G., *Catalog of Cometary Orbits*, (3rd rev. ed.). Cambridge, MA: Center for Astrophysics, 1979. Paperback.

National Aeronautics and Space Administration, *The Study of Comets*, Vols. I and II, SP–393. Washington, DC: Government Printing Office, 1976.

Olivier, Charles P., *Comets*. Baltimore: Williams and Wilkin, 1930.

Rahe, Jurgen, Bertram Donn, and Karl Wurm, *Atlas of Cometary Forms: Structures Near the Nucleus*, NASA SP–198. Washington, DC. Paperback.

Richardson, Robert, *Getting Acquainted with Comets*. New York: McGraw-Hill, 1967.

Richter, Nikolaus, et al., *The Nature of Comets*. London: Methuen, 1963.

Roth, Gunter D., *Astronomy: A Handbook*. Cambridge, MA: Sky Publishing Corp., 1975. Paperback.

Smithsonian Astrophysical Observatory, *Star Atlas*. Cambridge, MA: MIT Press, 1970.

——, *Star Catalog*. Cambridge, MA: MIT Press, 1966.

Vehrenberg, Hans, and D. Blank, *Handbook of the Constellations*. Düsseldorf: Privately published, 1970.

Watson, Fletcher G., *Between the Planets*. Cambridge, MA: Harvard University Press, 1956.

Whipple, Fred L., *The Collected Contributions of Fred L. Whipple*, Vol. I. Cambridge, MA: Smithsonian-Harvard Center for Astrophysics, 1972.

5

AMATEUR STUDIES OF THE SUN

The great turbulence of our sun, which is responsible for life-giving heat and light, goes largely unnoticed except when very sophisticated professional equipment is used to view it. Consequently, many amateur astronomers are disappointed by their first look at the sun and turn their attention to other celestial objects that afford more spectacular views. The solar observer has one distinct advantage, however. His or her work can be done during daylight hours.

Even the most casual observer notices at first glance that the sun's surface is not smooth and uniform but rather pocked with dark spots and faculae, bright areas associated with solar eruptions. Ancient Chinese astronomers, long before the advent of the telescope, noted naked-eye sunspots as the sun rose or approached the western horizon at sunset. The observer who uses a telescope can see even granular surface detail during moments of exceptionally steady seeing. The granular appearance is associated with rapidly rising convection currents in the sun's photosphere.

The most rewarding information that the amateur astronomer can obtain through study of the sun concerns the development and intensity of the sun's activity as measured through the most obvious of solar features—the sunspots. First telescopically seen by Galileo over 300 years ago, sunspots are magnetic anomalies on the sun's visible surface, cooler by several thousand degrees than the surrounding bright photosphere. They appear dark to us only because of the considerably brighter and hotter nature of the photosphere that they are in. If a sunspot could be isolated in a darkened room it would appear to be blindingly bright.

The sunspots are responsible for ionospheric disturbances on earth, resulting in momentary and frequent communications disruption. They also cause the spectacular aurorae borealis (the northern lights) and the aurorae australis (the southern lights). These spots, some of which attain areas hundreds of times the area of the earth, are now thought to be responsible for the earth's weather patterns and even its long-range climatic changes. Yet, the phenomena of the sunspot cycle and the patterns of a sunspot's development are not fully understood. Therefore, conscientious study by professional and amateur astronomers is needed. After all, the sun is the only star in our universe that is suitably placed for our study. It is a very important star from both a scientific and a humanistic viewpoint.

THE SUNSPOT CYCLE

The number of sunspots visible on any given day varies. The sunspots develop, grow, enlarge, and increase in number as if multiplying in the photosphere. The daily variation is somewhat unpredictable, but there are definite cyclic changes in the amount of solar activity that remain constant. The sun undergoes cyclic variations every 11 years; each cycle is fairly predictable when compared to the preceding cycle. It takes about 4.5 years for the sun to go from minimum activity to maximum activity but it takes slightly longer (about 6.5 years) to decline from this maximum back down to minimum. During minimum activity, days pass when not a single spot can be detected, even with high-resolution telescopes. During the maximum part of the sunspot cycle, as many as 20 large groups of sunspots, with a total of more than 300 spots, can be seen. In addition to this easy-to-follow 11-year sunspot cycle, the sun has a longer and more intense cycle of twice the "normal" period, or 22 years. The 11-year cycle is superimposed over the pattern of 22 years. Solar researchers are now investigating the likelihood that there exist even longer and more pronounced cycles on the order of slightly less than 200 years, and perhaps one even as long as 10,000 years. It is these long-term cycles that seem to affect general climatic trends on earth.

METHODS FOR OBSERVING THE SUN

PLEASE NOTE: *Before beginning* **any** *method of observing the sun, make sure all finderscopes and auxiliary scopes are capped at the objective end to block the light of the sun from entering these scopes. Crosshairs in finders will burn or melt in less than 5 seconds.*

Eyepiece Filters

Many small telescopes are supplied with dense filters that attach either to the top or the bottom of an eyepiece. **These filters should never be used, and they are illegal to sell in many states.** Such eyepiece filters are merely welder's glass and do not block the harmful wavelengths of light (infrared and ultraviolet). They have been responsible for severe retinal burns and for glass fragments lodged in the eye. The intensity of the sun is so great that these filters will eventually heat up and crack from the intense heat. Not only can these dangerous items cause eye damage, but damage to the telescope optics—particularly eyepiece elements—is highly possible or even likely. DISCARD ANY EYEPIECE FILTER FOR THE SUN THAT IS SUPPLIED WITH A TELESCOPE. **Never** use such a filter for any purpose. Discard it and find an alternative safe way to view the sun.

Solar Projection

Projecting the image of the sun through a telescope is a safe and widely used method of solar observing. The method has the following advantages:

- ☆ Almost all telescopes are capable of solar projection.
- ☆ It is the most economical method available to the amateur.
- ☆ The method is perfectly safe.
- ☆ Large groups of observers can view a projected image simultaneously.

One may employ varying degrees of sophistication when using solar projection. The simplest way is merely to hold a white card about 12 inches from the eyepiece of an unfiltered telescope. The telescope is focused in the usual manner until the sun's image on the white card is as sharp as possible. Moving the card away from

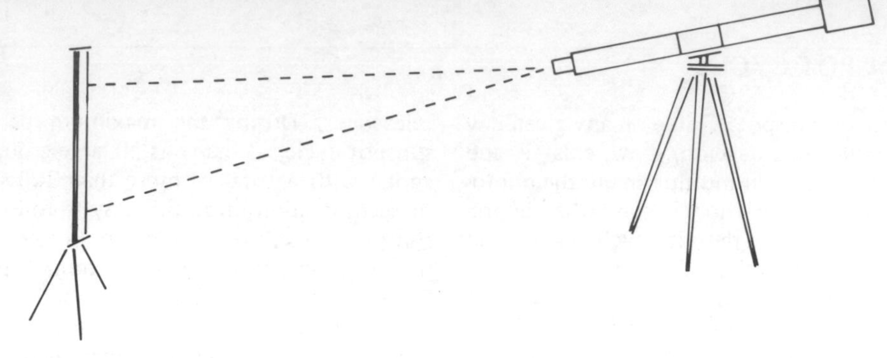

FIGURE 5-1. Projecting onto a movie screen.

the eyepiece results in an enlargement of the image, much as a slide projector is placed at varying distances from the screen according to the amount of enlargement desired. An even better method is to set up a small projection screen, or a large white card with a firm support, as shown in Figure 5-1, on which the image of the sun can be reflected. This provides a fixed distance between the eyepiece and the screen so that a constant focus is maintained. The farther away the screen is placed from the eyepiece, the larger the image is. However, the brightness of the solar disk decreases in proportion to the square of the distance from the screen to the eyepiece. A point will be reached, therefore, at which the image is no longer bright enough to be effectively used. Place the screen at a distance that allows adequate contrast but still close enough to provide an image large enough to enable the viewers to discern fine details.

When you project an image of the sun, block the white screen from any *direct* sunlight that is not focused through the eyepiece. When using a small screen, a simple card with a hole through which the eyepiece can project is fitted around the eyepiece drawtube, as shown in Figure 5-2. This arrangement has the effect of casting a shadow on the projection screen so the maximum contrast can be achieved. More sunspots and faculae are visible with such a shade than without. If a large projection screen is used, such as that illustrated in Figure 5-1, a shade is likely to be

FIGURE 5-2. Using a shade to block direct light.

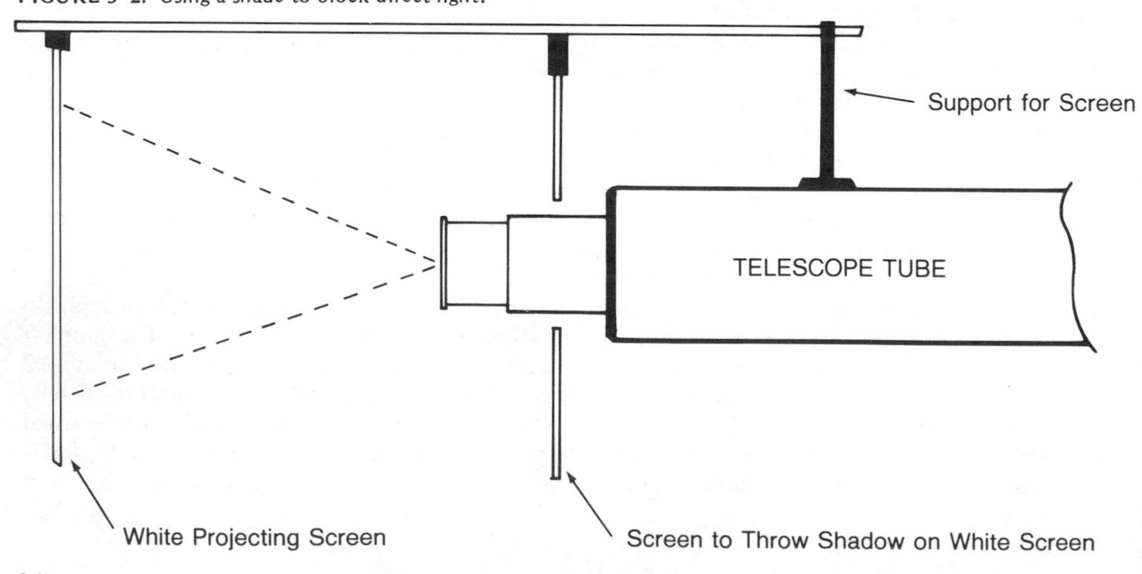

impractical. In such cases, a right-angle prism can be used to divert the image away at a right angle from direct sunlight, thus allowing only the light coming from the telescope eyepiece to be directed to the screen. The screen can be adjusted slightly to cast a shadow for itself. This prevents stray light from the sun from striking the screen. The stray light would tend to lessen contrast.

Although any type of telescope can be used for solar projection, the catadioptic, or compound telescope is not recommended for such a method. Heat buildup within the closed tubes of this mirror–lens system can damage delicate parts within the tube and melt certain cements used in the manufacture of the telescope. Any telescope used for projection should be periodically turned away from the sun for a minute or two to prevent excessive internal heating of both the telescope and the eyepiece. Turning the instrument away once every 10 minutes is usually adequate.

One final word concerning projecting the sun. Because of the intensification of the sun's light closer to the focal plane, any eyepiece used for projection will heat up. I recommend that you use only noncemented, inexpensive eyepieces (e.g., Kellner) for projection. Eyepieces with plastic fittings should be avoided, too, because they are subject to eventual melting. When you use the projection method of solar observation, it is wise to choose one eyepiece (usually about 10x per inch of telescope aperture) that you use for this purpose only, and use no others.

The Herschel Wedge

The Herschel wedge is a unique device, specifically designed for safe viewing of the sun. As shown in Figure 5-3, the wedge is simply a modified right-angle prism that allows about 95% of the sunlight entering the telescope tube to be deflected through the prism (or mirror) and out of the system. Thus, only about 5% of the sun's rays enter the eyepiece, resulting in minimal exposure of the eyepiece optics to heat. Even 5% of the sun's light is not safe to view with the eye, and some type of additional filtration must be used to reduce the amount of light further. Neutral density filters that transmit only 0.01 or 0.001% of the light can be used at the base of the eyepiece for this purpose.

As in projection viewing, the telescope equipped with the Herschel wedge should periodically (at least every 10 minutes) be turned away from the sun because the light entering the telescope is unfiltered and its intense heat can damage components within the telescope over prolonged periods.

Herschel wedges should not be used with Schmidt Cassegrain or Maksutov catadioptic systems, because heat will slowly build in the closed-tube systems of such telescopes.

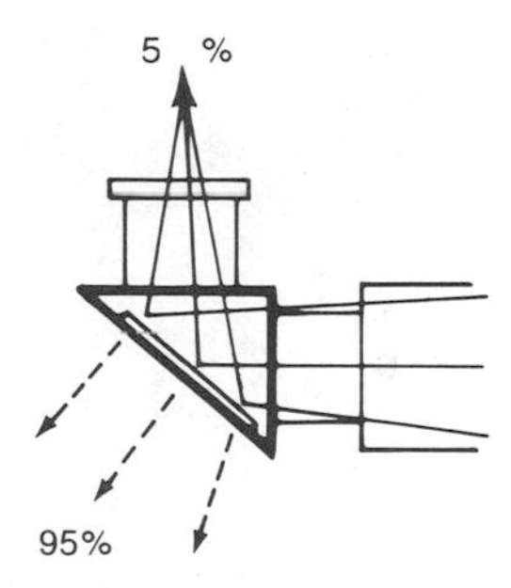

FIGURE 5-3. Diagram of the Herschel wedge.

A Newtonian reflector, provided its optics are well aligned, can be equipped with a Herschel wedge in place of the secondary mirror. Many observers who use a reflector have an arrangement that allows them to remove the secondary mirror quickly so that the Herschel wedge can be slipped into place with no loss of collimation. However, the use of a Herschel wedge in the focuser, as it would be used in a refractor, is difficult, if not impossible, because there is not enough inward travel of the focuser to accommodate the distance to focus (of the eyepiece) added by the light path through the wedge. Simply stated, the focuser cannot be turned inward enough with a Herschel wedge attached to the focuser to reach sharp focus in a Newtonian telescope.

Full-aperture and Off-axis Solar Filters

It would be ideal for the sun's light to be filtered *before* entering the telescope optics, thus eliminating the danger of internal heat buildup. However, at the present time, the supply of large

FIGURE 5-4. Telescope equipped with an off-axis solar filter. (Photograph by Brian Sherrod and Richard Wakefield)

filters of the quali.y necessary for solar work is almost nonexistent, except those supplied with certain catadioptic telescopes and made particularly for use in those telescopes.

Numerous suppliers stock filters made from a Mylar film on which a coating of aluminum has been put. At first glance, such filters appear to be aluminum foil with great strength and elasticity. The manufacturers and suppliers claim that such material performs as well as quality glass filters. After using a number of Mylar filters, I have found them totally useless for serious solar work, although they may be advantageous for the beginning observer who is unable to acquire a higher quality filter. The transmission of the coated Mylar filters, both for visual and photographic use, is off toward the blue end of the spectrum, in the region that provides little photospheric detail.

Quality glass filters, on the other hand, are ground plane-parallel on both surfaces and coated by means of a process that distributes a microscopically even layer of coating. The coating blocks, or reflects, most of the infrared and ultraviolet light in such a way as to allow the visible light transmitted to be as near white (as if no filter were used) as possible, thereby maintaining maximum clarity and definition.

In many observing situations, a full aperture (in which the full diameter of the telescope objective is used) may be less an advantage than an off-axis filter (which has a small hole located off to one side of the true center of the optical path) of smaller diameter (see Figure 5-4.) For an 8-inch (20-cm) telescope, a 3-inch (80-mm) off-axis filter is the minimum size; the proportion is maintained for telescopes of other diameters. Because solar observing is done during the day, it is more susceptible to interference by air currents rising rapidly from the heated ground and dissipating into the atmosphere, particularly after noon each day. A telescope of 8 inches or larger is capable of actually resolving more of the air currents that might go unnoticed in scopes having smaller apertures. Thus, a smaller filter on a large telescope enables you to see finer detail on the sun because the small filter is less likely to see the air currents visible in the large instrument. In most cases, the Herschel wedge provides details equaling or surpassing those provided in full-aperture solar filters, although the Herschel wedge is not practical for use in the modern catadioptic telescope.

An Inexpensive Solar Telescope

For many practical reasons a 3-inch refractor is an ideal telescope for solar observing. It is inexpensive, yet has exceptionally good resolution, and the size is optimum for maximum clarity of the sun's detail. However, if your telescope is located where conditions are conducive to using a larger aperture, and you exercise care in placing the instrument where it can be used to its full potential, you can use somewhat larger optics.

You can make an ideal solar telescope rather inexpensively if you have some parts that could be made into a Newtonian telescope. In the design of the solar scope, the primary mirror, fully finished and made parabolic, is left uncoated so that most of the light passes directly through the glass, rather than being reflected. Only a very small fraction of the total light is reflected off of the highly polished surface back to the eyepiece. However, as with a Herschel wedge, further filtration must be made to reach an intensity low enough for both comfort and safety. With this system, neutral density filters that transmit 0.01 and 0.001% of the sun's light are recommended.

Location of Telescope for Solar Observing

As important as the telescope itself is the location of the observing site from which you make your observations. If you mount your telescope on wood or pavement, you will receive highly distorted images as the heat escapes from these surfaces. If you locate your telescope in such a way that you must peer over the tops of houses or any structure that is losing heat, that also will result in poor views.

For best results, position your telescope in a location where the amount of heat lost versus the amount absorbed is minimal. Your best bet is to view from inside an enclosed building (with some substantial opening for the telescope, such as a dome), or better still, out in the open in a grass-covered field. Professional observatories, which constantly monitor the sun, are extremely conscious of the detrimental effects on the image of daytime heat. Studies have shown that the best conditions are found over large bodies of water. Consequently, one solar observatory—in an effort to achieve the highest resolution in solar photography—is located on a small man-made island in the middle of Big Bear Lake in California.

THE SUNSPOTS

When viewed through a properly filtered telescope, sunspots take many shapes, forms, and sizes. Each sunspot daily shows remarkable changes that should be consistently monitored by the amateur solar observer. Sunspots have two major components, although not every sunspot necessarily has both. The *umbra* is the very dark, often black inner portion of a complex sunspot group. It is considered the nucleus of sunspot activity. When you count individual spots, count the number of umbra present. In other words, every sunspot has an umbra. Surrounding many umbra is a "halo" known as the *penumbra*. Some sunspots do not exhibit a pronounced penumbra, whereas others show a penumbra with finely structured lines emanating from the umbra. The penumbra is commonly associated with a complex spot group, or with a very large single spot. The changes in the penumbra in less than one day's time can be remarkable. See Figure 5-5.

Although single spots exist as entities in their own right, they often have definite associations with other spots; together these groups of spots are known as a *sunspot group*. Over several weeks' time you will see that the spots and groups are rotating at a rate consistent with the rotational period of the sun.

The sun rotates in a period of about 27 days. However, with experience you will soon recognize that some spots move faster across the sun's disk than do others, and that their motions are governed by the latitude in which they are positioned. A spot moving from east to west across the sun's equator (near 0° latitude) makes a trip around the sun in about 26.7 days, which is indicative of the sun's rotational rate at the

FIGURE 5-5. The physical appearance of a large sunspot.

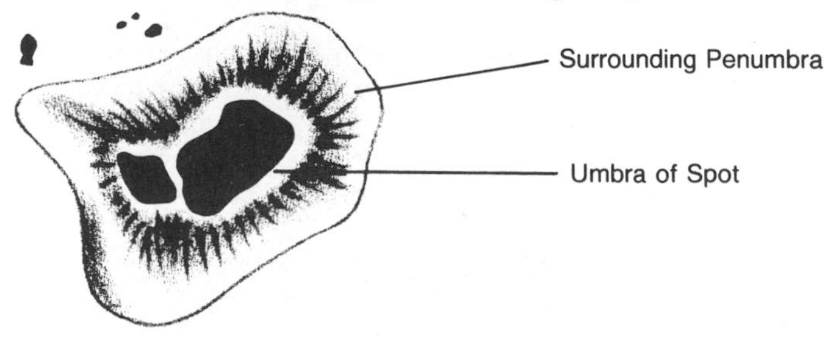

equator. A sunspot at about 40° latitude (north or south) requires a bit longer—29.3 days. Thus, the equatorial current of the sun—as it is on Jupiter—is considerably faster than the currents associated with the temperate and polar regions.

Sunspot Classification

So that some order and standardization might exist in solar astronomy as it does in other branches of the science, there is a very strict scheme for classification of sunspots and their groups. If you are a serious observer of the sun, you should master the nomenclature and classification as quickly as possible to add greater depth and understanding to your studies. Figure 5-6 gives the accepted classes of sunspots that normally might be seen on the surface of the sun. Most groups are easily recognizable. As spots near or appear at the sun's limb (edge), however, foreshortening causes spots and groups to become distorted in their appearance. In addition, larger telescopes used under very stable seeing conditions reveal groups in greater complexity than do smaller instruments.

THE ZURICH/WOLF NUMBER

Making daily sunspot counts is a simple and valuable endeavor for the amateur astronomer, but it is not quite so easy as it might first appear. Not only is it necessary to count the number of spots daily, but you must also count the groups that appear on the sun. R. Wolf proposed that the spot numbers (n) and groups (g) be combined in such a way that the groups represent 10 times (10 ×) the intensity of a single spot. Remember that the sunspots are important to astronomers as an *indicator* of the relative amount of activity of the sun as a whole. Sunspot groups represent areas that, in Wolf's opinion, should on an average be 10 times more active than areas in which single groups are seen. The Relative Spot Number (RSN), or *Zurich number*, as it is popularly termed, is represented simply by first noting the total number of groups and then multiplying that number by 10, and next adding the total number of spots to that figure. Or, shown mathematically:

$$RSN = (g \times 10) + n,$$

where

g is the number of groups, and n is the number of individual sunspots.

As you can see, the RSN is not just a simple sunspot count but an intensity index that measures the relative amount of solar activity day to day by the one phenomenon we can witness—the sunspots. It is important to count single isolated spots as a group if they appear not be associated with a group already counted. In addition, they must be counted as individual spots. All spots within any group must also be counted individually as part of the total number of spots (n). In other words, if one group consisting of 14 spots is seen, you record that group (g) as 1 and the spot number (n) as 14.

As an example, suppose that on a given day two groups of spots appear on the sun. Examining each group with high power reveals that one group contains 8 spots and the other has 4 spots. Thus:

the number of groups, $g = 2$
and the number of spots, $n = 12$

Thus,

$$RSN = (2 \times 10) + 12$$

or,

$$RSN = 32.$$

As previously mentioned, there is an obvious problem when comparing the RSN derived from an 8-inch (20-cm) telescope and the RSN from a 2.4-inch (6-cm) telescope. Not only is telescope *aperture* a determining factor when counting sunspots, but observer experience, the method used for the counting, and the observing conditions also affect the estimate.

Because of inconsistencies among observers,

FIGURE 5-6. Sunspot classification and description.

Class	Description	Example
A	A small spot, or spot group with no surrounding penumbra	
B	Similar to A, but spots showing definite association with one another, or they are symmetrically patterned (bipolar), but with no surrounding penumbra.	
C	A bipolar group in which the largest members are surrounded by one penumbra.	
D	A bipolar group in which major spots exhibit a penumbra.	
E	Very large bipolar group, larger than 10° across; the major spots exhibit very complex penumbra, between which are smaller spots many of which (or all) exhibit penumbra.	
F	The largest bipolar groups, 15° or larger, normally surrounded by complex penumbra and still showing random small spots.	
G	Similar to F but having no random spots.	
H	A large spot surrounded by penumbra with small random spots nearby; larger than 2.5°.	
J	A single spot (polar) with a penumbra.	

some factor must be applied to everyone's RSN number to bring his or her determination into agreement with a standard telescope and standard method. This "accepted value" is based on a telescope used at the Zurich Observatory in Switzerland. This telescope, used for over a century for sunspot counts, has an objective diameter of 3.15 inches (8 cm) and a focal length of 43 inches (110 cm). Because this telescope sets the worldwide standard, a factor must be adopted by each observer to bring his or her estimates to within the values determined daily by this instrument. This factor (k), once determined, is multiplied by the value already determined, or:

$$\text{RSN} = [(g \times 10) + n] \times k$$

A keen-eyed observer using a 3-inch (8-cm) refractor might be safely assumed to arrive at an RSN virtually the same as that determined by the Zurich telescope for the same day. Therefore, no factor would be necessary, assuming that the observer uses a filtered telescope, and not projection. Not using a factor gives the same results as multiplying the original value by 1; thus, this observer's k factor would simply be 1. It is safe for the beginner to use this value ($k = 1$) until a correction for this value is obviously needed. Remember that these estimates are *relative* values, and as long as the observer's results are consistent (estimates made with the same instrument, in the same location, and about the same time each day), no correction is necessary at first, because the values each observer obtains are combined with others—some higher and some lower—and averaged for a mean value.

A good way to check your estimates against those of other observers is to compare published values in astronomical journals against those you have made. *Sky & Telescope* magazine publishes daily sunspot counts in each monthly issue that are derived from the results obtained by observers in the American Association of Variable Star Observers Solar Division, and by the Zurich Observatory. The daily counts for a particular month appear in the journal three months later. For example, June values are published in the September issue. Most observers, when comparing their values to those published, are surprised to find that their values are too high. Therefore, the k factor must be adjusted downward (by less than 1), usually to about 0.8 or 0.7. In summary, this is how the count is made:

1. Count the number of groups (g) and multiply that number by 10.
2. Count all the spots visible, including those seen within the groups, and add this value (n) to the total.
3. Multiply the total obtained thus far by your k factor.

CHECKLIST FOR DETERMINING ZURICH NUMBERS

1. Always use the same telescope, the same method of observation, the same location, and observe at about the same time each day.
2. Starting with low power, scan the full disk of the sun during moments of steady seeing for unusual activity, and count the number of sunspot *groups*. Remember, a single, isolated spot constitutes a group and must be counted as both a group *and* a spot. If two large groups appear to be in any way connected physically (by faculae, or a line of tiny sunspots, or perhaps by a faint penumbra), consider such an area only *one* group, not two.
3. When the seeing is exceptionally steady, using higher power, scan very carefully for tiny, isolated spots that may have been missed at lower power. Small groups are particularly easy to overlook when near either of the sun's limbs.
4. Keeping the telescope at the higher power, begin counting the number of spots in each group. Write down—do not trust to memory—the total per group. If a large spot is very

complex, exercise care when counting. If the spot appears to be divided, say into three separate areas, make sure that each area is distinctly separated from the other two before counting it as a spot by itself. If it is connected in any way, it is still just one big spot. A clear line must divide the dark spot before it is counted two or more.

5. Add the totals of all groups and spots together, and derive the RSN for that day using the formula:

   $$\text{RSN} = [(g \times 10) + n] \times k$$

   where

   g = number of groups on the solar disk

   n = total number of spots including those in groups

   k = correction factor for observer and telescope.

6. Enter all information regarding the derived Relative Sunspot Number into a daily log book.
7. Classify all groups according to the descriptions given in Figure 5-6.
8. At the end of each month, send copies of all observations immediately to:

   Solar Division
   American Association of Variable
   Star Observers
   187 Concord Avenue
   Cambridge, MA 02138

   Your values will be averaged with others made on the same dates to derive the published daily Relative Sunspot Numbers for each day of the month, and they will serve as a basis on which professional astronomers can continue to probe the secrets of the sun.

CONCLUSION

Remember, *apathy* can be the amateur astronomer's worst enemy. Never assume that you can miss a day of observing because "others are probably observing the same thing anyway." Amateur astronomers don't have to go out and observe, but they do simply because they are dedicated to the projects they study. Always assume that your observations might be the only ones. Without them, no record of one day in the life of our sun would exist.

FIGURE 5-7. Simplified versions of various sunspot counts (k = 1).

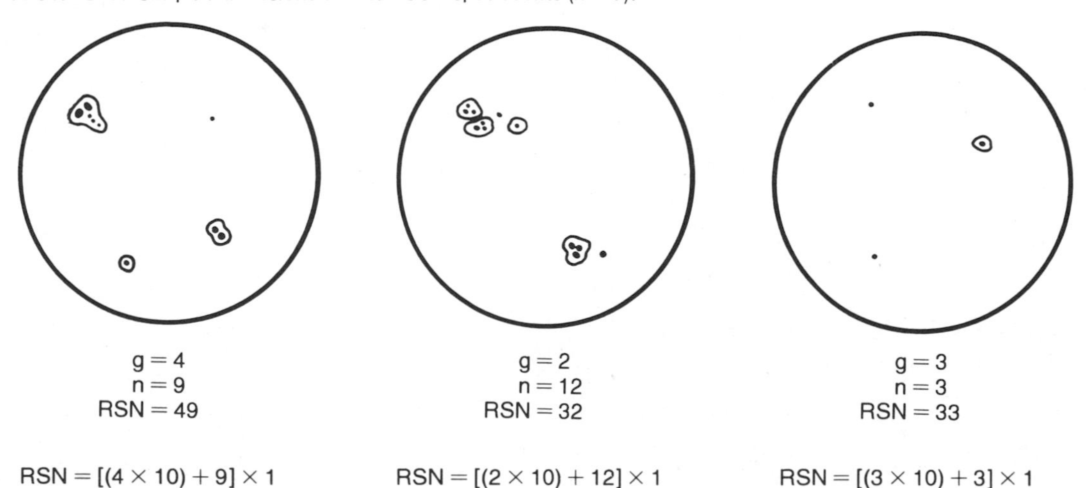

BIBLIOGRAPHY

Abell, George, *Realm of the Universe*. New York: Holt, Rinehart & Winston, 1976. Paperback.

Abetti, Giorgio, *The Sun: Its Phenomena and Physical Features*. London: Crossby Lockwood, 1938.

Abetti, Giorgio, *The Sun*. London: Faber and Faber, 1957.

Bray, R.J. and R.E. Loughhead, *Sunspots*. New York: Dover Publications, 1979. Paperback.

Ellison, M.A., *The Sun and Its Influence* (2nd ed.). London: Routledge and Kegan Paul, 1959.

Fredrick, L.W., and R.H. Baker, *An Introduction to Astronomy* (8th ed.). New York: D. Van Nostrand, 1974. Paperback.

Gamow, George, *A Star Called the Sun*. New York: Viking Press, 1964.

Kuiper, G.P., ed., *The Sun*. Chicago: University of Chicago Press, 1961.

Menzel, Donald, *Our Sun*. Cambridge, MA: Harvard University Press, 1959.

Moore, Patrick, *Astronomy Facts and Feats*, pp. 16–31. London: Guinness Superlatives, 1979.

Muirden, James, *The Amateur Astronomers' Handbook*, pp. 100–128. New York: 1974. Thomas Y. Crowell, 1974.

Roth, Gunter D., *Astronomy: A Handbook*. Cambridge, MA: Sky Publishing Corp., 1975.

Smith, H.J., and E. Smith, *Solar Flares*. New York: Macmillan, 1963.

Waldmeir, M., *The Sunspot Activity in the Years 1610–1960*. Zurich: Schultess, 1961.

6

LUNAR TOPOGRAPHY

Most observers familiar with the sky have grown somewhat blasé about the most appealing of all celestial objects—the earth's moon. Generally, it is the moon that attracts the largest crowds at star parties, and it is the moon that is first sought by the amateur with a new telescope. The moon is bright and easy to find. However, most observers gradually turn away from the moon as time goes on because the surface seems to be static and nothing new is expected. Most observers turn their attention from the moon much too fast, passing up the interesting studies that can be carried out concerning the lunar surface.

The study of lunar topography (i.e., surface features) demonstrates principles of earth geology. The measurement of lunar features requires some knowledge of geometry and trigonometry. And studies of the age and origin of many of the moon's features allow an observer to probe into the intriguing fields of cosmogony (the origin and nature of the universe) and the origin of the solar system.

THE LUNAR FEATURES

This section describes the common lunar features that are the focus of the projects considered in later sections of this chapter. Many of the features have somewhat controversial origins, but the most commonly accepted explanations are given here. I urge you to learn the correct nomenclature as well as the key identification features, so that you can properly identify those features.

Meteoric Craters

Most cratering on the lunar surface is a result of meteorite impact. Hence, the term *impact crater* is to be preferred to *meteoric crater* when referring to these features.

The size of impact craters depends on two factors:

1. *Size of meteorite*. Very simply, larger meteorites produce larger craters.
2. *Velocity of meteorite*. Because the force of impact is greater for a meteor traveling at a high rate of speed, we would expect a crater produced by a meteoroid at 20 mps to be larger than one produced at 15 mps, if both were of the same mass.

FIGURE 6-1. The 16-day-old gibbous moon. Along the terminator, shadows fill craters while direct illumination reveals sharp contrasts of rayed craters scattered throughout the dark maria near the western limb. (Photograph by the author).

Most impact craters on the moon are the result of its bombardment by meteoric material long ago, possibly as much as 4 billion years in the past. The large impact craters appear to have occurred during that time; the chance today for similar craters to be produced is very slim since there is no evidence of the survival of the cloud of large bodies which bombarded the surface. All bodies of the solar system were subject to the same bombardment at this time. Whether the meteoric material was excess planetary material left over after the origin of the planets that was gradually pulled in by gravity, or whether it was a huge swarm of particles traveling through space is not known. However, it appears that all the planets accumulated their masses through *accretion*, a process through which mass is gradually accumulated in a body by slow gravitational pull on whatever material is close enough to be subject to the gravity of the original mass. As more and more additional mass is accumulated, the gravitational influence of the object increases, thus extending its reach farther into space for additional mass. Eventually, all mass involved with the formation of planets by a particular star (in the earth's case, our sun) is accreted, and the process ends. The particles left at the end of this process might have caused 4 billion years ago the craters we see today.

The impact craters are generally quite large by comparison to other lunar features. The walls of small impact craters are well defined and (when new) quite sharp at their edges. If the crater has a smooth floor and almost vertical walls, its origin is probably meteoric.

Many impact craters have a sharp mountain peak in the center caused by volcanic action shortly after impact. Upon entering the lunar surface, the meteor actually melts the moon's crustal material, allowing lava to erupt at the point of entry. The velocity and size of the meteorite again determine the height of the mountain peak, due to the amount of penetration and thus the amount of lava produced.

In summary, the features that aid in identifying an impact crater are:

- ☆ The crater has a flat floor that at times is covered by lunar dust.
- ☆ Impact craters are generally quite large, particularly the oldest ones.
- ☆ The walls of impact craters are very sharp, sometimes even vertical.
- ☆ A sharp mountain peak can often be seen in the center, a result of volcanic activity caused by the impact.

Volcanic Craters

Some small craters are thought to be the result of volcanism other than that caused by impact. There is some skepticism about this theory, but there are indications that some craters were formed in this way. A volcanic crater has a sharp outer wall (i.e., it drops sharply to the moon's surface) and a gently sloping inner wall. Many craters that could possibly be volcanic appear to be quite new, as indicated by the top edges of the walls, which are sharp and uneroded by solar wind. Volcanic craters are generally less than 25 miles in diameter. If you wish to look for such craters, check for the following identification features:

- ☆ Less than 25 miles in diameter.
- ☆ Characteristic sharp outer wall, with inner wall sloping.
- ☆ Very little cosmic erosion.

A crater caused by a small, high-velocity meteorite entering the crust on a path perpendicular to the plane of impact could cause many of these same characteristics. However, there seems not to be much scattered debris surrounding these tiny craters, as would be expected if they were caused by impact, and as is typical of impact craters.

Collapsed Craters

It is difficult to identify collapsed craters, and they are very rare on the lunar surface. Caused by direct crustal collapse, these craters are often quite large and poorly defined. There apparently are pockets beneath the lunar crust that were once filled with material (molten?) that escaped, leaving empty spaces subject to collapse.

Lunar Seas

The lunar seas (maria) evolved as a result of ancient lava flows, hence the very smooth appearance of these features. However, the actual lava beds are now covered with a layer of lava

FIGURE 6-2. The crater Copernicus, the result of meteoric impact. Notice the bright albedo of this fresh meteoric crater. (Photograph by the author).

FIGURE 6-3. Lunar seas appear as dark areas on the moon's surface because they are less reflective of sunlight than features in surrounding areas. Notice the encroachment of the earth's shadow during a partial lunar eclipse. (Photograph by the author).

dust as much as 5 to 7 miles thick that has been caused by the erosive effects of the solar wind, as well as accumulated meteoric material. It is not certain whether the lava flows originated *after* the great meteoric bombardment, or before. It seems as though the moon would be somewhat more cratered than it is (even with the earth's "shielding effect") if it had indeed gone through the bombardment. It is more likely that the lava flow was the *result* of the bombardment, which produced the magma that now forms the lunar seas.

The seas are quite easy to identify, and anyone who has ever seen "the man in the moon" has observed the several lunar seas that comprise that familiar face. The seas are much darker than the surrounding highlands because of their less reflective surface, and they appear somewhat gray in color photographs.

Lunar Mountains

Any large irregular feature that cannot be classified as a crater or sea is generally classified as a mountain range. Many such mountains can be identified, but do not confuse lunar mountains with the very steep walls of large impact craters in areas where cratering is frequent.

Lunar mountains are formed in much the same way as the mountains of earth, that is by early geological changes during the moon's formative period. Generally, all lunar mountains are a result of primordial cooling of the moon's interior after its formation over 4 billion years ago. Tremendous pressures surged upward from the interior, pushing the crust upward. There are three different forms of mountain building:

- ☆ *Block*. The process by which a complete section of crust is pushed up, resulting in a range having material at the top that is identical with that of the moon's crust. The top is flat, like a plateau, and the sides are sharp, having originally been vertical.
- ☆ *Folded*. As in block mountains, a section of crust is pushed upward, but lateral shifts of the crust (also caused by internal pressure) result in the range being folded, one layer over the other.
- ☆ *Dome*. A small section of lunar crust that is especially weak is pushed up by internal pressures. Because the section is relatively small, the resulting mountain is isolated and steep. Piton is a good example of a lunar dome mountain.

Learn to identify all mountain ranges and peaks by their type of origin. This practice aids you in identifying the processes that make for the type of terrain in a particular lunar location.

Lunar Scarps

Lunar scarps are perhaps the most interesting of all lunar features. Many areas on the lunar surface seem to have been "scraped" by a meteor that hit the surface tangentially. Scarps are normally found between two craters, in mountain ranges, or on any other extended high surface on the moon. If two craters seem to blend together without any wall between them, it is likely that they were scraped by a passing meteoroid. The best example of a lunar scarp is the Alpine Valley, near the moon's north pole.

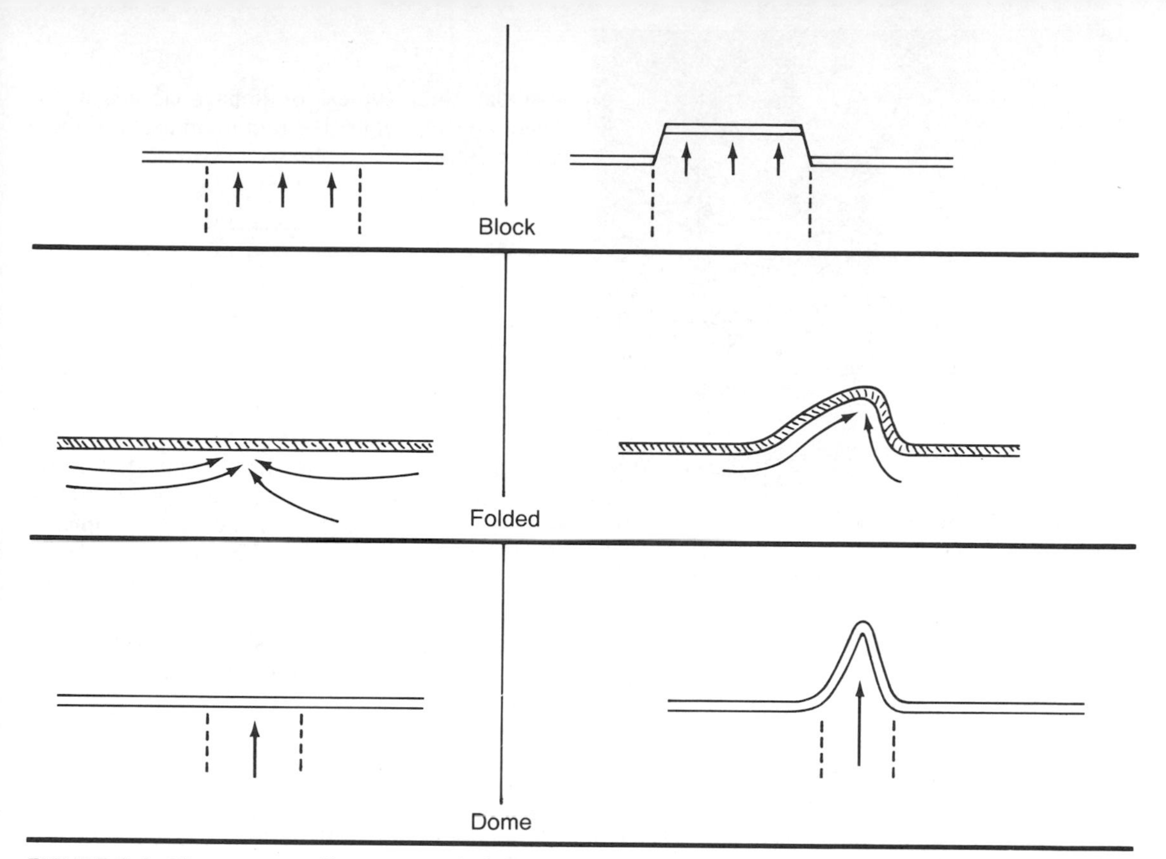

FIGURE 6-4. The processes of lunar mountain building.

FIGURE 6-5. The lunar Apennine mountain range rises high above the surrounding plains, casting dark shadows on the landscape below. (Photography by the author).

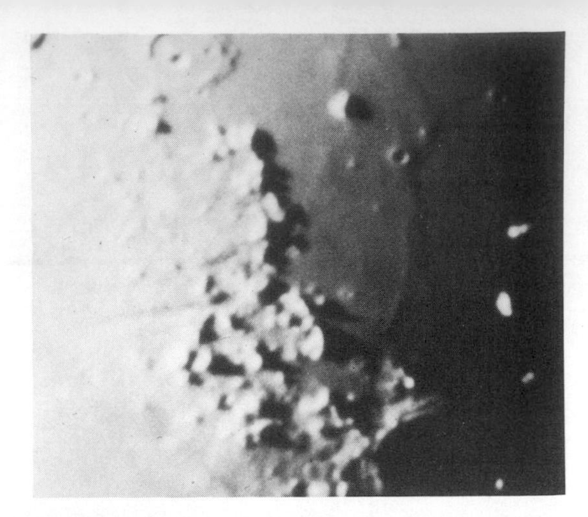

FIGURE 6-6. The lunar scarp, Alpine Valley (left), is the result of a tangent encounter of the moon with a passing meteoric particle. (Photograph by the author).

Lunar Rills

Very little is known about the origin of the lunar rills, but presently they are thought to be dried beds of water flow. They appear to the eye, and on photographs, as tiny channels cutting across flat areas of the moon, usually coming from or crossing a tiny but deep crater that has no walls (i.e., rims raised above the lunar surface).

One possible explanation for the lunar rills is that they are dried river beds left from water released from beneath the lunar crust at the moment of meteoric impact. The water beneath the surface is assumed to have been frozen until melted by the heat of impact. Although a lot of the water that may have been released would have been vaporized by the sudden exposure to the vacuum of space, it has been experimentally shown that some part of the water released to a vacuum will freeze in a fraction of a second and act as a cap under which more water can flow without rapid evaporation. Such a lunar "river" could last for no more than 100 years before the capping ice dissipated into space as vapor.

If this theory of the origin of the lunar rills is accurate, then it is possible that they can still be formed by the impact of any future meteorite. The craters from which the rills seem to originate are always quite deep relative to their diameter. Therefore, one cannot see if these craters have the flat floor typical of impact craters, nor is there any sign of central mountain peaks in these tiny craters.

Lunar Rays

Lunar rays are curious markings that are possibly one of the most controversial of all lunar features. The rays are bright extensions that seem to radiate from many large impact craters, indicating a powerful impact. The problem in determining their origin is whether the rays are actually cracks in the crust resulting from impact, merely material splashed from the impact, or a combination of the two. Currently, however, it appears that they are cracks that have become filled with lunar dust to a possible depth of 1 to 2 miles. During full moon, the rays emanating from the bright crater Tycho can easily be seen extending across as much as a third of the visible surface of the moon.

FIGURE 6-7. A network of lunar rills in the vicinity of Triesnecker (the bright crater, below center) and Hyginus (to the upper right of Triesnecker). (Photograph by the author).

FIGURE 6-8. Photograph showing the crater of Tycho with its bright rays. (Photograph by Thomas Koed).

FIGURE 6-9. An area of the moon that exhibits raised areas—domes—within maria. The most pronounced domes appear midway from top to center in this photograph. (Photograph by the author).

Lunar Domes

Recently, there have been many discoveries of lunar "domes" that resemble pressure domes associated with geyser activity on earth. The appearance of the domes leads to speculation that they originate from "hot spots" located within the crust of the moon. There is some question, however, as to whether molten magma exists beneath the lunar surface now. If not, then the domes are a result of past heating and pressure that now are permanent and static. You should make every effort to identify the domes now known and search for others not yet identified. If the domes are not static in their activity, then it seems reasonable to assume that they may be the source of some transient activity. Hence, you should actively monitor the domes for some type of change (swelling, "smoke," bright flashes, etc.).

Luna Incognita

Both the United States and the Soviet Union have sent numerous spacecraft to the moon, and extensive photography and general lunar mapping were carried out by almost every spaceship that landed on or orbited our satellite. With such numerous visits as these, you might expect that there is no further frontier on the moon. However, there is one region ideally angled to earth that was missed by the mapping cameras of the spaceships. This portion of the moon, known as the Luna Incognita, consists of over 270,000 sq km near the moon's southern pole: Amateur drawings and photographs of this region, concentrating on mountain ranges and other protrusions, are urgently needed to complete our mapping of the moon. Observations throughout a period of a month will allow you to observe the various features (mountain ranges, craters, and valleys) in changing relief because of the varying degrees and angles of sunlight.

The most efficient way by which the actual extents and heights of the features can be determined is through precise timings of the passage of the southern edge of the moon across the light of a distant star, which is known as a *grazing occultation*. These occultations will be discussed in depth in the Chapter 7.

Summary

All the moon's features previously discussed are important in understanding the origin of the moon, and hence the origin of the planetary system. Even if the moon seems "old hat" to you now, devote some time to the projects given in the remainder of this chapter to familiarize yourself with the principles of planetary astronomy. Take time to identify the features, both by origin and name; determine the sizes of some of the prominent features; and try your hand at determining the height of some high mountains and crater rims. These activities will not make you famous, but doing them will help you gain information and practice that you must have before you can understand planetary processes.

ESTIMATING THE AGE OF LUNAR FEATURES

The moon is thought to have originated about 4 billion years ago. It is unlikely that many of the original features of the original moon remain. There is a simple method by which the relative ages of features can be determined, using the details provided by impact craters. The mountains of the moon are probably the oldest of all lunar features, and the maria the second oldest. It is unlikely that either feature is changing its physical appearance a great deal, and therefore they can tell us little. However, the craters can help us estimate the age of features that they interrupt. The following are guidelines for estimating the age of lunar features.

1. *Crater size*. Generally speaking, the larger a crater is, the older it is because the meteoric material has been greatly reduced since the great bombardment 4 billion years ago. There are some very large craters in the southern hemisphere of the moon that are possibly 4 billion years old. Use these large craters as a model to help you estimate the size of other craters.

2. *Crater depth*. Craters that appear to be deep and rounded on the floor are generally young craters. Those in which there appears to be a great deal of dust, making a smooth floor, are thought to be very old. Also, many flat-floored craters contain lava from the magma that made the maria. Hence, they must be older than the maria themselves. Notice that many of the very large craters are also very shallow, or full.

3. *Crater walls*. Because of erosion caused by solar wind, crater walls become worn as they age. Hence, it is easy to identify relative ages by noting the degree of rim erosion on the crater wall. Very sharp rims that descend sharply to the lunar floor are younger than those that are rounded and worn. Notice also the amount of "slumping" on the inside wall of the crater. If it appears that the wall has collapsed, or that there has been a landslide in the wall, assume that the crater is moderately old, possibly 1.5 to 3 billion years.

4. *Secondary cratering*. Notice that many large craters appear to have smaller craters within them or in their walls. Some of those smaller craters may even contain craters. This is called

FIGURE 6–10. A photograph that shows varying depths of lunar craters. (Photograph by the author).

FIGURE 6–11. Comparisons of crater walls. Notice that the large crater near the top has an eroded and deteriorated wall whereas the smaller crater near the bottom maintains a sharp, well-delineated wall. (Photograph by the author).

FIGURE 6–12. Notice that the smooth crater Plato (upper photograph) is free of secondary cratering. Clavius (lower photograph), on the other hand, has within it a multitude of minor craters, which is characteristic of the craters near the moon's south pole. (Photograph by the author).

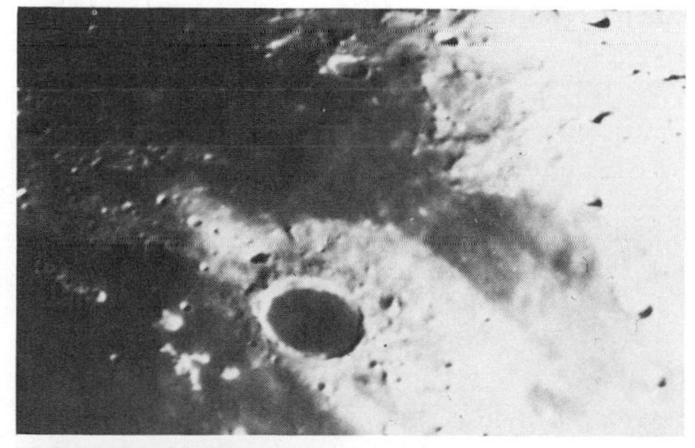

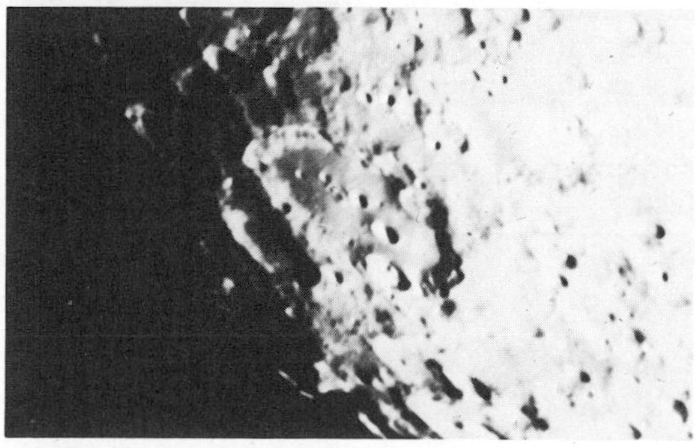

secondary cratering and is a good way to judge the relative age of craters. Because cratering is somewhat rare now, it must be assumed that a crater in which many smaller craters are found must have accumulated those craters slowly over a long period of time, and that hence, they must be quite old.

5. *The albedo of craters*. The *albedo* is simply the amount of reflectivity of a surface feature. The brighter a feature appears to us, the more sunlight it is reflecting and the higher the albedo. Older craters contain more hard lava and dust than do new ones, and consequently they are darker and have a lower albedo. Many old craters are gray; new ones are bright, and sometimes white.

6. *Rays from craters*. Craters from which rays originate are relatively new. It is thought that they are the newest of all the larger craters.

7. *Rills.* It is thought that the tiny craters with rills are very new because the total amount of erosion of them that we see is small compared to that of other features. The rill structure associated with them would not last long because of the erosion caused by the force of the solar wind.

FIGURE 6-13a. The crater Eratosthenes, located just west of Mount Apenninus, appears to be an astronomically recent addition to the moon's features. (Photograph by the author).

FIGURE 6-13b. The crater Ptolemaeus is representative of many of the older large craters on the moon's surface. (Photograph by the author).

TELESCOPIC EXAMINATION OF SELECTED LUNAR FEATURES

Craters. Locate the lunar craters listed in the table below, and sketch the area seen in the telescopic field of view. Be particularly aware of the mountain in the center of each of these craters, and represent in your sketch the shadow cast by these mountains as accurately as possible.

Crater	*Longitude*	*Latitude*
Petavius	—	-25°
Delambre	—	- 2°
Agrippa	+10°	—
Eratosthenes	-11°	—
Hansteen	—	-11°

Determine those longitudes and latitudes not in the tables either by consulting your lunar atlas, or by measuring on a high-quality photograph. You may wish to take the photograph yourself.

Another crater of interest is Alphonsus, similar in many ways to those listed in the table. However, Alphonsus has from time to time shown traces of fog and strange illumination, as though the volcanic processes associated with it are still active.

The craters listed in the following table have many rays, with lines of material originating at the crater wall. Locate these craters, and sketch the ray pattern seen in the telescope. If the rays seem to be concentrated on one side or the other, it might indicate the direction of the angle of impact of the meteor.

Crater	*Longitude*	*Latitude*
Pickering	+48°	–
Bettinus	–	-63°
Copernicus	–	+10°
Reiner	-56°	–

Some craters, such as those listed below, seem to be flooded by lava, or material similar to that of the lunar seas. Thus, the craters must be older than the maria themselves, since the crater existed prior to being flooded by the material that flooded the maria.

Crater	*Longitude*	*Latitude*
Plato	-10°	+52°
Francastorius	+33°	-21°
Letronne	-42°	-12°

Observe all features, particularly the craters, when the terminator (the dividing line between dark and light on the moon) is near the crater. On the other hand, you can best observe the lunar rays when the moon is full because the rays do not appear during all its phases.

Scarps. Locate the Alpine Valley, near the north pole of the moon, and examine this feature with high power. Running through the middle of this large rift is a well-defined rill, resembling a tiny wandering river bed. Sketch the view of the region of the Alpine Valley as seen in your telescope.

The straight wall. There is an interesting collapsed feature at longitude -8°, latitude -22°, known as the *straight wall*. It is difficult to say if one side of this feature collapsed, or if the other half uplifted.

This feature appears as a thin very black line near first and third quarter phases of the moon, probably somewhat better during 1st quarter. Look for any irregularities along this wall and sketch the field as it is seen, being careful to note the day and phase of the moon.

Naked-eye moon. Using a blank about 4 inches in diameter, sketch the moon as it appears to the naked eye when full. To enhance contrast, use sunglasses. For best results, do not examine a map of the moon, and try not to have any preconceived notions of what you expect to see. Such drawing gives you good insight into your ability to make consistent and reliable planetary drawings.

Locating Features by Age

Locate in the telescope, draw, and identify by name two each of the following features:

- ☆ Very old "original" craters.
- ☆ Old craters, but not original.
- ☆ Younger craters, or suspected new ones.

Locating Other Features

Locate the following types of features on the moon, using a telescope at relatively high magnification. You can find the proper names for these features in books, or you can take the names from Figure 6-14.

- ☆ Block mountain.
- ☆ Dome mountain.
- ☆ Folded mountain.
- ☆ Scarp.
- ☆ Network of rills.
- ☆ Lunar sea.
- ☆ Collapsed crater.
- ☆ Impact crater.
- ☆ Possible volcanic crater.
- ☆ Ray with no crater.

Determining the Heights of Lunar Features

Relative heights and depths on the lunar surface are quite familiar to the observer who has viewed the moon's disk near the terminator as shadows are cast by crater rims and mountains to fill the

FIGURE 6–14. First-quarter moon. North is at top, as seen by the naked eye. (Lick Observatory photograph)

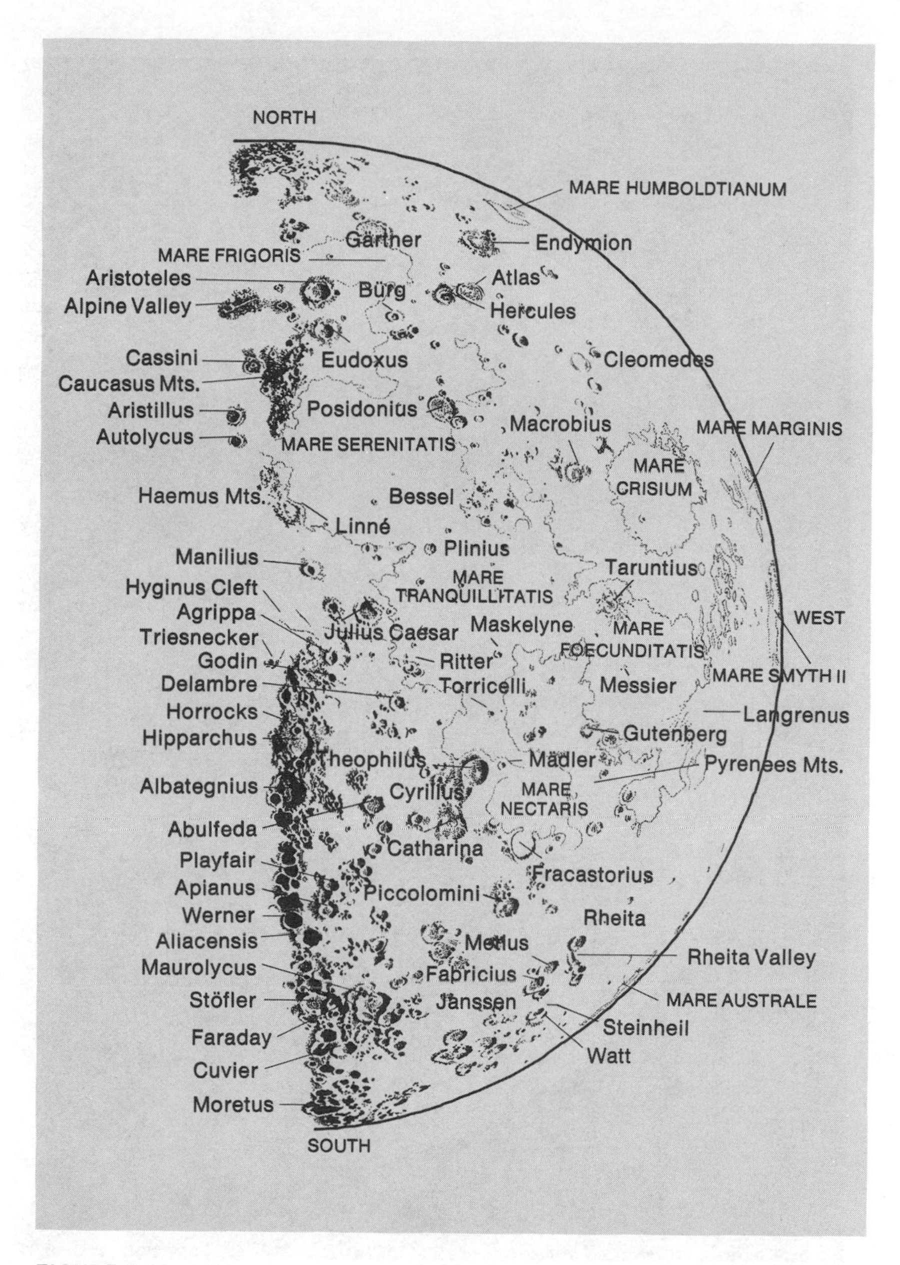

FIGURE 6–15. Key to the first-quarter moon. (From *Dynamic Astronomy*, 3rd ed., by Robert T. Dixon. By permission of the author, Prentice-Hall, Inc., and Herschel Wartik, Inc., New York).

FIGURE 6-16. Third-quarter moon. North is at top, as seen by the naked eye. (Lick Observatory photograph)

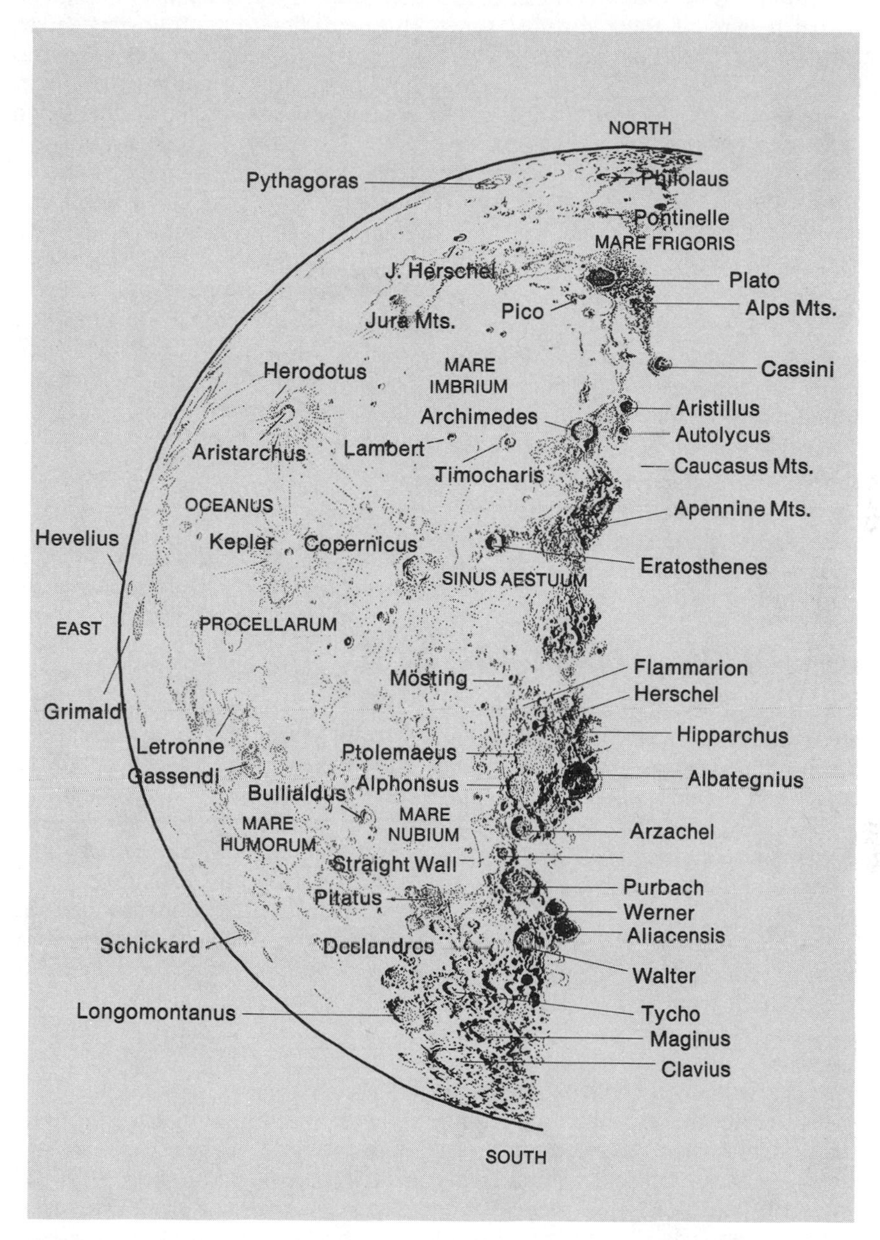

FIGURE 6-17. Key to the third-quarter moon. (From *Dynamic Astronomy*, 3rd ed., by Robert T. Dixon. By permission of the author, Prentice-Hall, Inc. and Herschel Wartik, Inc.)

valleys below. Methods by which you can determine the actual heights of these features, some methods being more refined than others, follow:

☆ *visual determinations*—By merely using the eye and some known values, the estimations of the shadow lengths of features can be made and subsequently reduced to corresponding lunar heights.

☆ *filar micrometer measurements*—Perhaps the most precise of all filar micrometer measurements. The measurement of a crater of known size is made as a standard. The shadow of the new feature is then measured and subsequently reduced trigonometrically to the actual height.

☆ *photographic measurements*—If done with exacting care, this method can be nearly as precise as using the filar micrometer. It has the convenience that it can be done at leisure and in the daytime.

Determining the Length of the Shadow

The first step in the reduction of heights of lunar features is simply to measure the size of the shadow cast by that feature. Using any of the steps just described should yield a proper value. To ensure the greatest accuracies attainable, follow the guidelines in the next several paragraphs when making those measurements.

Estimating with the eye. To estimate by eye is the least accurate of all methods because substantial bias and error on the part of the observer can occur. Nonetheless, if no other method is available, it is suitable. Estimating by eye merely requires that you compare the extent of a lunar feature to the extent of something known, usually a prominent crater. Crater sizes (in kilometers) are then used to determine what fractional part of that size is the shadow. For example, if a crater extends 100 km (as in Figure 6-10), and a shadow appears to be about half that distance in its total expanse, then the shadow is approximately 50 km long.

The drift method. As in comet observing, the length of a lunar shadow can be determined by allowing it to drift across a reticle in the telescope eyepiece and timing the duration from the preceding to the following end of the shadow. Because the moon's eastward motion causes the moon to drift slightly eastward in a telescope while you are tracking, the shadow will drift slowly across the crosshair at a rate very near 1 kps. To find the shadow's length by the drift method, leave the clock drive on, and with a stopwatch time in seconds the interval required for a shadow to drift its entire length. Then determine how long it would take for the entire lunar disk, centered at the lunar equator, to drift across. Knowing the exact diameter of the moon (3475.6 km), it is a simple matter to convert using the formula:

$$x = \frac{3475.6 \times b}{a}$$

where x is the value sought for the length of the shadow, b is the drift time for the shadow, in seconds, and a is the drift time for the lunar disk.

Measuring with the filar micrometer. Good bifilar micrometers are again coming into amateur hands. The most precise determinations of lunar size may be made using such an instrument, and the method is quick and straightforward. Using a crater of known size as a standard, make a measurement of the crater and record the thread value (the reading on the graduated micrometer scale). The shadow is similarly measured and recorded. Then it is a simple matter to compare proportionately the values:

$$x = \frac{C \times b}{a}$$

where x is the actual length of the shadow in kilometers, C is the diameter of the standard crater (also in kilometers), b is the thread value determined for the shadow by using the micrometer, and a is the thread value determined for the diameter of the standard crater.

Photographic measurements. Photographic measurements are almost identical to the measurement of shadow length using the filar micrometer, and they are quite simple. Merely measure, using an accurate millimeter scale or ruler, the diameter of the standard crater and the length of the shadow cast by the feature whose height you wish to determine. Once the measurements are made, the formula is identical to that used for the filar micrometer measurements. The values correspond to the symbols as:

C = the diameter of the crater in kilometers

b = the size in millimeters of the shadow

a = the size in millimeters of the crater

Once the value for the length of the shadow has been determined, it is necessary to compute the position of the feature you wish to measure in values known as Eta and Xi coordinates. This is done simply by determining from a good lunar map, the exact latitude and longitude on the moon of the shadow to be measured. You must arrive at an accuracy near 0.001 lunar radii. To determine the coordinates, simply compute from the latitude and longitude as follows:

$$\text{Eta} = \sin B$$

$$\text{Xi} = \sin L \times \cos B$$

where B is the latitude as determined for the shadow and L is its longitude. Modern pocket calculators have programmed functions that yield these values quickly.

Computing the height of the lunar feature. Now the height of the lunar feature can be determined, using a bit of mathematics. Again, the procedure can be greatly accelerated if a calculator with standard trigonometric functions is used. You will have to get many of the determinations of the variables from the *American Ephemeris and Nautical Almanac* because the earth-moon-sun positions change daily and even hourly.

The variables associated with the computations of lunar heights are as follows:

- F - angle between the earth and sun, as would be measured from the moon's center.
- A - height of the sun, in degrees, as seen from the lunar feature.
- L' - earth's selenographic longitude (from *American Ephemeris*).
- B' - earth's selenographic latitude (from *American Ephemeris*).
- C - sun's selenographic colongitude (from *American Ephemeris*).
- B'' - sun's selenographic latitude (from *American Ephemeris*).
- L - feature's longitude on the moon.
- B - feature's latitude on the moon.
- D - shadow length, expressed in decimals of lunar radii.

Obtain all the values L', B', C, and B'' for the date of the determination from the *American Ephemeris and Nautical Almanac*. The values *Eta* and *Xi* are now converted into lunar longitude and latitude for use in the final formula, and they yield the variables B and L. To convert to lunar longitude and latitude, use the following formulas (a calculator will expedite the results):

$$B = \text{latitude} = \sin^{-1} Eta$$

$$L = \sin \text{longitude} = Xi/\cos \text{latitude}$$

The *height of the sun* over the feature to be measured is expressed in degrees, and you can obtain it by using the following formula:

$$\sin A = (\sin B)(\sin B'') + (\cos B)(\cos B'') \times \sin (C + L).$$

This value is sin A. Compute the final value for the sun's height from that value.

The *angle between the earth and sun,* as seen from the moon's center (F), is found by using the following equation:

$$\cos F = (\sin B')(\sin B'') + (\cos B')(\cos B'') \times \sin (C + L')$$

The value F is subsequently obtained from its cosine value.

And finally, calculate the value for the height of the feature. The resulting value is in decimals of lunar radii, which you than convert to kilometer values. Determine height by using the following formula:

$$H = D (\sin A)(\csc F) - 0.5 (D^2)(\csc^2 F)(\cos^2 A).$$

BIBLIOGRAPHY

Abell, George, *Realm of the Universe* pp. 155–165. New York: Holt, Rinehart & Winston, 1976. Paperback.

Alter, Dismore, *Pictorial Guide to the Moon* (3rd rev. ed.). New York: Thomas Y. Crowell, 1979. Paperback.

Arthur, D.W.G., and A.P. Agnieray, *Lunar Designations and Positions*. Tucson: University of Arizona Press, 1964. Paperback.

Baldwin, R.B., *The Face of the Moon*. Chicago: University of Chicago Press, 1949.

Fredrick, L.W., and R.H. Baker, *An Introduction to Astronomy* (8th ed.), pp. 94–117. New York: D. Van Nostrand, 1974. Paperback.

Guest, John, *The Earth and Its Satellite*. New York: McKay, 1971.

Gurnis, Michael. "Endogenetic Craters on the Floors of Large Lunar Craters," *Journal A.L.P.O.*, Vol. 27, Nos. 1–2 (1979), 7–14.

Hartmann, William K., *Moons and Planets*. New York: Bogden and Quigley, 1972.

Harvard College Observatory, Annals of, *A Photographic Atlas of the Moon*, Vol. 51. Cambridge, MA, 1903.

Jamieson, Harry D., "Lunar Heights Measurements Made Easy," *Journal ALPO*, Vol. 26, Nos. 11–12 (1977), 235–240.

Match, Thomas A., *Geology of the Moon: A Stratigraphic View*. Princeton, N.J.: Princeton University Press, 1970.

Menzel, Donald H., *A Field Guide to the Stars and Planets*. Boston: Houghton Mifflin, 1964. Paperback.

Moore, Patrick, *A New Guide to the Moon*. New York: Nortons, 1977.

——, *A Survey of the Moon.* New York: Norton, 1963.

Moore, Patrick, and Peter Cattermole, *The Craters of the Moon*. London: Lutterworth Press, 1967.

Muirden, James, *The Amateur Astronomers' Handbook*, pp. 77–99. New York: Thomas Y. Crowell, 1974.

NASA. *The Moon as Viewed by Lunar Orbiter*, SP-200. Washington, D.C.: Government Printing Office, 1970.

National Geographic Society. *Map of the Moon*. Washington, D.C.

Roth, Gunter D. *Astronomy: A Handbook*, pp. 264–288. Cambridge, MA: Sky Publishing Corp., 1975. Paperback.

Rukl, Antonin, and John Gribbin, *The Amateur Astronomer*. Prague, Czech.: Mayflower-Octopus Books, 1979.

Rackman, Thomas. *The Moon in Focus*. New York: Pergamon Press, 1968.

7

LUNAR OCCULTATIONS

Star occultations occur when the moon in its eastward path about the earth passes in front of stars and eclipses them. The precise timing of the occultation—that moment when the star seems to blink out behind the lunar limb—supplies vital information regarding the earth-moon orbit, and any changes in the velocity or distance of that orbit, corrections for sidereal time, and the possibility of discovering double stars not resolvable by even the greatest telescopes of earth. These *lunar occultations* of stars by the moon are intriguing in that every amateur, even those with the smallest telescope, can obtain the accurate timings necessary for those determinations. Not only that; occultations give the advanced amateur, the person who has all sorts of sophisticated instrumentation for dark-sky observing, something to do when the moon is out and many stars cannot be seen.

Timing an occultation is quite simple, yet you must exercise the greatest care when making and reducing the observations. Basically, you note—using WWV time signals—the exact time, to the nearest 0.1 sec, when a star disappears behind the moon. It is also necessary to know, of course, which star it is that has been occulted and which stars in the path of the moon are potential candidates for occultation.

Instrumentation For Total Lunar Occultations

Any firmly mounted telescope will suffice for timing lunar occultations. It is necessary that the telescope be as rigid as possible, however, because the distraction of wind and the vibration from any other cause can prevent an otherwise accurate timing. The larger the instrument used for occultation work, the more stars that can be timed on any evening, because the larger instrument makes it possible to see fainter stars. However, stars at the very limits of perception in *any* telescope should not be timed because atmospheric conditions can cause spurious blinking and inaccurate timings.

As high a magnification as possible can be used for your occultation work, but do not use a magnification so great as to cause the image of the star or the edge of the moon to seem to be shimmering as a result of currents of air. In many cases, it is advantageous for you to use excess power to shield most of the field of view from the intense glare of the moon. Many times you will be able to eliminate entirely the illuminated portion of the moon from the telescopic field of view by merely resorting to higher magnification and and a smaller field. Higher magnification also results in dimmer images, but if the star is relatively bright in your telescope, it will not substantially impair your timing if high magnifications are used on a steady night. It is a great advantage that higher magnification reduces (inversely proportional to the square of magnification used) the intensity of the illuminated lunar disk as well. This is particularly useful when timing occultations near full moon, when the glare from the lunar disk could otherwise prevent you from discerning the dark limb.

Tape Recorder

Record all occultation work—your notes and the precise timing—on a tape recorder. A portable recorder using cassette tape is the best unit to use for lunar occultations. Be aware, however, that if you use a tape recorder in very cold weather, great variations in the speed of the driving mechanism of the tape recorder can, and frequently do, occur, which results in imprecise reductions of the exact time. To prevent such problems, you can take several steps. First, if the tape recorder is small, you can store it inside a jacket, against your body. Second, if the recorder can be battery-operated, never operate it on battery power when the temperature is below freezing. The power from the batteries fluctuates wildly in such cold and will affect the speed of the tape recorder. Third, the most efficient way you can ensure the reliability of your recorder's speed is to use a second tape recorder in case some fluctuation does occur in the first one. That is, use them simultaneously so that you have two sets of data.

The tape recorder stores the sound of the WWV radio transmission, which gives you accurate time tones from which you can reduce your data. In addition, the recorder should also pick up your voice. Before each timing, check the recorder for good operation by testing it with a 5 sec voice message, which you then replay. During observation, make a record on the tape recorder of changes in air turbulence, the wind's shaking the telescope, and any other factors that could affect the quality of your observations, including, of course, the exact time of your signal that occultation has taken place. Further use of the tape recorder and its efficiency in timing lunar occultations is discussed in the following sections.

The Stopwatch

Regardless of what method is used, it is necessary that you have a way to time to 0.1 sec accuracy the exact moment of the occultation. A good sports stopwatch with a sweep hand can be used. One method of timing the occultation at the telescope requires the use of the stopwatch, although there are several problems in this method—as well as any stopwatch use—as will be discussed in the following pages.

The Shortwave Radio

The shortwave radio, tuned to WWV time standard signals, must be used for all occultation timings. The frequencies of these signals are 5, 10, 15, and 20 MHz, although not all of them can be received at any one location.

A common problem with using the WWV broadcasts is the fading in and out of the stations, usually at the most inopportune time, right before the occultation. Because of varying locations, hillsides, and external interference, many locations can pick up only one station at any given time. Some nights the signal from 5 MHz can be quite distinct; on other nights it is unrecognizable. Therefore, whatever receiver you choose should have tunable bandwidths to cover all frequencies. In the event that the radio fades away just prior to the occultation, the timing can still be reduced by measuring from the last recognizable minute announcement if you are using a tape recorder. Or, if you are using a stopwatch, you can measure, while the stopwatch is still running, to the next available recognizable time mark after the occultation.

NAVAL OBSERVATORY PREDICTIONS OF OCCULTATIONS

On any given night, an observer using an 8-inch (20-cm) telescope can see about eight stars occulted. And when the moon passes through the star-rich regions of the Milky Way, twice that number may be seen. Usually, the difficulty in timing the star's disappearance behind the moon is not nearly so great as determining exactly which star it was that was occulted. If you are a serious observer, you will need to know when to expect these events so that you can schedule your observing time. Such predictions are available through:

The United States Naval Observatory
Occultation Timing Division
Washington, D.C. 20390

The computer-generated predictions available from the Naval Observatory are based on your location so that you can achieve maximum accuracy. Consequently, you must have a permanent observing location and know its exact latitude and longitude to within at least 25 feet. We have discussed the best method for the amateur to make such determinations in Chapter 2.

Not all observers should request the computer predictions for occultations. Only if you seriously wish to undertake regular timings of occultations should you write for them. If you are casually interested in observing only the most spectacular occultations, you can obtain predictions for the United States and Canada through the publication *Sky & Telescope* free of charge by merely requesting them. Most of the major occultations for each year are predicted in this nicely printed booklet. The predictions are prepared by the Nautical Almanac Office of the Royal Greenwich Observatory in England. The address to which to write for the booklet is:

Sky & Telescope
49–50–51 Bay State Rd.
Cambridge, MA 02138

Amateur astronomers using the U.S. Naval Observatory prediction sheets will find there predictions for a wealth of faint stars for which timings are badly needed. If you request the predictions, you will be asked to fill out a questionnaire regarding your observing experienc, telescope, observing location, and serious intent. The Observatory personnel then determine the degree of difficulty that they feel you can handle. For example, very faint stars will be left off a beginner's list, as well as many timings that could be made only near the time of full moon. If you are an experienced observer, however, you can set your own limits within reason. The USNO computer predictions are in code for the ease of processing information. You must understand the correct interpretation of the code in order to use these excellent tables to their fullest. Each star is coded as described in the following section.

Explanation of Format

On the USNO's list of possible occultations for your area, there will be some typed in data concerning the observer—you. Most of the data were supplied by you on the original questionnaire that you submitted to the USNO, and the data are now stored in the computer. Figure 7–1 shows a page of an observer's computer-generated predictions. First is given the year on which the predictions are based, and then comes the page number. Following these is a list of options that either you or the observatory personnel handling your request for predictions have determined. The first code is the *Observability Limit Code*, which is set by the type of instrumentation you have and your expertise in occultation timing. This information helps the computer to select the events that are printed for each night. Only events rated at the difficulty level chosen, or those easier, are printed. The second code is a number designating your status as it concerns mailing further predictions. If the number is 0, it shows that you have been inactive and should be contacted to determine if you are still interested in the occultation work. Otherwise, the number 1, 2, 3, or 4 indicates in which quarter of this year that next year's predictions should be mailed to you, so that if you need early information, you will get it in time to plan your program of observation. The third code designates several options available to you. The number 1 indicates that you have chosen a photoelectric option and that you probably have access to such equipment. The number 2 tells you that any graze occultations within a set radius of your station have been so designated in the list supplied to you. The number 4 sets extended limits for grazes, if you so choose. The number 8 indicates that the printed data are condensed, and therefore they do not include all the information that is included on a standard form. The condensed form is used primarily by observers who are somewhat erratic in their observations, and those who wish to know about only the most notable occultations. The last code number indicates the number of printouts of the prediction sheets that the observer has been issued at his or her request.

Across the top of the printed sheet is the address code, which begins with the letter A and contains a total of five characters that contain your name and mailing address. Once this address code is assigned to you, it is permanent even if you should move to another address. For example, my own observing code has always been AE405. Following the address code is the station code, which begins with the letter S followed by five digits that identify the station name (regardless of the observer) and identifies its geodetic coordinates on the globe.

Across the second line of the page headings are the coordinates of the station (in latitude and longitude and elevation), the location of the station (by city or town), and the *identification of the observing site*, which might include the names of all the observers using it, the name of the institution or observatory, or any specific instruments that might have been mentioned in your response to the original questionnaire.

In summary, on just the top two title lines of the USNO predictions are the following:

1. The year for the predictions.
2. All chosen options, including:
   a. difficulty of observations
   b. mailing priority
   c. photoelectric capability
   d. graze observer
   e. number of prediction copies issued.
3. The address code of the observer.
4. The station code of the observing site.
5. The latitude, longitude, and elevation of the observing site.
6. The city in which the site is located.
7. Specific information about the site.

Column Headings

The pages of the computer prediction sheets consist of vertical columns, each of which gives specific information about the star that will be occulted. Each column is headed by an abbreviated code translated according to the following:

DAY TIME MONTH (UT). This column gives the day and the time in Universal Time to the nearest second of when the occultation event is predicted for the observer's location.

P. The letter P stands for "phenomenon," and tells what type of occultation will take place. If it is predicted that the star will disappear, a D appears in this column. A reappearance of a star from behind the moon's disk is signified by R. The letter M signifies that there is a graze near the observer's location, but the occultation will not be visible from his station.

D. The D column gives information about any *multiple* star that might be occulted. If the star is known or suspected to be double, certain letters—N, S, P, F—are used. The letters mean, respectively, that the prediction is for only one component—the *n*orth, the *s*outh, the *p*receding, or the *f*ollowing component. If the letter A appears, it refers to the Aitken catalog of double stars. The letter C signifies stars in the Innes catalog. The letter B means that the star is triple. The letter D signifies that the predicted time of occultation is that for the primary star only; E denotes the timing for the secondary component. Other designations (of no interest to amateur observers) may appear in this column.

AC. The number in the AC column represents

```
U.S. NAVAL OBSERVATORY TOTAL OCCULTATION PREDICTIONS FOR 1980          PAGE 50          OPTICNS -- 0,0,1,1          ACCRESS CODE AE405
COORDS., LCCATION, IDENT.-- W 92 14 17.4, +34 48 44.0, +0136M -- NO.LITTLE ROCK, ARK.-- C14ILZ+7IQP/SHERROD    -- STATION CODE SE492

DAY TIME-UT P AC  USNO    C  MAX SP  PCT ELG SN MN CA     PA  CNTCT    WA LONG LAT R-RATE DIST     DM      SAO     HA     DECL.  RT. ASC.
    H  M  S  D   REF NC V   MAG     SNLT       AL AL             ANGLE        LIB LIB AS/S         KM REF NC  REF NO  O / // O / // H M  S
DECEMBER                                                       DECEMBER      ********************************************** DECEMBER
26/06 31 27/R  5 X15432 73  8.8 GO  77- 123    37 85S 284.6-178.5 262.3 6.5-2.2 .4068 387443+13 2232 099038 -533418 122306 101625.6
26/07 23 06/R  6 X15470 73  8.4 GO  77- 123    47 69N 310.6 157.9 288.3 6.4-2.2 .3502 386811+13 2239 099057 -405952 122315 101757.9
26/07 43 59/R  3 X15474 83  8.7 G5  77- 123    50 47S 247.4-137.8 225.1 6.3-2.2 .2730 386606+12 2190 099059 -355040 120442 101817.7
26/07 43 35/RF 9 X15475 81 10.0 G5  77- 123    50 43S 242.9-133.3 220.5 6.3-2.2 .2527 386610+12 2190 000000 -355739 120339 101819.2
26/09 28 17/R  3 X15526 83  9.0     77- 122    65 53S 253.3-138.7 230.9 6.0-2.1 .2575 386102+12 2195 099079 -101609 115220 102034.8
26/09 30 17/R  3 X15528 71 10.3 G5  77- 122    66 79N 301.0 173.7 278.5 6.0-2.1 .3403 386102+12 2197 000000  -94750 120427 102043.7
26/10 10 19/R  3 X15561 71 10.3 KO  77- 122    67 53N 327.0 149.1 304.5 5.8-2.1 .2939 386170+12 2204 000000    -409 120335 102153.9
26/11 51 56/R  8 X15594 93  8.9     76- 121    57 41S 241.8-124.6 219.3 5.5-2.0 .2079 387003+12 2208 099101  245628 112707 102342.8
26/13 34 25/R  2   1529 72  6.6 G5  76- 121  3 39 77N 303.1 172.6 280.5 5.2-1.9 .4151 388711+12 2211 099132  500154 112449 102611.6
27/05 03 38/R  2 X16614 64  8.5 F5  69- 112     8 84S 285.6 179.8 261.8 6.0-3.4 .4942 394913+09 2445 118637 -861033  84956 110243.8
27/07 29 37/R  6 X16727 63  9.0 KO  68- 111    37 79S 280.7-169.6 256.8 5.7-3.5 .3842 392453+09 2451 118674 -503641  82854 110649.6
27/08 30 33/R  3 X16771 61 10.4 KO  68- 111    48 64N 317.4 156.7 293.5 5.6-3.5 .3312 391730+09 2459 000000 -354533  82829 110833.7
27/09 17 42/R  4 X16799 85  7.9 K5  68- 110    55 25S 226.7-110.3 202.8 5.4-3.5 .1196 391361+08 2465 118702 -241135  75942 110931.2
27/09 39 28/R  9 X16805 71 10.6 K5  68- 110    58 53S 254.7-137.4 230.8 5.3-3.5 .2500 391252+00 0000 000000 -184550  80246 110943.5
27/09 52 49/R  4 X16815 71 10.5 GO  68- 110    60 50S 251.8-134.0 227.9 5.3-3.4 .2344 391206+08 2466 000000 -153017  75956 111000.5
27/09 58 50/R  2 X16816 63  9.2 GO  68- 110    60 88S 290.2-172.1 266.2 5.3-3.4 .3335 391191+08 2467 118705 -140138  80901 111008.5
27/09 35 19/R  2 X16822 74  8.6 AO  68- 110    58 36N 345.8 131.3 321.8 5.3-3.5 .2249 391269+09 2465 118709 -195724  82214 111022.9
27/10 06 22/R  3 X16844 73  8.9 G5  67- 110    61 31N 351.0 127.3 327.0 5.2-3.5 .2037 391176+08 2471 118718 -122019  81736 111106.0
27/11 53 40/RP 3 X16874 61 11.0 GO  67- 110    60 87S 289.5-169.3 265.5 4.9-3.3 .3369 391541+08 2473 000000  141140  74948 111224.0
27/11 56 07/RF 3 X16877 61 10.3 K5  67- 110    60 89N 292.6-172.4 268.6 4.8-3.3 .3405 391562+08 2473 000000  144741  75009 111228.1
27/11 32 33/R  3 X16880 74  8.2 KO  67- 110    62 40N 341.6 138.6 317.6 4.9-3.4 .2540 391384+08 2475 118729   85051  80249 111237.6
28/05 27 12/R  2 X17740 71  9.3 KO  59- 101     2 41S 244.3-136.4 220.0 5.0-4.6 .3640 399685+05 2550 000000 -910806  43607 115006.5
28/06 21 56/R  2 X17771 74  9.1 F5  59- 100    13 43N 339.4 130.2 315.0 4.9-4.6 .2976 398621+05 2554 119095 -775744  44809 115219.7
28/07 21 13/R  2 X17799 61 10.7 GO  59- 100    24 76S 279.0-166.8 254.6 4.8-4.6 .4072 397535+00 0000 000000 -632115  42711 115318.9
28/07 30 36/R  3 X17804 71 10.8 GO  59- 100    26 40S 243.0-130.4 218.7 4.8-4.6 .2671 397374+04 2542 000000 -610445  41632 115339.3
28/08 54 13/R  2 X17844 65  8.3 A2  58-  99    42 68S 271.1-154.6 246.8 4.6-4.6 .3296 396148+04 2549 119122 -403442  41013 115530.9
28/08 47 17/R  3 X17854 73  9.3 A3  58-  99    40 42N 340.4 135.7 316.1 4.6-4.6 .2635 396234+05 2558 119127 -422723  42516 115601.9
28/09 42 34/R  3 X17863 62 10.2 G5  58-  99    49 66S 268.5-149.9 244.2 4.4-4.6 .2990 395655+04 2550 000000 -284246  40135 115631.8
28/09 56 52/R  3 X17870 63  9.5 GO  58-  99    51 84N 299.1-179.9 274.8 4.4-4.6 .3413 395544+04 2552 000000 -251319  40656 115656.7
28/10 27 16/R  3 X17884 64  9.1 F8  58-  99    55 71N 312.3 168.0 288.0 4.3-4.5 .3268 395369+04 2554 000000 -174725  40439 115740.5
28/10 17 30/D  2   1733 61  5.2 AO  58-  99    54-69S 133.6 -13.5 109.2 4.3-4.6-.3271 395416+04 2556 119156 -203344  34542 115857.9
28/11 46 53/R  2   1733 69  5.2 AO  58-  99    59 86S 288.8-166.9 264.4 4.0-4.5 .3198 395308+04 2556 119156   15043  34542 115857.9
28/12 34 31/R  2 X17939 66  7.9 F5  57-  98 -8 57 61N 321.5 160.3 297.2 3.8-4.4 .3146 395557+04 2560 119169  132841  34420 120011.1
29/06 45 15/R  2 X18639 63  8.6 F5  49-  89     7 56S 258.7-147.5 235.0 3.8-5.5 .3981 402161+01 2728 119510 -822506   1923 123733.1
29/07 45 30/R  2 X18672 63 10.1 KO  49-  89    18 62N 321.0 152.6 297.3 3.6-5.5 .3809 400972+01 2736 000000 -675010   2331 123934.3
29/07 51 09/R  2 X18674 64  9.7 KO  49-  89    19 65N 318.5 155.3 294.8 3.6-5.5 .3863 400865+01 2737 000000 -662701   2208 123941.1
29/08 53 40/R 10 X18704 86  8.2 F5  49-  88    31 12S 214.9 -98.1 191.3 3.5-5.5 .0547 399778+00 2972 138923 -510801  -1209 124118.4
29/09 35 55/R  3 X18711 63 10.0 F5  49-  88    38 72S 275.4-156.8 251.8 3.4-5.5 .3344 399172+00 2973 000000 -404253   -548 124148.8
29/09 30 49/RP 3 X18719 62 10.5 G   49-  88    38 55N 327.7 150.6 304.0 3.4-5.5 .3190 399239+00 2976 000000 -420443    620 124209.9
29/09 31 33/R  3 X18720 65  9.0 F8  49-  88    38 56N 327.6 150.7 304.0 3.4-5.5 .3191 399229+00 2976 119551 -415357    611 124210.8
```

FIGURE 7-1. Computer-generated predictions.

the degree of accuracy assigned by the computer to the predicted time of occultation. The estimated degree of accuracy is obtained in part by data supplied in past years by amateurs. The number is the total number of seconds to be added or subtracted because of uncertainties about the contour of the moon's edge, the position of the star, or the conditions of observability.

USNO REF NO. The number in this column is the star number, as referred to in the catalogs of the U.S. Naval Observatory. A series of four numbers indicates that the star is given the Zodiacal Catalog (ZC) number. Numbers preceded by Z or SZ and consisting of five digits are taken from the Smithsonian Star Catalog.

V. This column gives the value of the event in terms of needed data. Easy occultations that hundreds of people might be recording do not have as high a value as does a very faint star that will be occulted on the illuminated edge of the third-quarter moon. The scale is from 1 to 9, with 9 having the most value and generally being the most difficult.

O. This column gives the observability code, which is a factor in determining the value given to each occultation. The code is on a scale of 1 to 9, with 9 being the easiest events, such as a bright star occulted on the dark limb of the first-quarter moon. For example, a value of 1 may indicate that the event will occur near full moon, or that a faint star may reappear on a bright edge. Your option code, determined by your expertise and equipment, helps the USNO determine the difficulty level of events put on your prediction list.

MAX MAG. The magnitude of the star is given in this column. Variable stars are generally noted as such, and the magnitude supplied is the highest predicted magnitude for each star.

SP. The spectral type of the star is given in the SP column. The color of spectral stars will greatly affect the accuracy of your timing, particularly if the event takes place on an illuminated limb.

PCT SNLT. This column given the percentage of solar illumination on the moon's disk as seen from earth at the predicted time of occultation. If that time is precisely the time of the full moon, that value will be 100. A plus sign (+) indicates the waxing phase of the moon; a minus sign (-) indicates waning phase.

ELG. Under this column are values that give the elongations of the moon, in degrees from the sun, up to 180° (near full moon). If the value is exactly 180°, a lunar eclipse will take place; if the value is exactly 0°, a solar eclipse will occur.

SN AL. This column gives the altitude of the sun from the nearest horizon to the moon on the date of a predicted occultation. When the sun is -12° from the horizon, astronomical twilight begins. If your code is high, then some bright stars predicted for daylight occultation will be included. In such cases, the sun's altitude is a positive (+) number.

MN AL. The moon's altitude above the horizon is also a factor in the observability and the timing reliability of many occultations. This column expresses that altitude relative to the nearest (east or west) horizon, measured in degrees up to 90°.

MN AZ. This column provides the value for the moon's azimuth, as measured from due north toward east. Due south is 180° and west is 270°.

CA. This value is important in sighting stars of low magnitude, or those making a reappearance. The value is simply the angle from the closest cusp to the position at which the occultation will occur. It is identified also by either N, S, E, W—the north, south, east, or west cusp respectively. If the event takes place on the dark limb, the cusp angle is given as a positive number; if it occurs on the lighted limb, the cusp angle is negative.

PA. Much like the column for cusp angle, this column provides further means for locating the point of a predicted event. However, this is a value measured from the north pole of the moon eastward. An event that takes place exactly on the east limb of the moon has a position angle of 90°.

VA. This is another measurement of position of the event, little used by the amateur. Numbers in this column reflect the vertex angle, as measured at the center of the moon's disk from the zenith, eastward to the star.

CNTCT ANGLE. This column gives values for the angle at which the star is occulted, all the values being in relation to an occultation at the middle of the moon's disk. The angle is based on the curvature of the moon's edge. If the occultation takes place precisely on the middle of the moon's limb (east to west, i.e., along its equator), then the contact angle is 0°. Any deviation from that is measured in degrees either north (+) or south (-). A graze occultation has a value of either +90° or -90°, whereas a reappearance would be just the opposite of a disappearance, or 180°.

WA. This value is the Watts angle as measured from the north axial pole of the moon eastward. It is a precise measurement for use with special programs and charts.

LONG LIB. This column gives the values for the moon's longitudinal libration for the date of occultation.

LAT LIB. This column provides the libration of the moon in latitude for the date.

A B C (M/O M/O S/K). This column gives correction values if your data deviate from the recorded position. It allows the time for an event to be recomputed. The values indicate minutes of time for every degree of west longitude or north latitude change (M/O) and indicate seconds of time for each added kilometer of altitude.

DM REF NO. This is the number of the star for the predicted event, as given by the standard reference catalog *Bonner Durchmusterung*.

SAO REF NO. This is the number of the star as given in the Smithsonian Astrophysical Observatory *star catalog*. If this line is blank for any given event, it indicates that the star is not in that catalog.

HA (O / //). This is the hour angle for the moon at the time of an event, measured from the meridian to the east (negative) or west (positive) to the star.

DECL (O / //). This column supplies the declination of the star to be occulted.

RT ASC (H M S). The final column gives the known right ascension for the current epoch of the star for the predicted event.

ACCURATE TIMING OF AN OCCULTATION

We briefly mentioned in the section on instrumentation some of the methods by which accurate determinations of the precise time of an occultation can be made. We now explore the advantages and pitfalls of each method.

No matter what method is used, the accurate counting of time is essential. The only way by which such precise timing can be made is through audio signals supplied by WWV time standard stations each minute. All that is necessary for the reception of these signals is a shortwave radio. The allowable margin of error in the timing of occultations is only 0.5 sec, which would not rate highly with observers who are able to determine visually the time of occultation to within 0.01 sec. In most cases, amateurs who have spent about a year timing occultations can achieve an accuracy of about 0.03 sec, and sometimes better, in easy events. To split seconds in this way, you must have a good stopwatch with a sweep second hand. Regardless of what actual timing method is employed, the stopwatch is a part of every method. Stopwatches can be obtained with the following variations:

- ☆ *one minute*—The sweep hand requires 1 min or 60 sec to make one revolution.
- ☆ *ten second*—The sweep hand makes one complete revolution in 10 sec and provides better accuracy.
- ☆ *one second*—The sweep hand can record to 0.01 sec. The watch is electronic, and although expensive, is the best choice for accurate timing.

Some stopwatches are made with combinations of two or more metals so that contraction in extremely cold weather does not affect the accuracy of the sweep hand. Such watches are preferred over the cheaper monometallic types. Most stopwatches come with a data sheet explaining its type and proper use.

Occasionally, a stopwatch must be tested for accuracy against the time tick of a WWV radio. It is best to run the stopwatch for 1 hour and determine its rate of inaccuracy to within 1 min. You should also check your stopwatch under observing conditions, including the heat and moisture of a summer day and the bitter cold of winter. At the time of an important event, you should know any peculiarities of your watch. Occultation time is not the time to begin testing your stopwatch.

With the WWV radio signal tuned in, your stopwatch wound and checked, and your tape recorder (if you use one) loaded with a fresh cassette tape, you can use one of three methods for timing occultation events: the *stopwatch method*, the *tape recorder method*, or the *eye-ear method*.

The Stopwatch Method

The first, and most widely used, method of occultation timing is the stopwatch method. Although the method is used by about two out of every three observers, it is the method, in my opinion, that gives the least accurate reading because of the nature of the timing.

To use the stopwatch method, monitor the star as the moon's limb approaches. At the instant of occultation (or reappearance), depress the button of the watch as quickly as possible. Leave the watch running until you hear the next voice signal that announces the minute on the WWV radio. At that instant—when you hear the tone that signals the start of the minute *following* the occultation—stop the watch. Subtract the time that elapsed between the occultation and the stopping of the watch from the hour and minute signified by the tone to obtain the real value for the occultation time. For example, an observer notes an occultation of a star and starts the stopwatch; at the tone of "6 hours 48 minutes coordinated Universal Time" and the following tone which indicates the precise instant of that minute, the watch is stopped. The observer then notes the difference in elapsed time on the stopwatch, which is 31.8 sec and subtracts that value from 6:48:00.0, or:

6:48:00.0	next time-tone after event
:31.8	elapsed time
6:47:28.2	actual time of occultation

The several problems inherent in using this method can be avoided simply by using one of the alternative methods. The first and foremost problem is that one has only one chance to note the event. If you cannot get the watch started instantly, or if you start it a little too quickly in anticipation, the timing will be inaccurate. There is no way you can go back and time the occultation again. You get only one chance.

The second problem, as has been previously mentioned, is that the accuracy of the stopwatch must be consistent and its deviation well known. For example, every stopwatch, no matter how expensive or cheap, can have one or more of the following problems:

- ☆ *improper zeroing*—At the resetting of many stopwatches, the sweep hand cannot be reset exactly to zero. You can easily take this defect into account and correct for it.
- ☆ *delayed starting*—As watches get older, they often develop a lag when the button is first depressed. This lag can amount to as much as 2 sec. At first there is a lag, followed by a jump, which tends to accelerate the watch for a second or two. If the delay and the jump are predictable and constant, you can account for them and make the appropriate correction.
- ☆ *weather sensitive*—As previously mentioned, many watches behave erratically when used in humid, hot weather or very cold weather. You cannot realistically correct for this idiosyncracy because the amount of fluctuation is variable, depending on the extremes of the weather conditions and the workings of the watch.
- ☆ *sweep irregularities*—Of four stopwatches that I have used for occultation determination, all had drifting of the sweep hand as it passed around the dial. Sometimes the hand appeared to accelerate; at other times it slowed a bit. These conditions are inherent in almost every watch to some degree. But many watches have considerably less of this type of defect, and those should be used.
- ☆ *gravity influences*—If a watch is held with its face perpendicular to the ground, it will most likely run at a different rate than if it were held with its face parallel to the ground. You must do some testing of the watch while listening to WWV broadcasts to deter-

mine which position is the most accurate. The watch should be used in that position exclusively.

☆ *winding*—Overwinding will often cause a watch to run irregularly. In other watches underwinding will cause irregular running. You must predetermine the degree of irregularity and correct for it.

The accuracy of the stopwatch cannot be overemphasized; it is the heart of every method used for timing lunar occultations. Testing a watch is a simple procedure that requires at most 5 to 10 minutes; it is imperative that you know the exact correction that your watch requires.

A third problem with using the stopwatch timing method is the evaluation of your own *personal equation*, or reaction time. This subject is discussed in detail later in this chapter. For now it is important to know that your personal equation is of greater importance when using a stopwatch than when using a tape recorder.

The Eye-and-Ear Method

Occultation timings were done before the advent of devices even as simple as a stopwatch or a ship's chronometer. It is remarkable that old timings made using only the 1-sec tick of a clock were often as accurate as timings made today using fancy watches that count in 0.1-sec ticks. The eye-and-ear method of timing occultations is the simplest—and perhaps the least efficient—of the three methods by which accuracies better than 1 sec can be attained. You merely listen to audible ticks of 1-sec intervals (from a WWV radio) and estimate the tenth of a second at which time the event occurs. This method requires "splitting hairs," but, nonetheless, you can achieve some degree of accuracy by using it.

It is said that the human brain can mentally affix an image of the second interval at which an event occurred, and within the second can "mark" the fractional moment at which the occultation took place. If this is so, then the brain retains an image that can be mentally referred to later and thought about, allowing some degree of advantage over the stopwatch method.

The eye-and-ear method has one clear advantage over the stopwatch and the tape recorder: The eye and the mind respond more quickly to stimuli than does the hand. In either of the other two methods, transfer from mind to hand of the thought of the event taking place must occur, and that transfer requires a certain interval of time. By using your eyes and brain (in which to fix the image of the event), you can greatly shorten that interval. However, most observers tend to estimate at about 0.25 to 0.33 sec when using this method. Also, it is difficult to estimate with confidence unless the event takes place very near the beginning or end of a second. The best solution to problems inherent in this method is practice so that you can build your confidence that the method *can* work.

When learning the eye-and-ear method, it is best to have some type of backup system in case your efficiency fails, or in case the event takes place while the tone of the audible 1-sec beat of the WWV radio resonates over the occultation event. It is interesting that only one out of every 20 occultation observers uses the eye-and-ear method, yet those who do use it rate it very highly and are quite skilled in its use. It may be necessary to have a knack for such a method.

The Tape Recorder Method

When the sound of the WWV radio signal and the observer's voice signal of the occurrence of an occultation are simultaneously recorded on a cassette tape, it is known as the *tape recorder method*. This method is my personal preference of the three.

To use the tape recorder method, merely tune in the frequency (5, 10, 15, or 20 MHz) that you can most clearly receive about 5 minutes before the occultation you wish to record. At the same time start a small tape recorder, whose accuracy you have previously checked, so that time signals can be recorded. If the transmission of the WWV signal fades right before an event (which often happens), you will at least have a record of your voice signals on the tape. At the time of occultation, note the event as quickly as you can respond. Many people say "T" the moment they observe the event. The sound of that letter is unmistakable, and it also has the advantage that it can be "gotten off the tongue" more quickly than most other letters. Other observers use what is known as a "cricket," a small toy that can be bought in novelty and dime stores. This little metal device makes a prominent click when its metal flap is depressed. Whether the hand is quicker than the tongue, I cannot say, but I prefer the voice signal. My personal estimates show that when I use the metal cricket, my efficiency drops by as much as 1 sec as compared to the efficiency I achieve by using the voice method.

After the event, you can listen at leisure to the recorded data and determine the actual time of occultation. The tape can be played over and over again, until you are confident that you have determined the proper time. To evaluate the precise 0.1 sec at which the event took place, use an accurate stopwatch with a sweep hand, much as is done in the stopwatch method. In this case, however, start the watch at precisely the moment of the minute tone *preceding* the event, and stop the watch when you hear the voice signal (or cricket click). Do this as many times as necessary to reach the greatest accuracy. I always replay the tape three times for an easy event and at least five times for one that is questionable. Then I determine an average or mean time from all of those times I have obtained. That is the time I report.

Some problems are inherent in the method, however, and must be eliminated if the method is to be reliable. First is the problem of recorder performance, as mentioned earlier. In very cold weather, tape recorders run erratically, either because of the mechanics of the recorder, or the batteries (if the unit is so powered). Careful analysis of the recorded data, even if the recorder runs too fast or too slow, provides the accurate time, however. The second common problem with using a recorder is the stopwatch. But because the tape can be replayed many times, the event can be determined many times, thus eliminating some of the drift problems inherent in all stopwatches. Likewise, the problem of inaccuracies caused by weather extremes can be eliminated if all the data are reduced using the stopwatch *indoors* while playing the tape. The other deficiencies of the stopwatch are dealt with as described in the previous discussion of the stopwatch method.

EVALUATION OF YOUR PERSONAL EQUATION

In timing every occultation, the time that you report must be corrected by some small amount to account for the delay between the actual time the event occurred and the time that you noted on your stopwatch or tape recorder. This correction is known as your *personal equation.*

Every time you report an occultation, it is necessary to state whether or not a personal equation has been applied. If you do not make such a statement, the persons reducing the data automatically apply a 0.3-sec reduction to your time. If you are new to occultation work, I advise you to use 0.3 sec as your personal equation. As your efficiency and practice increase, your personal equation should drop to as low as 0.1 sec but not lower.

Delay time can be caused by many factors. In the stopwatch method, there can be a delay when you first start the watch as well as a delay when you stop it. In the eye-and-ear method, the delay can be caused by your attempt to decide in which fraction of the second the occultation occurred. In the tape recorder method, the delay is a result of the time it takes your brain to cause you to respond with your voice to say that the occultation has taken place. The delay factors caused by using a stopwatch to reduce taped information is virtually eliminated when you replay the tape many times and take a mean of the three best times.

The personal equation, then, might be called *reaction time*—the time required for the mind to realize that the event has taken place, recognize what must be done, and respond (either with voice or hand). For each individual, the actual time at which an event took place may be determined as follows:

Actual time = Observed time *minus* reaction time

Some Easy Tests For Your Personal Equation

Many observers wish to know exactly how their reaction time compares to that of others. Remembering that the average personal equation for the beginning observer is 0.3 sec and the best obtainable by even the most experienced observer can be only 0.1 sec, it is easy to determine at the outset just what delay time you can expect if you are a beginner.

The quickest way to determine your personal equation is to experiment with a stopwatch. Check the watch beforehand for any defects, as previously described, and set the sweep hand to zero. Predetermine a point at which you will stop the watch, precisely on the second. Let us say, at exactly 10 sec you will stop the watch. Starting and stopping the watch many times will show you just what sort of delay might be experienced during an actual occultation. But there is one big problem with such a test: You will anticipate the coming of the 10-sec mark because you can see that point on the watch dial.

An alternate method that can be tried and that is not as subject to anticipation as the first is to have a second person who is also equipped with a stopwatch give a voice signal precisely at some time, at which time you start your watch. At the next nearest minute, stop your watch, and your helper will stop his or her watch, too. Your helper has the advantage of knowing how many seconds have elapsed from the first minute before he or she gives the voice signal. You are thus caught off guard.

A more efficient method is to start and stop your watch at precise minutes, as given over a WWV radio. Start the watch precisely at the tone of one whole minute, and stop it at the tone of the subsequent minute, without looking at the watch. After you have practiced this many times, try doing it *without* the radio signal. Start the watch and mentally count the seconds, as you did when you heard them on the radio. This time, however, you are imagining second beats in your mind and not hearing them with your ear. Try starting the watch and stopping it exactly 1 sec later without looking at the watch. Once you get quite proficient at this, let the watch run for what you think is 30 sec, mentally counting off each of the seconds. Again, do not look at the watch face. This is a test of your mind's ability to think in time and react physically to the passage of that time. After about an hour's practice, you should have developed some degree of proficiency at counting mentally. By noting the tenths of a second that you are consistently off when you estimate the passing of only 1 sec, you can determine your personal equation.

If none of these methods seems to provide a consistent value, try the following method. On the WWV radio signal broadcasts, every second is denoted by a tone except those during which the announcer's voice gives the coming minute—numbers 51 through 59. Start the watch as precisely on the minute tone as possible, and let it run until you hear 10 subsequent tones (i.e., until 10 sec have elapsed). Stop the watch precisely on the tenth tone. If you miss the tenth tone, start all over again until you stop the watch precisely on the tenth tone. Then read the watch. It should read less than 10 sec because of the delay (reaction time) that resulted in your starting the watch a bit after the minute tone. The difference between the time observed on the watch (less than 10 sec but not less than 9.7 sec) and 10 sec is your personal equation. After a little practice you will find that the initial values you obtain will be substantially reduced.

DETERMINING YOUR ACCURACY DURING AN EVENT

Your personal equation has little to do with your accuracy in timing a particular event. Your personal equation may be quite low, yet poor seeing conditions or the faintness of an occulted star might make the accuracy of your timing of any particular occultation very poor. However, if you have a low personal equation, you will usually achieve closer accuracy than someone who has a high personal equation.

The reported value of an event should be the

time reading, less your personal equation, less the amount of time by which you feel your reading was inaccurate. If you think you delayed 0.5 sec before indicating that the event had taken place, your accuracy is 0.5 sec. Subtract your personal equation from the value thus obtained. Do not subtract your accuracy estimate, which serves only as an indicator of how much your reported time could deviate. If you think you timed a particular event 0.2 sec too early, put +0.2 sec next to the event in your report. If it was timed 0.2 sec too late, then report it as -0.2 sec.

After it is reported, each estimated timing is evaluated by the U.S. Naval Observatory to determine whether to accept or reject it. If there is a great deviation between the predicted and the observed times, with no probable cause, then your observations will probably be rejected. However, if you indicate a probable error by including the determined accuracy, your report would probably then be within range of the predictions, and the observation would be accepted by the United States Naval Observatory.

TYPES OF LUNAR OCCULTATIONS

Lunar occultations of stars are of three basic types, each of which has its special problems and its own value. It would be ideal to be able to time the instant a star disappears behind the moon's limb and the instant (many minutes later) when that star reappears on the opposite limb. This is difficult, however, because one limb of the moon is usually dark, which makes for accuracy in timing, whereas the opposite limb is bright, making the reappearance of a star difficult or impossible to see.

Disappearance of a Star

When the moon is in waxing phases, from new to full, stars disappear on the darkened limb. In the waning phases, stars disappear on the lighted limb, and they are, thus, more difficult to view. When stars disappear on the dark limb, particularly at about the time of the first-quarter moon, occultations of even quite faint stars are easy to observe. The value of such observations is not as great as for other, more difficult events. However, such occurrences are excellent practice for the new observer of occultations.

For timing disappearances on the dark limb of the moon, use moderate powers (i.e., 100x to 200x) of your telescope. Move the image of the illuminated portion of the moon outside the field of view so as not to impair your view of the star as the lunar limb approaches. After full moon, when disappearances of stars occur on the lighted limb, the value of the timing of the occultations becomes more important because fewer observers are willing to tackle the challenge. For stars with high magnitudes, the timing can be simple if the star's color is quite unlike the moon's. Whitish, blue, or red stars are easy to time against the yellow color of the moon. However, G-type stars are frequently occulted, and their coloration is quite similar to the moon's, thus increasing the difficulty of the observation. It is of great help, up to a point, in timing disappearances on the bright limb if you use very high powers (i.e., 200x to 400x) of your telescope. Choose a magnification that clearly and steadily shows the star yet that shows only a small fraction of the moon's disk in the field of view.

Whether the moon is in its waxing or waning phase, begin looking at the field about 5 minutes before the occultation. First, locate the star. If it is a faint one, this may take some time. Next, choose a magnification suitable for the approaching event and insert it as quickly as possible. Use the remaining time to allow your eyes to become accustomed to the brilliance of the moon and the situation under which the occultation will take place.

Reappearance of a Star

The timing of the reappearance of a star from behind the lunar disk is a more valuable observation than the timing of a disappearance. During waxing phases of the moon, when its western limb is bright, the reappearances of all but the very brightest stars are quite difficult to time. Most are impossible. However, during the waning phases, the western limb is darkened, and the reappearances are spectacular.

Even under the most favorable circumstances (e.g., when a bright star reappears midway on the moon's darkened western limb), the timing of a star's reappearance is a formidable challenge. Because the event is instantaneous, there is no prior warning, other than the time predicted on your list from the USNO. You must have some preconceived idea of *where* the star will reappear. It is under these circumstances that the data on the computer-derived predictions are invaluable. Using the *position angle* and the *cusp angle* data as guides, you can use a simple reticle cross-hair eyepiece to determine from the moon's north pole approximately where the event will occur. Some observers have etched their reticle eyepieces to predict precisely where a reappearance will take place, using a movable web that can be centered at precisely the predicted position angle. However, a little practice with a normal cross-hair reticle will allow one to determine very closely the predicted location.

For example, if the star is to reappear on the western limb, midway from the north to south poles, the position angle would be 270° (0° is due north, 90° east, and 180° south). Therefore, a cross-hair eyepiece centered with the vertical hair running through the moon's north and south pole and the horizontal hair bisecting the moon halfway from pole to pole should have the westernmost horizontal cross hair nearly at 270°. Of course, not every occultation will be so neatly placed, but a little practice will enable you to interpolate between the 90° increments to within 5° of the predicted position.

In general, I have found that the accuracy rating of reappearances is much higher than disappearances. Reappearances should be reported as accurately and honestly as possible. If your eye is searching for the reappearance only 10° away from the actual point of emergence, a delay of as much as 1 sec is common.

The reappearances of stars, even very bright ones, are not usually included on prediction sheets while the moon is in its waxing phases.

Graze Occultations

A rare and special type of occultation is the *graze occultation*, which occurs when the northern or southern polar regions of the moon skim across a star. As the moon progresses steadily eastward, the star can be seen to flicker in and out of unseen valleys and behind unknown mountain ranges (see also Chapter 6). Because of the uncertainty of the nature of the moon's poles (particularly its south pole), these grazes are extremely valuable as a mapping tool. The sophisticated photography done by the earth's spacecraft during their visits to the moon notwithstanding, many areas are unknown to us, both in their character and altitude. From the graze occultations, the altitudes, shapes, and sizes of large lunar features can be determined.

The predictions supplied by the USNO will call attention to any graze occultations within your radius of coverage, generally to within 20 to 40 miles of your observing site. Stars within that range will be so marked, and details for determining the path of the star graze will be included, allowing you to draw on a geodetic survey map the graze line, along which the star will be multiply occulted.

Graze occultations are best observed by a team of observers, each separated from the other by about 0.1 mile perpendicular to the path along which the graze will be *central. Central graze* is not the path along which the graze is *total*, or totally occulted, nor is it a *miss*, where no occultation is seen. Rather, central graze is a fine line between those extremes, where the star appears to pass midway behind any projections on the lunar pole, making it blink on and off. Each observer finds a station that can easily be identified on the geodetic survey maps and that is very near 0.1 mile from the preceding station. The distance between the stations can be measured off in feet, along a highway or by a standard automobile odometer that has increments of 0.1 mile. Each observer's latitude and longitude, as well as altitude, must be determined to within 20 feet.

For recording graze occultations, you must use the tape recorder method because the events come in rapid succession. Time both the disappearances and reappearances. I prefer to use

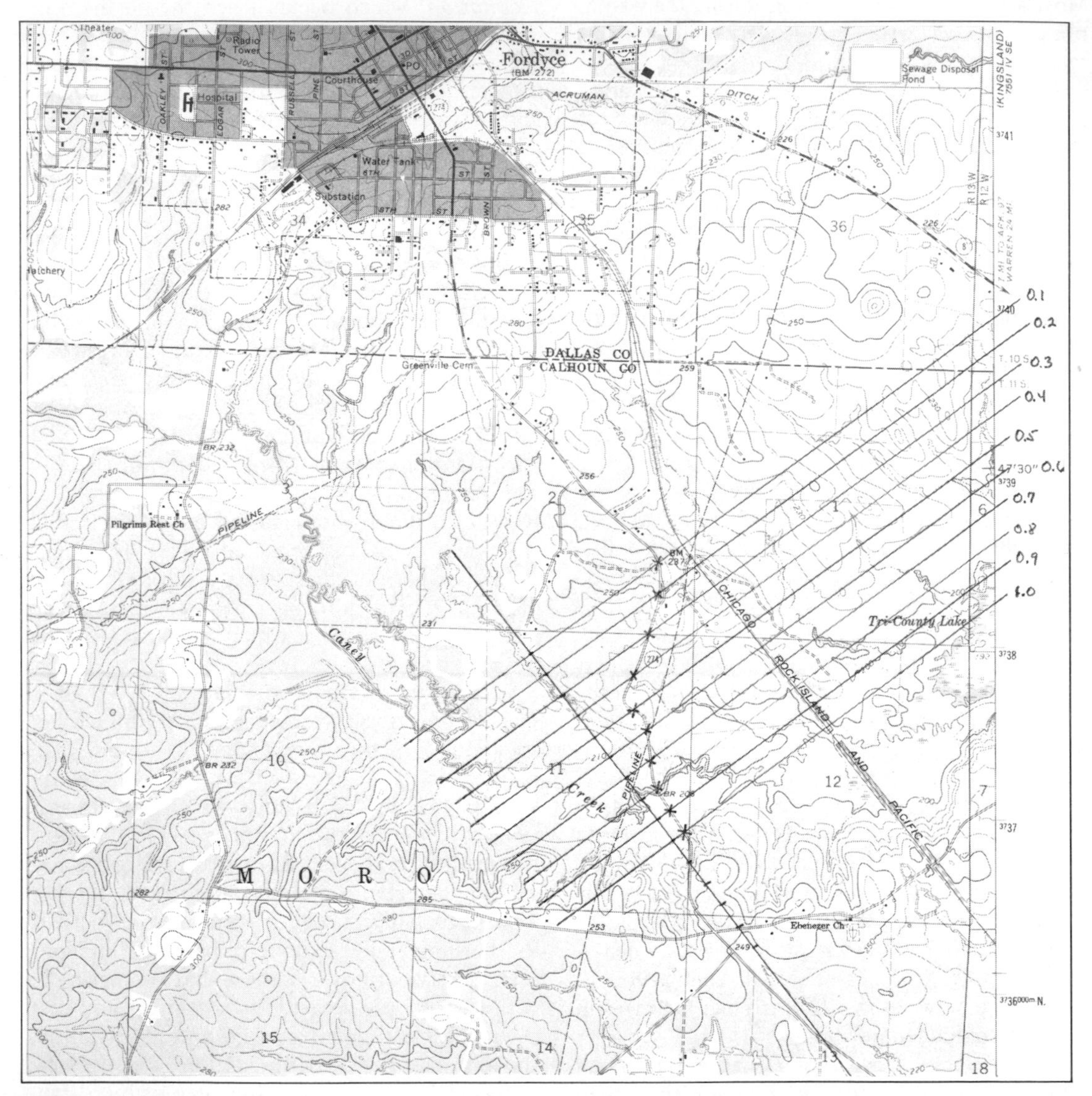

FIGURE 7-2. Copy of work map for observer placement along a highway in preparation for a graze occultation of a star by the limb of the moon. Every line must have at least one observer on it, regardless of his or her location laterally along the line. On this topographic map, the lines are separated by only 0.1 mile.

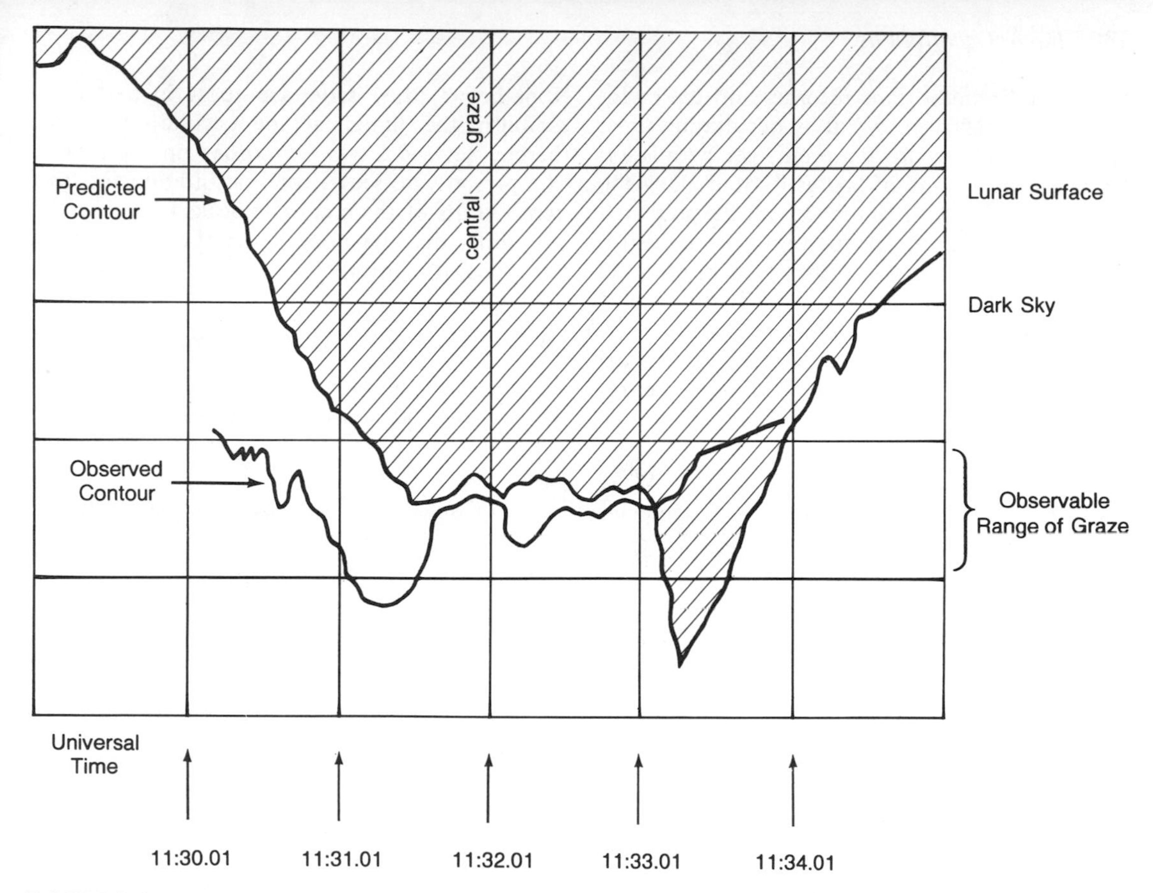

FIGURE 7-3. A relief map of the moon's south pole, determined by a graze occultation of the plant Venus, December 26, 1978.

the voice signals "T" for disappearances and "R" for reappearances because these letters can be spoken sharply, and they can be differentiated clearly on the final tape.

One word of caution concerning setting up for a graze occultation. Because the occultation takes place at night, it will be dark. Although setting up along a road can be dangerous, most roads provide the best perpendicular to the graze path, and they can be quickly located on the survey maps to determine latitude and longitude. Therefore setting up on a road, rather than on private property, is preferred. Nonetheless, it is imperative that each observer set up as far from the road way as possible because approaching traffic might not see the station in the dark of night. It helps to have an observing buddy who can help look out for traffic or wave a red flashlight to warn approaching automobiles of your location. One author recommends setting stations up along railroad tracks for convenience in determining position and spacing. Do not ever do this. Not only is every life along that track at stake, but the equipment as well. Should a fast train approach, you could easily make a simple jump to get off the tracks. But what of the equipment left behind?

Graze occultations can be exciting and rewarding when your map of the lunar pole is finally completed from the data of the observing team. Graze occultation timing is an excellent club or school project. Each team member should be given a checklist of items he or she is required to take—flashlights, tape recorders, shortwave radios, telescopes, and so on. Each team should have at least one extra set of batteries for both the tape recorder and the radio, should the originals play out by occultation time. About a week before the event, get permission from the occupants of houses located along the graze path to be on their property at the predicted time of occultation. The local sheriff should also be told. One does not wish to be interrupted right in the middle of a graze by flashing lights and angry policemen and homeowners.

And, in all fairness, be sure to inform the persons on the extreme north and south stations that they might not see a graze occultation. Nonetheless, they serve a purpose in that they establish the limits of the occultation. They thereby set the lowest elevations on the moon. The person who sees a total occultation sets the upper limits of the mountain peaks (the observer who notes that the moon does not cover the star, but misses it slightly).

Occultations with Multiple Events

The moon goes through cyclic periods that occasionally cause sequences of occultations of

FIGURE 7-4. Diagram of Hyades cluster as seen in binoculars to magnitude 10.0. The figures given in the illustration are magnitudes with decimals omitted.

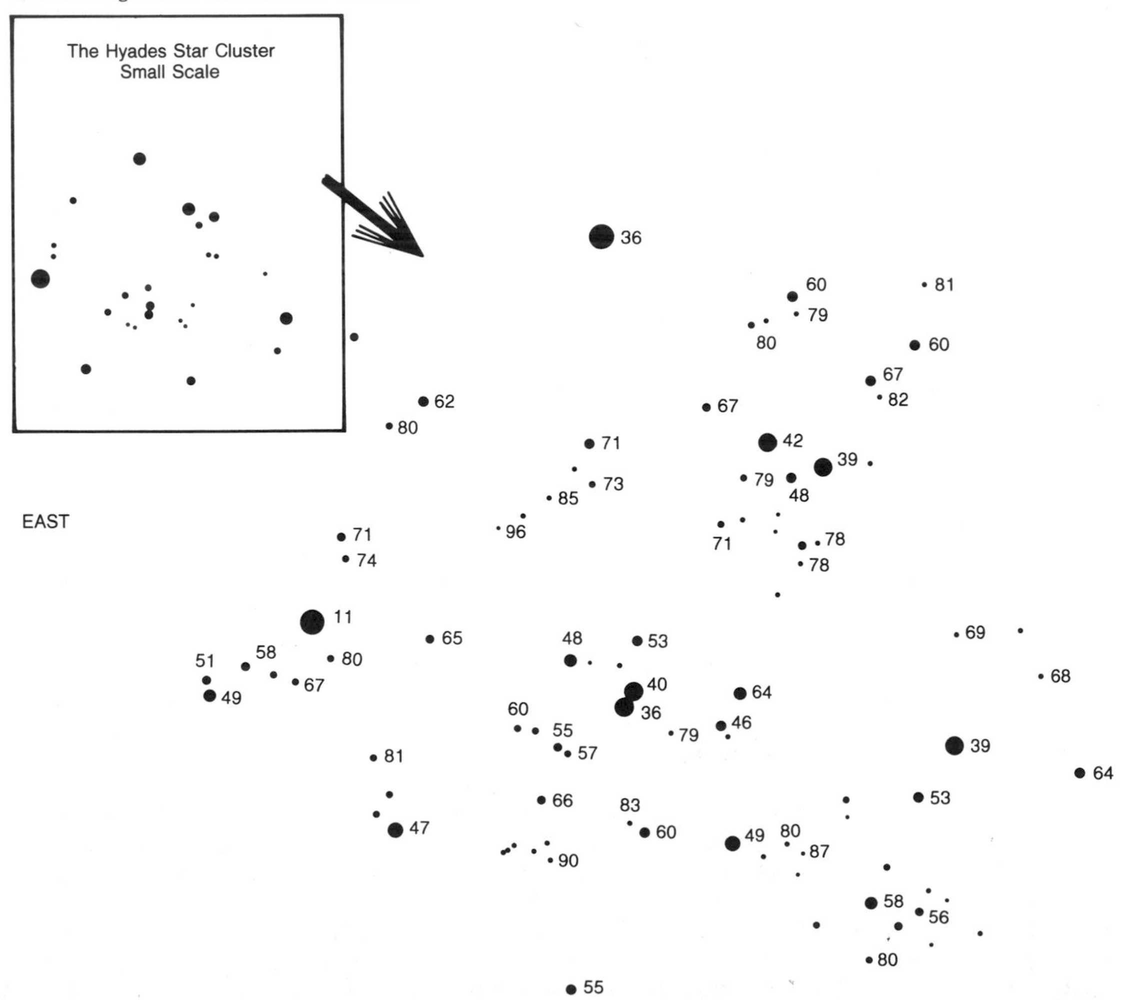

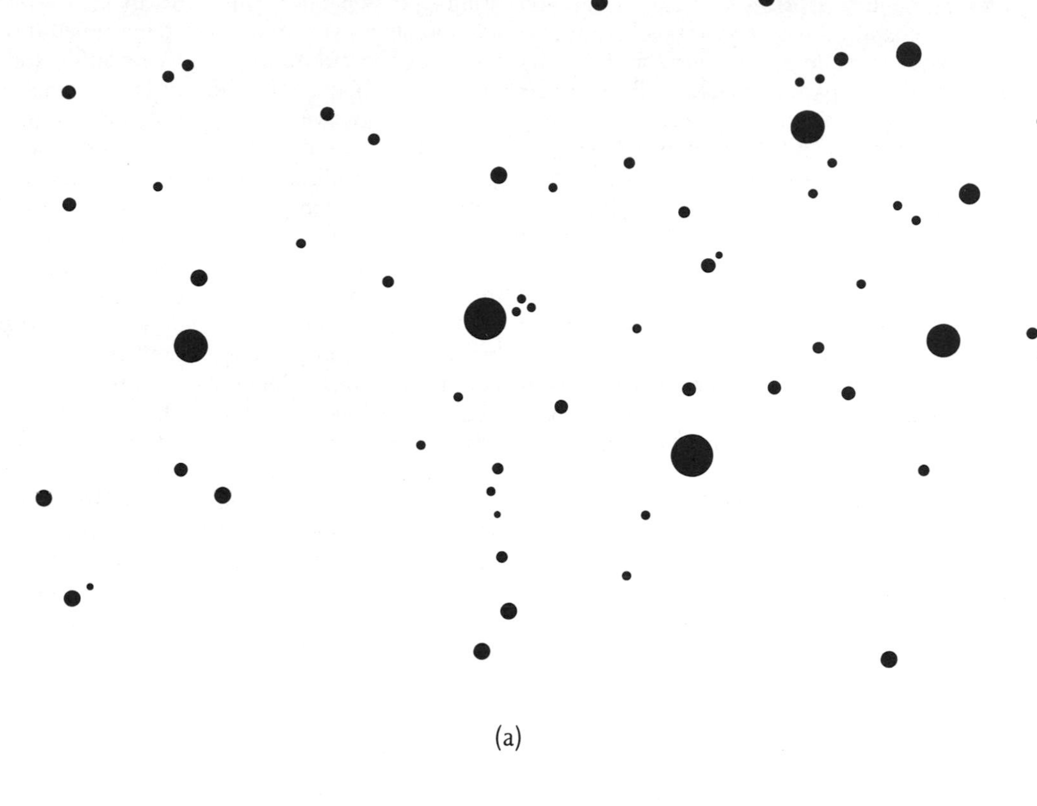

(a)

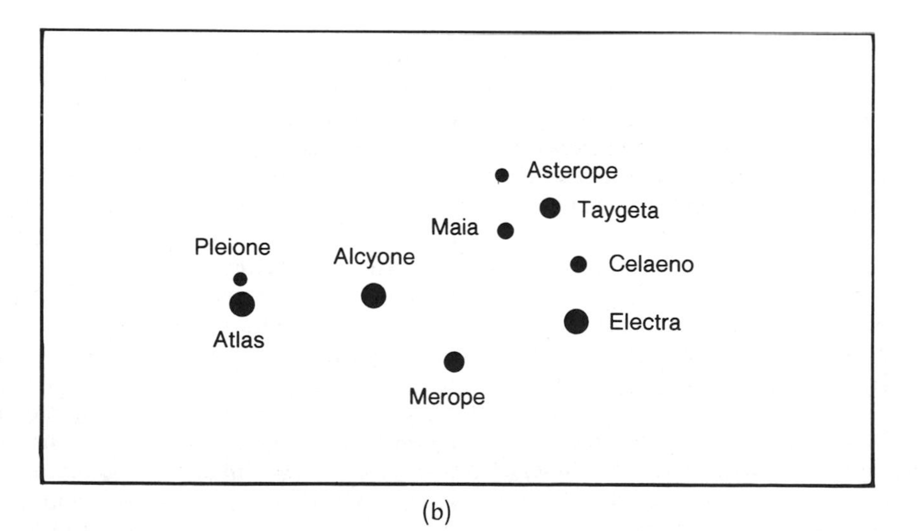

(b)

FIGURE 7–5. (a) The Pleiades cluster to magnitude 10.0. (b) Names of the nine brightest Pleiades.

dense star areas, such as the Pleiades and the Hyades star clusters. It is necessary that you know beforehand of such occurrences because it usually requires at least an hour for the many stars in the moon's path to be occulted. Observers of such events can get really excited as they watch the moon encroaching on a dense group of stars and covering them up one at a time. The only suitable method to use in timing these events is the tape recorder method because occultations will probably occur in rapid succession. If USNO predictions are being used, say, for example, on a Pleiades occultation, some stars that are occulted might not be listed. These stars should be timed and reported, however, using their approximate positions in relation to the primary stars of the cluster if no positions can be determined.

The next series of Pleiades occultations begins in 1986. The Hyades in Taurus are frequently in the moon's path; some of its outer members are occulted each month.

Occultations of Double Stars

Many known double stars are occulted by the moon. The less that is known of a double star, the more valuable are observations of its occultation. The separation distances of some spectroscopic binary stars have been determined through total lunar occultations. Many new binary stars are the result of the resolution of known single stars that were observed during a total occultation.

During the occultation of a double star, the components blink out in rapid succession. The star often appears to dim quickly, then suddenly blink out, signifying that a double star may have been resolved. If the star is known to be double, the USNO predictions will indicate this fact in the column containing the double star code. If there is a letter next to a star prediction, then the event is double. The event is quite rapid, as one star blinks away followed by the other. You should try to make as accurate an estimate of the time interval between the two as possible, although the accuracy will be low. Some stars of great separation show two components in the telescope, whereas other close pairs appear as only one star. Remember that *any* star being occulted could be a potential double star. Always take notice of any star that seems to fade before occultation.

PROBLEMS OF THE OCCULTATION OBSERVER

Unlike many other projects the amateur astronomer can undertake, the observation of occultations is a relaxing and rewarding experience. With a little patience, a dedicated observer can build up quite a record in the occultation business. However, as with all aspects of amateur observing, you will night after night contend with certain unpleasant things in order to make the observations meaningful. As if mosquitos in summer and the bitter cold of winter were not enough, other factors affect not only your attitude but also the reliability of your estimates. The following are only some of these factors:

1. *Seeing conditions.*—Bright stars twinkle in the sky; they flash with varying brightness. Faint stars twinkle as well, but the alternating brightness often makes them fall below the threshold limit of the observer and the telescope. Such a condition causes several false alarms before the actual event. During times of great atmospheric turbulence (particularly when the star is near near the horizon), the degree of inaccuracy increases and should be accounted for in your final report.

2. *Haze and clouds.*—You can frequently start occultation work in spite of heavy haze and clouds. Occasionally, clouds cause some false "occultations," but the cause of most of these is obvious. Sometimes, however, no cloud is visible, but one will momentarily cover a faint star. If the star reappears just before actual occultation, nothing is lost. However, if it is obscured only seconds before the event, you might not be able to see it reappear. You might thus be deluded into thinking that the star's disappearance behind the cloud was the actual occultation. Your timing would be for the cloud, and not the lunar limb. Likewise, the faint clouds and dense patches of haze might cause a star to suddenly "appear" from behind the moon, even though the true reappearance had occurred the moment prior.

3. *Preconception.*—Bias, or preconception, can make you somewhat lazy. Because you know from the predictions just when the occultation is to occur, and if you are uncertain about your timing, it is tempting to slant it toward what was actually predicted. Make every attempt to avoid "remembering" when the occultation is predicted to occur, and time only what you see.

After a month of recording star occultations, you may have accumulated data that no one else has. Each observer's estimates are important, and all data are welcome by those who use them. Twice a year send your observations to:

International Lunar Occultation Centre
Astronomical Division
Hydrographic Department
Tsukiji-5, Chuo-ku
Tokyo, 104 Japan

Membership and reports from occultation observers are also welcome by the following organization, comprised chiefly of amateur astronomers:

International Occultation Timing Association
(IOTA)
Post Office Box 596
Tinley Park, Illinois 60477

BIBLIOGRAPHY

Jones, Aubrey, *Mathematical Astronomy with a Pocket Calculator*, pp. 235–250. New York: John Wiley, 1978.

Link, F., *Eclipse Phenomena in Astronomy*. New York: Springer-Verlag, 1969.

Muirden, James, *The Amateur Astronomers' Handbook*. New York: Thomas Y. Crowell, 1974.

Roth, Gunter D., *Astronomy: A Handbook*, pp. 313–324. Cambridge, MA: Sky Publishing Corp., 1975.

8

MARS: THE RED PLANET

Since the Viking and Mariner spacecraft visited the vicinity of the mysterious planet Mars, some of the wonder and mystique have disappeared from our attitudes toward the red planet. We have now closely examined the surface, and it appears that there are no creatures there. Indeed, not even the basic organics necessary for life exist in the Martian environment. Yet it appears that Mars at one time was quite active both geologically and meteorologically, considerably more so than at present.

Near the beginning of the twentieth century, astronomer Percival Lowell had the theory, based largely on the ideas of Italian astronomer Schiaparelli, that Mars was a dying world and the creatures that inhabited the planet were starving from the lack of water. Lowell's theory was that, knowing the Martian spring and summer would result in melting the Martian polar caps, the inhabitants built a vast network of canals to take the life-giving moisture of those caps into the arid equatorial regions of Mars.

We now know that Mars does not have canals and that there are no inhabitants who wait eagerly for the advent of spring. Yet Lowell was correct about one point of his theory—Mars is, indeed, a dying world. The processes we see today on Mars are the result of phenomena that occurred only yesterday in the life of our solar system; many of the processes continue to this day, only on a considerably smaller scale. Massive volcanoes not yet eroded by the winds of Mars still loom on the Martian surface. And still fresh canyons appear as the result of the activity in the recent past.

AMATEUR STUDIES OF MARS

Because of the many uncertainties that still exist regarding the visible phenomena of the Martian surface and atmosphere, Mars continues to appeal to the amateur astronomer. In addition, we can rationalize in our human way that the spacecraft may have missed something. Why do the maria, or dark areas, of the Martian plains appear to darken when the polar caps melt? What causes the mysterious "blue clearing" and the circulation of the Martian clouds? We still do not clearly understand why the darker features change shape and size erratically as seen from earth. Spacecraft have provided little additional knowledge of these phenomena.

Mars comes into opposition with the earth every 2 years and 50 days on an average, yet each two successive oppositions are unique because the orbit of Mars is considerably more eccentric than that of earth. Favorable oppositions of Mars occur approximately every 2 years, either in January or February, or in late summer. Those that occur in January or February are known as *aphelic oppositions*, and they show Mars as a very small disk of only about 13″ arc. The oppositions that occur in summer, known as *perihelic oppositions*, bring Mars closer to the earth and sun so that the disk is twice the size as it is in the aphelic oppositions. From the United States, Mars is seen quite high in the sky during aphelic oppositions and very low in the south during perihelic oppositions. Consequently, the conditions for observing the planet are improved during the winter months even though the disk is considerably smaller than during the summer oppositions.

Observations that the amateur astronomer can pursue are badly needed for our better understanding of Mars. Areas of study within the scope of nonprofessional equipment should include the following:

- ☆ Studies of the Martian polar regions and the caps.
- ☆ Long-term programs examining the Martian atmospheric phenomena.
- ☆ Patrol observations for brightenings on the Martian maria or plains.

FIGURE 8-1. A map of Mars, drawn in Mercator projection, showing major features visible in amateur instruments. See key to features in Table 8-2.

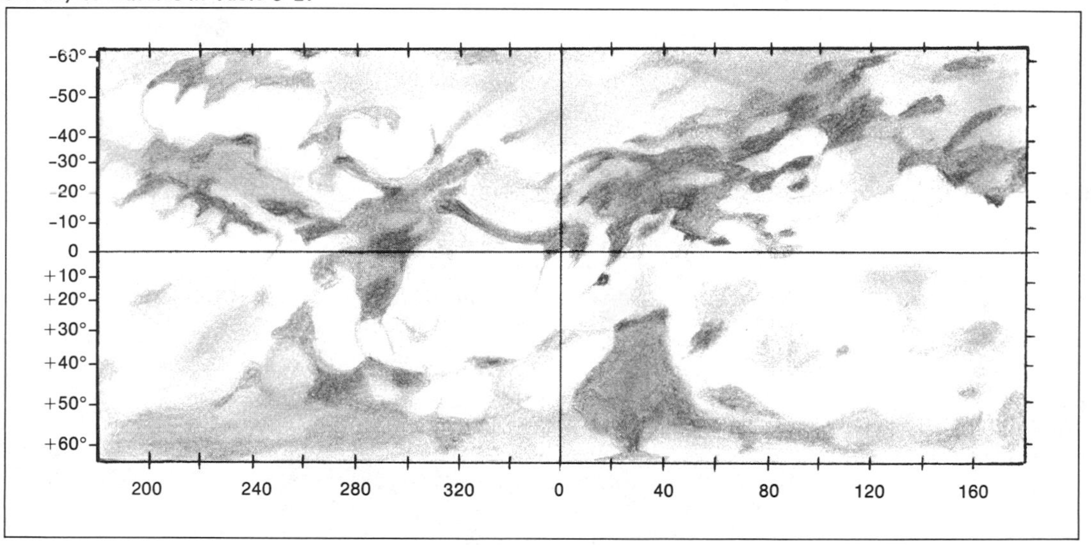

TABLE 8-1. Martian physical data.

Distance from the sun	228 million (km) ave.
Orbital velocity	24.1 km/sec
Length of Martian year	687 earth days
Equatorial diameter	6800 km
Maximum size of polar caps	2200 km
Mass (X earth)	0.11
Albedo	0.15
Maximum magnitude	-2.8
Maximum apparent diameter	26″
Atmospheric constituents	N, H_2O, CO_2, Ar

TABLE 8-2. Features of Mars.

Feature	Longitude	Latitude	Feature	Longitude	Latitude
Acidalium Mare	30°	+45°	Hellas	290°	-40°
Aeolis	215°	- 5°	Hellespontica Depressio	340°	- 6°
Aeria	310°	+10°	Hellespontus	325°	-50°
Aetheria	230°	+40°	Hesperia	240°	-20°
Aethiopis	230°	+10°	Hiddekel	345°	+15°
Amazonis	140°	0°	Hyperboreus Lacus	60°	+75°
Amenthes	250°	+ 5°	Iapigia	295°	-20°
Aonius Sinus	105°	-45°	Iscaria	130°	-40°
Arabia	330°	+20°	Isidis Regio	280°	+20°
Araxes	115°	-25°	Ismenius Lacus	330°	+40°
Arcadia	100°	+45°	Jamuna	40°	+10°
Argyre	25°	-45°	Juventae Fons	63°	- 5°
Arnon	335°	+48°	Laestrygon	200°	0°
Aurorae Sinus	50°	-15°	Lemuria	200°	+70°
Ausonia	250°	-40°	Libya	270°	0°
Australe Mare	40°	-65°	Lunae Lacus	65°	+15°
Baltia	50°	+60°	Margaritifer Sinus	30°	- 2°
Boreum Mare	90°	+50°	Memnonia	160°	- 2°
Boreosyrtis	290°	+55°	Meroe	285°	+35°
Candor	90°	+10°	Meridiani Sinus	0°	- 5°
Casius	260°	+40°	Moab	350°	+20°
Cebrenia	210°	+50°	Moeris Lacus	270°	+ 8°
Cecropia	320°	+60°	Nectar	72°	-28°
Ceraunius	95°	+20°	Neith Regio	275°	+35°
Cerberus	205°	+15°	Nepenthes	260°	+20°
Chalce	0°	-50°	Nereidum Fretum	55°	-45°
Chersonesus	260°	-50°	Niliacus Lacus	30°	+30°
Chronium Mare	210°	-58°	Nilokeras	60°	+25°
Chryse	30°	+10°	Nilosyrtis	290°	+42°
Chrysokeras	110°	-50°	Noachis	330°	-45°
Cimmerium Mare	220°	-20°	Ogygis Regio	65°	-45°
Claritas	110°	-35°	Olympia	200°	+80°
Copais Palus	280°	+55°	Olympus Mons	133°	+18°
Coprates	65°	-15°	Ophir	68°	- 8°
Cyclopia	230°	- 5°	Ortygia	0°	+60°
Cydonia	0°	+40°	Oxia Palus	18°	+ 8°
Deltoton Sinus	305°	- 4°	Oxus	10°	+20°
Deucalionis Regio	345°	-12°	Panchaia	200°	+60°
Deuteronilus	0°	+35°	Pandorae Fretum	340°	-25°
Diacria	180°	+50°	Phaethontis	155°	-50°
Dioscuria	320°	+50°	Phison	320°	+20°
Edom	345°	0°	Phlegra	190°	+30°
Electris	190°	-45°	Phoenicis Lacus	110°	-12°
Elysium	215°	+30°	Phrixi Regio	70°	-40°
Eridania	220°	-45°	Promethei Sinus	280°	-65°
Erythraeum Mare	40°	-25°	Propontis	185°	+45°
Eunostos	220°	+22°	Protei Regio	50°	-23°
Euphrates	335°	+20°	Protonilus	315°	+42°
Gehon	0°	+15°	Pyrrhae Regio	38°	-15°
Hadriacum Mare	270°	-40°			

TABLE 8–2. (cont'd)

Feature	*Longitude*	*Latitude*	*Feature*	*Longitude*	*Latitude*
Sabaeus Sinus	340°	- 8°	Thyle II	230°	-70°
Scandia	150°	+60°	Thymiamata	10°	+10°
Serpentis Mare	320°	-30°	Tithonius Lacus	85°	- 5°
Sinai	62°	-25°	Tractus Albus	80°	+30°
Sirenum Mare	155°	-30°	Trinacria	268°	-25°
Sithonius Lacus	245°	+45°	Trivium Charontis	198°	+20°
Solis Lacus	85°	-35°	Tyrrhenum Mare	255°	-20°
Styx	200°	+30°	Uchronia	260°	+70°
Syria	100°	-20°	Umbra	290°	+50°
Syrtis Major	298°	+10°	Utopia	250°	+50°
Tanais	70°	+50°	Vulcani Pelagus	15°	-35°
Tempe	70°	+40°	Xanthe	50°	+10°
Thaumasia	75°	-30°	Yaonis Regio	320°	-40°
Thoth	256°	+30°	Zephyria	190°	-12°
Thyle I	180°	-70°			

☆ Studies of seasonal changes, including correlations with the melting of the polar caps.

☆ Visual and photographic studies of surface features.

Whatever program you undertake, it is better that you do not attempt to accomplish all objectives listed above. Concentrate fully on one or two closely related studies, such as the correlation of the appearance of the polar regions in conjunction with predicted seasonal changes. There is no substitute for dedicated systematic study of the planet. Try to overcome the urge to put too many irons in the fire, and restrict your observations to a well-developed program. Basic physical knowledge of Mars is summarized in Table 8–1.

TELESCOPES FOR OBSERVING THE MARTIAN SURFACE

The telescope required to observe the Martian phenomena depends on the type of program you plan. Obviously, if your objective is to patrol the Martian environment photographically, the telescope requirements would be different from those of a telescope used to monitor for Martian clouds, using selected color filters.

For serious study of Mars, you should use at least a 4-inch refractor or a 6-inch reflector. The best telescope for an amateur to use to make best use of resolution, light gathering, and—perhaps above all—budget, is a 10-inch Newtonian reflector. A 10-inch aperture is just about the size limit that has power capable of resolving the currents of air in our atmosphere. Anything larger would not be effective on nights of average air steadiness. Smaller telescopes can often actually be more efficiently used on nights of poor seeing conditions than can larger telescopes.

One prime requisite for your Mars telescope is that it be motor driven and have the capacity for accurate tracking. Because Mars shows at best a disk less than half that of Jupiter, higher powers (perhaps 70x per inch of aperture) are necessary for observing its disk. The fine detail present on the Martian surface requires your best concentration. The necessity for making constant adjustments to a nontracking telescope will eventually frustrate even the most patient observer.

Magnification

The disk of Mars is quite small, even when the planet is closest to us, and thus very high magnification is required to view its subtle changes. The optimum magnifications, based on tests with a range of telescopes, all at the same location operating under the same conditions, are recommended as follows:

Aperture	*Magnification*
6 inch (15 cm)	300x
8 inch (20 cm)	400x
10 inch (25 cm)	450x
12 inch (30 cm)	500x
14 inch (35 cm)	550x
16 inch (40 cm)	600x

On nights with very poor seeing conditions, observations of scientific quality are impractical. If the steadiness of the air changes rapidly and if there are some moments of average steadiness, magnifications of about half the amounts given above would permit some patrol work. The efficiency and that reliability of fine detail that are discerned at lower magnifications are not as great as with higher magnifications.

Use eyepieces of the highest quality, preferably of orthoscopic design, no matter what type of telescope is used. Top quality eyepieces permit the finest color correction, improved eye relief, and the best resolution as opposed to that obtainable by using cheap, inexpensive makes.

Filters for Observing Mars

Observation of Mars, more than any other planet, requires good filters. They often make the difference between being able to see some subtle marking and not being able to see it. A good set of optical glass Wratten (or equivalent) filters is a must for the serious observer, both to cut down unnecessary glare from the Martian disk and to accentuate fine detail. It is this detail that shows the start of many of the large-scale changes on Mars that are of scientific interest to the amateur.

Although the optical glass filters are superior, gelatin sheet filters may be cut and mounted in the standard cardboard frames available for mounting 35-mm color-slide film. Gelatin filters offer only the advantage of being cheaper than the glass. They will wear out and get scratched and stained when fine optical glass will not. Gelatin filters are usually simply held by hand (requiring the use of an otherwise free hand) between the eyepiece and your eye. Of course, the film should not be touched, because fingerprints are difficult to clean from the soft gelatin material. Glass filters are normally manufactured by commercial firms to thread directly onto the base of standard eyepieces. If the eyepiece is not threaded, the glass filter too may be simply held by hand.

Table 8–3 lists the various filters of interest to the amateur astronomer. Notice that not all filters can be used to any advantage when observing Mars whereas several others can be of considerable help.

Most drawings of Mars should be made using either the orange (No. 21) or the red (No. 25A) filter. It is helpful when observing Mars to run through quick overviews with all the filters to check for various features that might be visible through one filter but not through another. However, after you have made these checks (say, for the blue clearing), insert either the orange or red filter for critical observations and drawings. It is best to use the orange filter for photography because the red filter is so dense that a factor of up to 8 times normal is necessary for the proper exposure.

Auxiliary Equipment for Observing Mars

Little else is needed for the systematic observation of Mars except a good telescope and steady skies. By using a few auxiliary items, your study of Mars can be greatly expanded. Some are:

☆ *filar micrometer* This device allows you to determine accurately the placement in latitude of Martian features as well as to determine precisely the changes in size of prominent areas (e.g., the diminishing polar cap as the Martian summer approaches). The

TABLE 8-3. Types of glass filters and their uses.

Wratten Filter Number	*Color*	*Peak Transmission*[a]	*Use to Observe*
47	Violet	470	Not particularly good for observing Mars; can possibly use for heavy blue clearing.
80A	Blue	486.5	Use every night to search for blue clearing and to look for high (H_2O) clouds in atmosphere.
58	Green	538	Use for observations at the "melt line" around polar caps and searching for yellow dust storms.
12	Yellow	583	Not much advantage when used for Mars; might accentuate the outline of polar caps and white clouds.
21	Orange	593	Very good filter for Mars because it reduces intensity of the reddish coloration and allows observers to note very fine delineation on the red plains; in addition it accentuates the maria, showing mottling within them.
25A	Red	617.2	Much the same advantages as for No. 21, but with much better color differentiation between red plains and dark maria. Red filter is quite dense and can be used to advantage only with lower powers and steady seeing.

[a]In nanometers.

filar micrometer is becoming available from commercial sources. The use of each instrument is dependent on the method by which it was made. Most measurements made with such an instrument are made in relation to the total disk size. For each date the *American Ephemeris and Nautical Almanac* gives a precise angular size of the Martian disk (in seconds of arc). Measure the Martian feature in proportion to the disk size for that date.

☆ *camera* Photography of the Martian disk is discussed briefly later in this chapter. It is covered in detail in Chapter 14.

☆ *observing forms* See Appendix IX.

☆ *accurate time* A watch accurate to 1 minute of WWV radio transmission from shortwave radio is necessary. This aids in the accurate determination of Martian longitudes of various features, and can be used in indirect determination of the east-west extent of a feature.

☆ *The American Ephemeris and Nautical Almanac* gives physical data for each date throughout the year, such as longitude of central meridian for 00h 00m Universal Time, phase angle of Mars, size, magnitude, and so on.

OBSERVATION OF THE POLAR CAPS

Like earth, Mars has two polar caps, although neither is as densely deposited on the Martian poles as on those of earth. A tilt of up to 24° of Mars as it approaches and recedes from the earth allows you to view one pole well, although the northern cap is easily visible only during the aphelic oppositions. Thus it is somewhat more difficult to study.

Prior to the Viking spacecraft visits to Mars, it was predicted that the polar caps of Mars were probably composed of frozen carbon dioxide (CO_2)—or "dry ice"—with traces of water as well. This hypothesis was first made by astronomers R.B. Leighton and B.C. Murray in 1966, and it was confirmed partially by the 1969 visit to Mars by the Mariner spacecraft. Beginning with the spectroscope's use in planetary astronomy, scientists have been at odds as to the nature of the caps, whether they were water ice or carbon dioxide ice. As recently as 1952 photoelectric infrared studies by G.P. Kuiper suggested that the caps were composed predominantly of water vapor deposited thinly on the Martian poles, with little or no CO_2 being present. By contrast, the Mariner flyby in 1969 revealed a cap composed predominantly of frozen CO_2, and having little or no water vapor. More recent trips by the Viking probes, which landed on the surface of Mars, proved both schools of thought to be correct. As Mars recedes from the sun and the Martian autumn and winter approach, the little water vapor present in the atmosphere is slowly deposited at the poles, because the freezing point of H_2O is higher than CO_2. Then, as Mars rapidly cools below the freezing, or sublimation, point of CO_2 (-110°F), the CO_2 is also deposited over the thin sheet of water ice.

Therefore, neither Kuiper nor the Mariner craft was wrong. If Kuiper's observations were made during the Martian early spring or late autumn, he would have seen an ice cap, composed chiefly of water, from which the CO_2 had sublimated because of the higher temperature. Similarly if the temperature was below -110°F, the Mariner craft may have recorded a cap composed of CO_2 that covered any sign of the water ice beneath.

Observations of the North Polar Cap

The northern polar cap of Mars apparently never completely disappears during the aphelic summers. E.C. Slipher, of the Lowell Observatory, determined that the cap never shrinks below an average width of 6°, usually greater, and may reach a maximum expanse of 72°. During the advent of autumn on Mars, a haze stemming from the polar regions develops, and any further shrinking of the northern cap ceases. Even during late summer, when one might expect the maximum recession of the cap, you can sometimes see that it stops shrinking and actually increases in size. Interestingly, the growth of the cap is always associated with the reappearance of the haze in the northern polar region. The unpredictable nature of this cap can perhaps be explained by the following three factors:

- ☆ *The topography of the northern polar region* may allow for greater deposits of ice or CO_2 in the higher altitudes. Exposure to Martian winds may also increase the effective chill.
- ☆ *Average temperatures of the polar regions,* depending on planet-wide meteorological conditions, may vary during certain seasons, and the distribution of temperature gradients may change from Martian year to year.
- ☆ *The density of the deposits* at the caps may also vary in the amount of mass of either the CO_2 or H_2O vapor.

Further study of the degree of melting of the northern polar cap are important and desirable, and the effect and cause or the polar haze associated with this cap are equally important.

Observations of the South Polar Cap

During perihelic oppositions of Mars, the large south polar cap is visible to observers on earth. The south polar cap seems to go through more changes than its northern counterpart. During approaches to Martian summer at perihelic oppositions, the tilt of Mars' axis of rotation can change from 5° to 24°, giving excellent views of the cap. And the closer proximity to the earth during these close approaches helps astronomers make more detailed studies of the southern cap.

Considerable meteorology is associated with the thawing and melting of the southern ice cap on Mars. It is this defrosting that enables Mars to warm considerably as a result of decreased reflection of sunlight from the large cap. The many cloud formations that are discussed in following pages are the result, either directly or indirectly, of the thawing of this large cap. As spring on Mars advances, the southern polar cap appears to split into two segments, the result of rapid thawing of a feature known as Novus Mons, or the Mountains of Mitchell. This large mountain range protrudes high above the cap remnant and is the first visible delineation of the cap seen at each opposition. Around the edge of the southern cap, you can also often see finely detailed rifts that emanate from the melting cap and stretch delicately into the peripheral plains. The meteorological conditions in the Novus Mons area are of particular importance to astronomers.

Procedures for Observing Martian Polar Caps

Both the northern and southern polar caps are studied similarly, although each cap has its own characteristics. Five areas of detailed studies enable the amateur astronomer to provide important patrol information, as follows:

1. Examine the caps carefully for such meteorological phenomena as the formation of clouds, mists, or fogs when the polar caps begin to melt and again when they begin to refreeze. Make examinations for such meteorological phenomena throughout Martian spring and midsummer, and again throughout the autumn when the cap is reforming. The cloud formation will usually be confined to the hemisphere in which the cap under observation is. So, during perihelic oppositions (every 13 years, roughly), the southern cap should instigate meteorological activity in the southern hemisphere. Likewise, at aphelic opposition, the northern cap will be the precursor of activity in the northern hemisphere. Predictable meteorological activity that might occur is the formation of the north polar haze during Martian summer and the formation of white clouds in the Hellas basin during July and August when Mars is at perihelic opposition. Green, yellow, and orange filters aid in the study of such developments.

2. Draw carefully, or measure on photographs, the size and shape of the polar cap as it begins melting. Continue your drawing and/or measuring until Mars is no longer visible. Measurements of the polar caps can be made in two ways: (1) by using a filar micrometer, and (2) by using a reticle eyepiece. I have heard of successful observers using the *drift method* (described in Chapter 4), which is convenient if you have neither a filar micrometer or an eyepiece.

☆ *the filar micrometer*—This instrument allows

you to make precise determinations of the extent of the polar cap. First take a measurement of the size of Mars from north pole to south pole to determine the micrometric extent for any given day. Then convert it into thread measurement per degree by finding the angular extent of Mars on that date from the *American Ephemeris and Nautical Almanac* (AENA). Then measure the polar cap width and convert it similarly. With the advent of micrometers available to the amateur astronomer at a reasonable cost, many contributions have been and continue to be made with this instrument.

☆ *a reticle eyepiece*—By using a graduated reticle inserted into an eyepiece whose field is well known (see Appendix IV), you can extrapolate those gradations of the reticle in fractions of the field of view. Consequently, by determining the angular extent of Mars' disk on the night of determination (using the *AENA*), you can compute the division-per-degree ratio of the planet's surface. Then it is a simple matter to determine proportionately the total angular extent of the width of the cap in relation to the 180° pole-to-pole extent of the planet.

Such measurements enable you to compile a "melting" or "freezing" curve that graphically displays the changes in size of a cap throughout its appearance. Because the amount of atmospheric activity is thought to be attributable to the degree of melting, the changes in size give you the percentage of the polar cap that is melting or freezing.

3. Examine the caps under high magnification during moments of very steady conditions to search for any significant features within the caps, such as Novus Mons in the southern cap. It is important to note the time and date when you first saw the feature, any changes in intensity and character, and the date when you last see it. You can obtain valuable information concerning the temperature, the terrain of the land, and the relation to atmospheric phenomena from the visibility of features of polar caps.

4. Measurements and drawings of the polar cap melt line are of interest to the amateur astronomer. This dark band surrounding the periphery of the polar cap develops as the cap melts in early Martian spring, and it is visible until summer. These bands are probably caused by the melting of the water ice and the dispersion of liquid from that portion of the cap. Because the amount of H_2O that may be locked up in a cap can vary depending on conditions previously described, the melt line also can vary. Studies comparing the rapidity of polar cap recession and the width of the melt line are valuable. Record the time when you first see the melt line and when it disappears.

STUDIES OF MARTIAN ATMOSPHERIC PHENOMENA

Mars is not a static world; it is a planet still alive both geologically and meteorologically. The atmosphere of this small planet is representative of the climate, the seasons, the chemical constituents of the air and ground, and—of course—the melting of the polar caps. There is no doubt that the seasonal variations in Mars' atmosphere are directly related to the melting and the sublimation of the polar caps as Martian spring progresses. As the atmosphere condenses at the poles, the formation of clouds and transient veils diminishes, and you can monitor these events readily.

Other than a telescope of moderate aperture, what is most needed for the studies of Martian atmospheric phenomena is a good working knowledge of the Martian terrain as it normally appears. Good maps of the Martian features aid you in quickly identifying transient clouds superimposed over major known areas.

Five types of atmospheric phenomena are suited for studies by amateur astronomers. Except for the efforts of these nonprofessional astronomers, very little investigation from earth-based observatories is being made into these phenomena of Martian meteorology, which are: (1) whitish blue and white clouds; (2) yellow clouds, (3) dust storms, (4) the blue clearing, and (5) the W-shaped clouds.

Whitish-Blue and White Clouds

The whitish-blue and the white clouds are attributed to near-surface fogs and mists or perhaps to actual deposits of frost in sheltered depressions on the Martian surface. The very white clouds appear to increase in number and the area they cover at about the time the polar caps melt, and they are not necessarily restricted to one hemisphere. Thus, the white clouds are thought to be seasonal occurences, beginning in Martian spring and ending in early autumn.

In addition to being seasonal occurrences, it is possible these whitish clouds are daily occurrences as well, forming in early morning on the Martian terminator and disappearing in the heat of day. It is important for amateurs to monitor for such changes to help establish the theory of the possible daily formation of low ground fogs and frost deposits. You should record the appearance of *any* cloud, whether the cloud was visible on the morning terminator or the evening terminator of Mars, or whether it was visible all day. The areas given in Table 8-4 are known to exhibit the seasonal forming of white clouds.

Most of the reported white clouds are confined to the midtemperate and equatorial latitudes, and they are more predominant in the southern hemisphere than in the northern. This may be misleading, because the true cause for such distribution may be the favorable circumstances under which we observe the southern hemisphere. Not only is the planet of greater angular size when its southern hemisphere is visible, but the angle of reflected sunlight from features such as the white clouds is considerably improved over that seen in oppositions of a northern tilt.

Your observation of the whitish clouds can be considerably improved if you use the Wratten No. 58 green filter for the maria or the Wratten No. 12 yellow filter for the plains, or their equivalents. These filters enhance the brightness of the clouds on the maria and plains.

TABLE 8-4. Areas of seasonal white clouds (in order of increasing longitude).

Feature	*Latitude*	*Longitude*
Aram	00°	15°
Sinai	-25°	62°
Ophir	-08°	68°
Thaumasia	-30°	75°
Tharsis	+02°	100°
Nix Lux	-08°	112°
Olympus Mons	+18°	133°
Memnonia	-20°	160°
Zephyria	-12°	190°
Elysium	+30°	215°
Isidis Regio	+20°	280°
Nelth Regio	+35°	275°
Nymphaeum	+08°	305°
Hammonis Cornu	-10°	315°
Deucalionis Regio	-12°	345°

The Yellow Clouds of Mars

Precursors of the great dust storms of Mars, the yellow clouds form readily during years when Mars is near perihelion and at the time of the Martian summer solstice. Because of the timing of such developments, it is speculated that the yellow clouds as well as the dust storms are raised by the rapid transfer of heat in the thin Martian atmosphere, which causes violent winds during this interval. The winds carry the clouds of dust and distribute it all across the planet.

Yellow clouds form quite rapidly, and sometimes they spread equally fast. They can turn the planet into a vague diffuse globe, with no sign of even the brightest or darkest feature beneath the blowing dust. Although the clouds appear quickly, they can take weeks, even months, to disappear. It is their longevity that allows amateur and professional astronomers to map the transient circulation patterns of Mars.

The origin of the dusty yellow clouds is highly localized, coming from the Serpentis-Noachis-Hellas basin area, centered at latitude -28°, longitude 320°. With few exceptions, these clouds, as well as the dust storms, occur quite low in the Martian atmosphere, skirting the landscape of Mars with dust-laden clouds. You will get best results if you monitor the yellow clouds and the extents of the dust storms using a Wratten No. 12 (yellow), No. 21 (orange), or No. 25 (red) filter.

The Martian Blue Clearing

If you view Mars in blue light, you will usually see very little detail. The planet appears to have a uniformly smooth surface with a bright polar cap. The farther you go into the blue region of the spectrum (toward the ultraviolet) in your viewing of Mars, the less you will be able to see. You might glimpse some detail if you use the Wratten No. 80A (medium blue) filter, but all detail vanishes if you use violet light (No. 47 filter). Because of the nature of the materials that make up the Martian landscape and portions of its atmosphere, more light is absorbed in the blue end of the spectrum, whereas most of the light of longer wavelengths—in yellow, orange, and red—is reflected.

A little-understood phenomenon occasionally occurs during which the Martian surface is very poorly delineated if you use filters of the longer wavelengths that you would normally use in viewing the planet. If you use a blue (No. 80A) or violet (No. 47) filter, however, the surface features again become visible. Such conditions, known as the *blue clearing*, may last for several days and are of great interest to astronomers. This clearing in blue light was originally suspected to be caused by a blocking layer of ultraviolet clouds, high in the Martian atmosphere. Recent studies, however, indicate that this might not be the case, or that another phenomenon—the albedo of various surface features, as well as the polarization of light from those features—might be a partial cause of the blue clearing.

To search for the blue clearing, merely look first with the blue filter on your telescope and then look without the aid of a blue filter. It is best to search for the clearing using the medium blue (No. 80A) filter. If you suspect the presence of the blue clearing move to the more dense No. 47 filter for further scrutiny. You should become suspicious of the presence of the blue clearing if, when viewed through white or unfiltered light, the planet looks like a featureless, orange disk.

Make observations of a blue clearing on every possible date, noting the location (whether the clearing is restricted to a small area or is planet-wide), the date and time you first saw the clearing, and the results of your filter observations. This unusual phenomenon is not usually restricted to small areas, as are some of the other atmospheric phenomena on Mars; rather, the clearing seems to affect the entire planet on most occasions.

The Curious W-shaped Clouds

The most curious of all Martian phenomena—the "W-shaped" clouds—form in the vicinity of massive volcanic peaks. Reported first by Earl Slipher of the Lowell Observatory, and confirmed later as a recurring phenomenon by Charles Capen in 1966, the clouds (usually quite large) apparently are associated with reflections in longer wavelengths of light. They are therefore best seen in the medium blue to violet range of the spectrum (Wratten No. 80A filter). It is possible that these unusual clouds occur less frequently in the southern hemisphere than in the northern. They are also seen more often in the Martian summer and early fall.

Because the W-shaped clouds move fairly rapidly, it is important that you record them on the date you first see them. Record also their motions across the planet relative to known features; if possible, longitudes on each date should be determined, and the latitudes of these clouds estimated. It is important that you record and report the point of origin of the cloud.

The W-shaped clouds form near the following features (in order of increasing longitude):

Feature near Origin	Latitude	Longitude
Ascraeus Lacus	+11°	104°
Pavonis Lacus	+01°	112°
Arsia Silva	-09°	120°
Olympus Mons	+18°	133°

It is interesting that all the features listed above

are large, volcanic peaks and are located near one another. Perhaps the origin of these clouds is not nearly as simple as might be explained by atmospheric conditions.

The High, Blue Clouds

High in the Martian atmosphere are thin clouds not visible to the naked eye, or even to the eye aided by a telescope. Photographs taken in far-violet or ultraviolet light reveal the existence of these small clouds; ultraviolet-sensitive film is often used on steady nights by amateurs who possess large instruments [i.e., 32 cm (12½ inches) and larger] capable of taking long photographic exposures at high magnification.

I emphasize that the would-be Mars observer should become familiar with the Martian surface as it appears *without* clouds, so that atmospheric phenomena can be recognized for what they are. All clouds and any other atmosphere-related features should be recorded as follows:

1. The date and time (UT) on a feature is first seen.
2. The location of the feature relative to known features on the planet.
3. The approximate size of the feature, in proportion to the total size of the apparent disk.
4. The longitude of the feature and an estimate of its latitude.
5. A disk drawing of the planet, showing the feature (optional).
6. Date on which the feature is last seen.
7. The amount and direction of the feature's drift on the planet.

Recording such information can be of great help to the professional planetary astronomer. It also will enable you, the amateur, to compile all your observations and derive a cloud map of Mars, showing the locations of the most frequent areas of cloud formation and their circulation after being formed.

PATROL OBSERVATION OF MARS

You can make patrol observations of the Martian surface either telescopically or photographically, but you should make them as consistently as possible. Basically, by setting up a patrol, you scan the surface on every available date so that you are sure to see any rapid changes. Report such changes immediately so that others may study them. Such changes include cloud formation, blue clearing, dust storms, changes in the size and shape of the maria, the appearance of bright spots on the plains or maria, new features in the polar regions, and so forth. Unless some new phenomenon is noted, you need not make drawings unless you have the time and the desire to do so.

THE SEASONAL CHANGES ON MARS

Because Mars has an axial tilt of 24.9°, it has a seasonal cycle much like that of the earth. On Mars, however, the seasons are more extreme and last twice as long. Some effects of the seasons on the Martian polar caps have already been discussed, and these basic changes cause most of the other seasonal changes on the planet. For example, the density of the atmosphere changes proportionately with the changes of the polar caps. The density increases as the caps melt and sublimate, and it decreases as they refreeze in Martian autumn and winter. With the melting of the ice caps, the amount of sunlight reflected from the planet decreases, which causes more heat absorption by the planet's surface. This in turn, results in increased atmospheric convection.

From your standpoint as an amateur observer, the most notable seasonal changes, other than those of the caps themselves, occur in the maria, which are seen as dark regions. The changes were formerly thought to be caused by growing vegetation that made these regions darken as the Martian summer progressed lighten as autumn approached. We now know that these are related to the albedo, or the amount of sunlight absorbed compared to the amount reflected. The more light that is reflected, the brighter the feature appears to us.

The wave of darkening, as it has been termed, begins in midspring on Mars and continues until most of the polar ice cap is gone. It is more predominant in perihelic oppositions than in aphelic ones, possibly as a result of the differing densities of the north and south caps. As the caps melt or sublimate, the darkening progresses from pole to equator. An increase in atmospheric H_2O slowly disseminating from the pole and drifting toward the equator could possibly explain such sequential darkening.

Not only do the features darken, their size also often increases greatly over their size in the early spring, and they even change their shape and their position on the planet. Again, your familiarity with the normal appearance of the maria is essential. Only when you know how they are supposed to appear, can you detect the subtle seasonal changes that always occur. Table 8-5 is only a small sampling of areas that exhibit changes such as those described.

TABLE 8-5. Notable areas of seasonal changes.[a]

Feature	*Latitude*	*Longitude*
Margaritifer Sinus	-02°	30°
Hydrae Sinus	-02°	30°
Mare Australe	-65°	40°
Acidalius Fons/Tempe	+58°	60°
Nilokeras/Lunae Lacus	+25°	60°
Solis Lacus	-35°	85°
Candor/Tharsis	+10°	90°
Aonius Sinus	-45°	105°
Amenthes	+05°	250°
Thoana Palus	+35°	256°
Thoth	+30	256°
Nepenthes	+20°	260°
Moeris Lacus	+08°	270°
Antigones Fons/ Astaboras	+22°	290°
Syrtis Major	+10°	298°
Aeria	+10°	310°

[a]In order of increasing longitude.

You can determine the relative degree of darkening of the maria by using a scale of 1 to 10, noting the relative contrast between a feature and its surroundings, with 1 being the lowest contrast and 10 the highest contrast. All the maria can be better differentiated by using color filters; the orange (Wratten No. 21) or the red (No. 25) are the best for such studies.

TELESCOPIC AND PHOTOGRAPHIC RECORDS OF MARS

To compile a history of the appearance of Mars is a rewarding experience. You can compile the history by drawing on a standard disk form the appearance of Mars as you see it in the telescope, or you can photograph the planet. Whichever way is chosen, adhere to it. Do not switch back and forth to one or the other method.

Photographic records of the planets, particularly Mars, are somewhat less desirable than high-resolution drawings. With the aid of your 10- to 12-inch (25- to 30-cm) telescope, you can detect more visually than can professional astronomers making photographs through the largest telescopes on earth. This is so primarily because of the state of the earth's atmosphere when the photograph is taken. The high magnification required to discern detail on the tiny Martian disk also magnifies the unsteadiness of the air and requires long exposure time.

Photographing Mars

If you choose to record photographically Mars throughout its appearances, you must exercise great care in recording all pertinent information when the photograph is made, including the date, time (UT), film type, and exposure length. The great advantage of a photographic program is that the photo verifies itself, so to speak. A photograph of a W-shaped cloud would be of great value, for example, and would virtually substantiate itself even if observers elsewhere had not seen this transient feature.

For photographing Mars, choose a high-contrast, fine-grain, and red-sensitive film. The Kodak 2415 film, developed in Kodak HC-110 developer, dilution D, is best suited for photography of the red planet. For more information about computing effective focal length and basic planetary photography, refer to Chapter 14.

Visual Studies of Mars

Visual studies of Mars are better suited for the endeavors of the amateur astronomer than are photographic studies. Telescopic studies reveal low contrasts and subtle detail that cannot be recorded by the camera. You can record all the Martian phenomena thus far discussed quite accurately by making highly detailed drawings on which you draw and label precisely what you see in the eyepiece. During moments of steady seeing, you will be able to see extraordinary detail crisscrossing the Martian surface.

Your first efforts at drawing the tiny disk of Mars may seem somewhat comical, with strange abstract figures and geometrical patterns drawn on a white circle. With a little practice and the experience of only a few drawings behind you, however, the very low contrasts and subtle details begin to form, appearing much as they do in the map shown in Figure 8-1. Both Mars and earth have atmospheres, and it is only when both atmospheres are simultaneously steady that the telescope can penetrate to the Martian surface to view the fine network of detail. The trained Mars observer knows to wait for such moments.

Drawings of the Martian disk provide a permanent record of your impression of the planet. No drawing is without a certain amount of bias, showing features that result from some preconceived notion of what you *think* should be there. Draw only what you really see—not what you think you should see. If you are not familiar with the Martian features, particularly the very small ones, you might not record some fine linear marking that you saw only for a moment, simply because you do not trust what your eyes have seen.

When seeing conditions permit, attempt to make observations at least every other night, preferably at about the same time each night. Ideally, make two drawings each night, when possible, to allow for greater coverage of Martian longitude. These drawings can make a composite record of all changes seen over a long period as well as show the expansion of very large features. If you make a second drawing 4 hours after the first, almost 60° change in longitude on Mars will be shown as a result of the rotation of the planet.

Use a standard form, part of which is a circle 2 to 3 inches in diameter, as a standard for making visual observations of Mars. The form should have places where additional data can also be recorded, including the following:

1. The observer's name, the date, and time of the observation.
2. Telescope used, magnification, and any filters used.
3. The steadiness of the air on a scale of 1 to 5 (5 best, 1 worst).
4. Transparency of the sky (1 to 6, representing the magnitude of the faintest star seen to the naked eye).
5. Any transient or unusual features on the planet.
6. Martian longitude on the central meridian at the beginning of the drawing.

Make copies of each drawing as accurately as possible from the original, retaining the original and filing it in some systematic way. Send the copies to:

MARS SECTION
The Association of Lunar and Planetary Observers
Box 3 AZ
University Park, NM 88003

Appendix IX shows the standard observing form used by the Mars section of the Association of Lunar and Planetary Observers (ALPO).

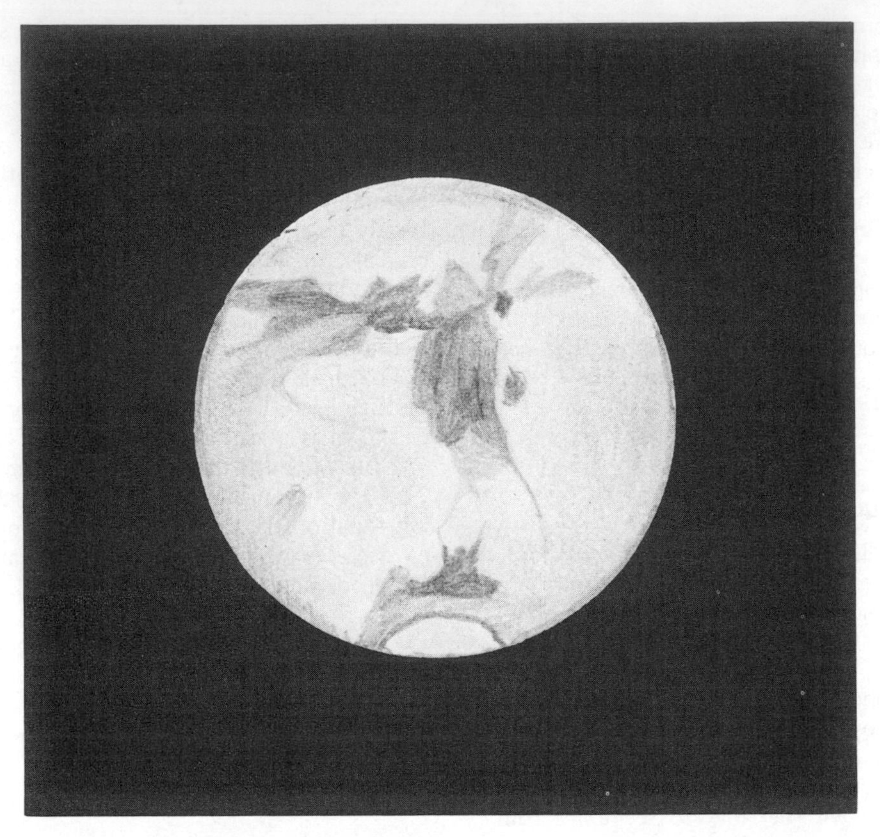

FIGURE 8-2. Drawing of Mars made by the author on February 25, 1980. The prominent feature, Syrtis Major, is seen near the central meridian. The north polar cap is at the bottom. In all astronomical drawings, *south* should be at the top.

For your drawings, it is best to follow a strict sequence in which features are recorded to provide accuracy and consistency in your records. The following sequence is recommended:

1. Draw the polar caps first. This allows for better accuracy in the placement of other features, and sets up a basic north-south orientation for you.
2. Draw all prominent detail in the center of the disk, using the maria as guides for small, indistinct detail.
3. Draw details on the preceding limb (eastern, or left on the form shown in Appendix IX). Notice that south is at the *top* on this form, as it should be in all drawings.
4. Sketch in all fine details, looking carefully for clouds, "canals," bright spots, and so on.
5. Add details seen in filters that were not seen without them, but be sure to note (by a small number) which were seen this way, and through what filter.

Adhering to this routine helps you to place features in their proper positions, rather than too far north, south, east, or west. After all the most prominent details have been drawn in their approximate positions, then draw the finer details on the drawing in relation to the obvious features. A normal Mars drawing should require a full hour. You can take the time to wait for moments of optimum steadiness in order to discern the finest detail.

DETERMINING LONGITUDE ON MARS

To make your observations more scientifically valuable, it is necessary that you determine the central meridian of the Martian globe at the time of any particular observation. For each drawing made, or photograph taken, determine the longitude of the central meridian. The central meridian (CM) is merely an imaginary line passing from the north pole to the south pole of the planet, perpendicular to the equator. Record the central meridian at the time when the drawing was begun, and again when the drawing was finished. If it took you one hour to render the disk drawing, then 14.6° of longitude will have rotated past the CM. For a photograph, the CM recorded should be that at the instant the photograph was made.

Because the rotational rates of Mars and the earth are almost the same, the same face of Mars will be visible with only slight deviation for several successive nights. Over a period of 36 days, you can view an entire rotation of Mars through 360° of longitude if you make your observations each night at approximately the same time.

Another important advantage in determining Martian longitude is that it allows you to determine the correct longitudinal placement of features, as well as their angular size horizontally. The longitudes of Mars' surface features are determined by noting the time, to the nearest minute, at which that feature is exactly centered on the Martian disk (the central meridian would then cut symmetrically through the feature). To determine angular longitudinal expanse, one simply times the instant the preceding edge of the feature first crosses the CM, and the instant the following edge crosses some moments later. After determining the longitudes for each timing, the difference of the two is the angular extent in longitude of that feature.

The process of determining longitudinal placement of features on Mars is much easier than for those on Jupiter or Saturn, because of the much slower rate of rotation of Mars. Closer precision may be achieved if a vertical wire in a finely threaded cross-hair eyepiece is positioned to exactly coincide with a north-south line.

FIGURE 8-3. Drawing showing the central meridian line on Mars.

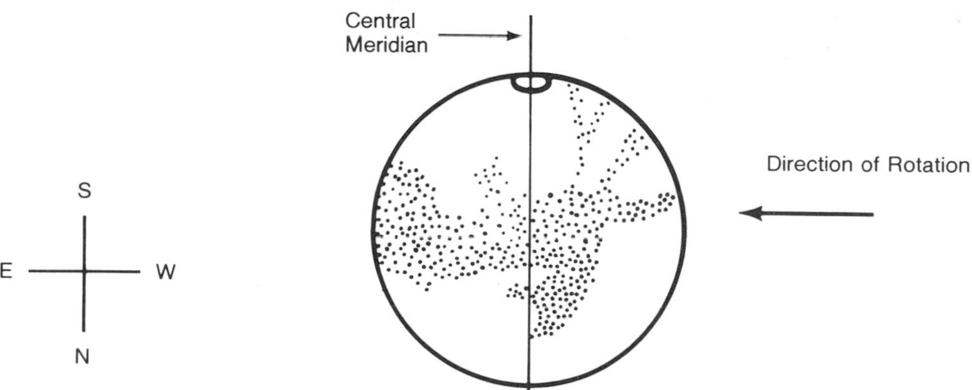

Reducing the Timings To Determine the Longitude

Every planet has 360° of longitude starting with 0° and progressing all the way through 359.9°. On Mars the 0° point–the starting point for all subsequent measures–was arbitrarily set at the center of a prominent feature, Sinus Meridiani, which is also located near the Martian equator (-05°). All subsequent markings are measured in increasing longitude westward from that point, as they are on earth.

MARS, 1978

EPHEMERIS FOR PHYSICAL OBSERVATIONS

FOR 0h UNIVERSAL TIME

Date	k	i	Defect of Illumination	Position Angle of Defect	Position Angle of Axis	Central Meridian Of Date	Central Meridian Of Following Date	Universal Time of Transit of Zero Meridian Of Date	Universal Time of Transit of Zero Meridian Of Following Date
		°	"	°	°	°	°	h m	h m
July 1	0.920	32.88	0.40	111.61	15.80	315.50	305.75	3 03.0	3 43.0
3	.921	32.64	0.39	111.75	16.49	296.00	286.25	4 23.1	5 03.2
5	.922	32.39	0.38	111.88	17.17	276.49	266.73	5 43.3	6 23.5
7	.923	32.14	0.38	112.01	17.86	256.97	247.21	7 03.6	7 43.7
9	.925	31.89	0.37	112.13	18.54	237.44	227.68	8 23.9	9 04.1
11	0.926	31.63	0.36	112.24	19.21	217.91	208.14	9 44.2	10 24.4
13	.927	31.37	0.35	112.35	19.89	198.37	188.59	11 04.6	11 44.8
15	.928	31.11	0.34	112.45	20.56	178.82	169.04	12 25.0	13 05.2
17	.929	30.84	0.33	112.54	21.23	159.26	149.48	13 45.4	14 25.6
19	.930	30.57	0.33	112.62	21.89	139.70	129.91	15 05.8	15 46.1
21	0.932	30.30	0.32	112.69	22.55	120.13	110.34	16 26.3	17 06.6
23	.933	30.03	0.31	112.76	23.20	100.55	90.76	17 46.8	18 27.1
25	.934	29.75	0.30	112.82	23.84	80.97	71.18	19 07.3	19 47.6
27	.935	29.47	0.30	112.87	24.48	61.38	51.59	20 27.9	21 08.2
29	.937	29.19	0.29	112.91	25.11	41.79	31.99	21 48.4	22 28.7
31	0.938	28.90	0.28	112.94	25.74	22.19	12.39	23 09.0	23 49.3
Aug. 2	.939	28.61	0.28	112.97	26.35	2.59	352.79	. . . . .	0 29.6
4	.940	28.32	0.27	112.98	26.96	342.99	333.18	1 10.0	1 50.3
6	.941	28.03	0.26	112.99	27.56	323.38	313.57	2 30.6	3 10.9
8	.943	27.73	0.26	112.99	28.14	303.76	293.96	3 51.2	4 31.6
10	0.944	27.43	0.25	112.98	28.72	284.15	274.34	5 11.9	5 52.2
12	.945	27.13	0.24	112.96	29.29	264.53	254.72	6 32.6	7 12.9
14	.946	26.83	0.24	112.93	29.84	244.91	235.10	7 53.3	8 33.6
16	.947	26.53	0.23	112.89	30.38	225.29	215.47	9 14.0	9 54.3
18	.949	26.22	0.22	112.84	30.91	205.66	195.85	10 34.6	11 15.0
20	0.950	25.91	0.22	112.78	31.43	186.04	176.22	11 55.3	12 35.7
22	.951	25.61	0.21	112.72	31.93	166.41	156.60	13 16.1	13 56.4
24	.952	25.29	0.21	112.64	32.42	146.78	136.97	14 36.8	15 17.1
26	.953	24.98	0.20	112.55	32.89	127.16	117.34	15 57.5	16 37.8
28	.954	24.67	0.19	112.46	33.34	107.53	97.72	17 18.2	17 58.5
30	0.956	24.35	0.19	112.35	33.78	87.90	78.09	18 38.9	19 19.2
Sept. 1	.957	24.03	0.18	112.24	34.21	68.28	58.46	19 59.6	20 39.9
3	.958	23.71	0.18	112.11	34.62	48.65	38.84	21 20.3	22 00.6
5	.959	23.39	0.17	111.97	35.01	29.03	19.21	22 41.0	23 21.3
7	.960	23.07	0.17	111.83	35.38	9.40	359.59	. . . . .	0 01.7
9	0.961	22.75	0.16	111.67	35.73	349.78	339.97	0 42.0	1 22.4
11	.962	22.42	0.16	111.50	36.06	330.16	320.35	2 02.7	2 43.0
13	.963	22.09	0.15	111.32	36.38	310.54	300.73	3 23.4	4 03.7
15	.964	21.77	0.15	111.14	36.67	290.92	281.12	4 44.0	5 24.4
17	.965	21.44	0.14	110.94	36.94	271.31	261.50	6 04.7	6 45.0
19	0.966	21.11	0.14	110.73	37.20	251.70	241.89	7 25.3	8 05.7
21	.967	20.78	0.13	110.51	37.43	232.09	222.29	8 46.0	9 26.3
23	.968	20.45	0.13	110.28	37.64	212.48	202.68	10 06.6	10 46.9
25	.970	20.11	0.12	110.04	37.82	192.88	183.08	11 27.2	12 07.5
27	.970	19.78	0.12	109.79	37.99	173.28	163.48	12 47.8	13 28.1
29	0.971	19.45	0.12	109.53	38.13	153.68	143.88	14 08.4	14 48.7
Oct. 1	0.972	19.11	0.11	109.25	38.25	134.08	124.29	15 28.9	16 09.2

FIGURE 8-4. Extract from The *American Ephemeris and Nautical Almanac,* which gives Martian longitudes.

The *American Ephemeris and Nautical Almanac* provides the longitudes (and other information for physical observations) of the central meridian for 00h 00m Universal Time for each date of the year, as shown in Figure 8-4. (To convert Central Standard Time, for example, into Universal Time, see the conversion chart in Appendix I.)

Not all observations conveniently take place at exactly 00h 00m UT, therefore some form of conversion is necessary to account for the amount of time that has transpired from 0 hours until the observation is made. Because Mars is rotating, as the earth is, 1 hour of rotation allows for 14.6° of longitudinal change. Therefore, for every hour, another 14°.6 increase in longitude accumulates. Table 8-6 provides the appropriate amount of correction, which should be *added* to the values given for 00h 00m in the *AENA* for the elapsed time. For example, suppose that on July 20, 1978 (Universal Date and Time), an observer begins a drawing at 02:12 UT. To determine the longitude visible on the central meridian when the drawing was begun, five steps are necessary:

1. Look in the *American Ephemeris and Nautical Almanac* (see Figure 8-4) for the Martian longitude for July 20, 1978, at 0 hr UT. This value is seen to be 129.9°.
2. The upper listing of Table 8-6 gives the correction values to be added to 129.9° at 10-minute intervals for 24 hours. Because the time of the observation was 02:12 UT, 2 hours and 12 minutes have elapsed since the reading given in the *AENA*. Using Table 8-6 (note arrow), we find the value for 02:10, which is the closest interval for the actual time of 02:12. That value shows that the Martian disk has rotated an additional 31.7°.
3. The lower box of Table 8-6 gives correction values for 1-minute intervals. Because our actual time was 02:12 (not 02:10), it is necessary to add an additional amount for the 2 minutes. The lower box designates this value as 0.5°.
4. Adding these values together provides the correct Martian longitude on the central meridian at 02:12 UT on July 20, 1978:

129.9°	(Step 1—Longitude for date from *AENA*)
31.7°	(Step 2—Increased longitude for hour)
+ 0.5°	(Step 3—Longitude correction for nearest minute)
= 162.1°	(Correct Longitude)

5. If the determined value, after all adding is done, is greater than 360°, then 360° must be subtracted from that total, because there are only 360° longitude on a planet.

Thus, at 02:12 UT on July 20, 1978, an observer could have seen features have a longitude of about 162.1° centered on the disk of Mars, and could easily relate the observations to features shown on published Mars maps for that given longitude.

TABLE 8–6. Change in Martian longitude vs. Universal Time.

UT	*Correction*	*UT*	*Correction*	*UT*	*Correction*	*UT*	*Correction*
00:00	0.0	06:00	87.7	12:00	175.5	18:00	263.1
10	2.4	10	90.2	10	177.9	10	265.5
20	4.9	20	92.6	20	180.3	20	268.0
30	7.3	30	95.0	30	182.8	30	270.4
40	9.7	40	97.5	40	185.2	40	272.8
50	12.2	50	99.9	50	187.6	50	275.3
01:00	14.6	07:00	102.4	13:00	190.1	19:00	277.8
10	17.1	10	104.8	10	192.5	10	280.2
20	19.5	20	107.2	20	195.0	20	282.7
30	21.9	30	109.7	30	197.4	30	285.1
40	24.4	40	112.1	40	199.8	40	287.5
50	26.8	50	114.5	50	202.3	50	290.0
02:00	29.2	08:00	117.0	14:00	204.7	20:00	292.4
→10	31.7	10	119.4	10	207.1	10	294.8
20	34.1	20	121.8	20	209.6	20	297.3
30	36.6	30	124.3	30	212.0	30	299.7
40	39.0	40	126.7	40	214.4	40	302.1
50	41.4	50	129.2	50	216.9	50	304.6
03:00	43.9	09:00	131.6	15:00	219.3	21:00	307.0
10	46.3	10	134.0	10	221.7	10	309.4
20	48.7	20	136.5	20	224.2	20	311.9
30	51.2	30	138.9	30	226.6	30	314.3
40	53.6	40	141.3	40	229.0	40	316.7
50	56.0	50	143.8	50	231.5	50	319.2
04:00	58.5	10:00	146.2	16:00	233.9	22:00	321.6
10	60.9	10	148.7	10	236.3	10	324.0
20	63.4	20	151.1	20	238.8	20	326.5
30	65.8	30	153.5	30	241.2	30	328.9
40	68.2	40	156.0	40	243.6	40	331.3
50	70.7	50	158.4	50	246.1	50	333.8
05:00	73.1	11:00	160.8	17:00	248.5	23:00	336.2
10	75.5	10	163.3	10	250.9	10	338.6
20	78.0	20	165.7	20	253.4	20	341.1
30	80.4	30	168.2	30	255.8	30	343.5
40	82.9	40	170.6	40	258.2	40	345.9
50	85.3	50	173.0	50	260.7	50	348.4

Minutes	*Correction*	*Minutes*	*Correction*	*Minutes*	*Correction*
01	0.2	04	1.2	07	1.7
02	0.5	05	1.4	08	1.9
03	0.7	06	1.7	09	2.2

THE VISIBILITY OF MARS

Unlike the planet Jupiter, Mars does not present a favorable apparition each year. Consequently, make the best use of the times when Mars is close and high in the sky. Even at opposition the apparent angular size of the Martian disk varies considerably, thus varying the consistency of your records. Nonetheless, each opposition of this mysterious planet is equally important because the earth–Mars–sun distance is apparently a factor in many of the meteorological phenomena we witness.

About 50 days after conjunction with the sun, Mars will appear in the morning sky low on the horizon. It steadily climbs higher each morning until it reaches opposition almost a year after first being spotted in morning twilight. On the date of opposition, Mars will appear on the celestial meridian about midnight and can be observed in the sky all night long. Because Mars is a superior planet (i.e., it is outside of the earth's orbit), it will pass through retrograde 47 days before opposition and continue until 47 days after opposition. You will notice this apparent reversal of Mars' path through the "loop" that Mars makes relative to the distant stars.

TABLE 8–7. Oppositions of Mars, 1982–1999.

Opposition Date	*Apparent Diameter*[a]	*Magnitude*	*Constellation*
March 31, 1982	14.7	-1.2	Virgo
May 11, 1984	17.5	-1.8	Libra
July 10, 1986	23.1	-2.4	Sagittarius
September 28, 1988	23.7	-2.6	Pisces
November 27, 1990	17.9	-1.7	Taurus
January 7, 1993	14.9	-1.2	Gemini
February 12, 1995	13.8	-1.0	Leo
March 17, 1997	14.2	-1.1	Virgo
April 24, 1999	16.2	-1.5	Virgo

[a] In seconds of arc.

BIBLIOGRAPHY

American Ephemeris and Nautical Almanac. Washington, DC: U.S. Government Printing Office. Published for each year.

Blunck, Jurgen, *Mars and Its Satellites*. New York: Exposition Press, 1977.

Burgess, Eric, *To the Red Planet*. New York: Columbia University Press, 1978.

Capen, Charles F., and V.W. Capen "Martian North Polar Cap, 1962–68," *Icarus*, Vol. 13, No. 1 (1970), 100–108.

——"Martian Yellow Clouds—Past and Present," *Sky & Telescope*, Vol. 41, No. 2 (1971).

Eastman Kodak Co., *Kodak Filters for Scientific and Technical Use*. Rochester, NY: Eastman Kodak Co., 1970.

Firsoff, V.A., *The Solar Planets*. New York: Crane, Russak, 1977.

Hartmann, William K., *Moons and Planets*. New York: Bogden and Quigly, 1972.

Harvard College Observatory, Annals of, Vol. 53,

Part 8: *Martian Meteorology*. Cambridge, MA, 1907.

Hoyt, William G., *Lowell and Mars*. Tucson: University of Arizona Press, 1976.

Kuiper, G.P., *The Atmospheres of the Earth and Planets*. Chicago: University of Chicago Press, 1952.

Kuiper, G.P., and B.M. Middlehurst, *Planets and Satellites*. Chicago: University of Chicago Press, 1961.

Leighton, R.B., and B.C. Murray, "Behavior of Carbon Dioxide and Other Volatiles on Mars," *Science*, Vol. 153 (1966), 136-147.

Lowell, Percival, *Mars* (reprint). Bernardston, MA: Paul W. Luther, 1978.

Moore, Patrick, *A Guide to Mars*. New York: Macmillan, 1956.

National Aeronautics and Space Administration, *Mars as Viewed From Mariner 9*, SP-329. Washington, DC: Government Printing Office, 1974.

——*The Book of Mars*, SP-179. Washington, DC: Government Printing Office, 1968.

Roth, Gunter D., *Astronomy: A Handbook*, pp. 364-373. Cambridge, MA: Sky Publishing Corp., 1975.

——*Handbook for Planet Observers*. London: Faber and Faber, 1970.

Slipher, E.C., *Mars: The Photographic Story*. Flagstaff, AZ: Northland Press, 1962.

Steeg et al., *Astrofilters for Observation and Astrophotography*. Oakland, Optica b/c Co., 1973.

Tull, R.G., "High Dispersion Spectroscopic Observations of Mars—IV: The Latitude Distribution of Water Vapor," *Icarus*, Vol. 13, No. 1 (1970), 43-57.

9

JUPITER: AN AMATEUR'S GUIDE TO RESEARCH PROJECTS

Of the nine major planets, Jupiter affords the greatest opportunity for amateur astronomical research. This chapter describes preliminary research activity that can be initiated and completed by the amateur, using modest equipment.

If you wish to be a serious observer of Jupiter, I urge you to apply for membership in the Association of Lunar and Planetary Observers (ALPO), a society devoted to nonprofessional organized research of the solar system. Another aid to you is the excellent book, *The Planet Jupiter*, by B.M. Peek. This book provides great assistance in the interpretation of data accumulated at the telescope. The *American Ephemeris and Nautical Almanac*, published by the U.S. Government Printing Office, is also indispensable for reducing timings and determinations of Jovian longitude. Any of these books can be ordered through a local book store; many advertisements in astronomy journals also offer them for sale.

Nomenclature of the Jovian belts and zones. Figure 9-1 gives the accepted abbreviations and nomenclature of the belts (dark) and zones (light) of Jupiter. Full names are given below the diagram. It is essential that you become familiar with the correct names of the features, as well as their approximate latitudes as measured from the equator (0°).

TABLE 9-1. Jovian physical data.

Distance from the sun	778 million km
Orbital velocity	13 km/sec
Length of year	4332.6 days
Equatorial diameter	142,800 km
Mass (X earth)	318
Rotation period (day)	9h 50m 30s (System I)
	9h 55m 40s (System II)
Albedo	0.4
Maximum magnitude	-2.7
Maximum apparent diameter	48" arc
Atmosphere constituents	H, He, NH_3, CH_4, H_2O

Rotational patterns of Jupiter's clouds. The visible cloud patterns of Jupiter are subject to at least two different major rates of rotation, known as *System I* and *System II*. System I has a faster rotational rate than System II and encompasses only the Equatorial Zone (EZ) and features within that zone. System II encompasses all latitudes north and south of the EZ, including the North Equatorial Belt (NEB) and the South Equatorial Belt (SEB). The rotational rates for these two systems are:

System I	9 hours 50 minutes 30 seconds (average)
System II	9 hours 55 minutes 40 seconds (average)

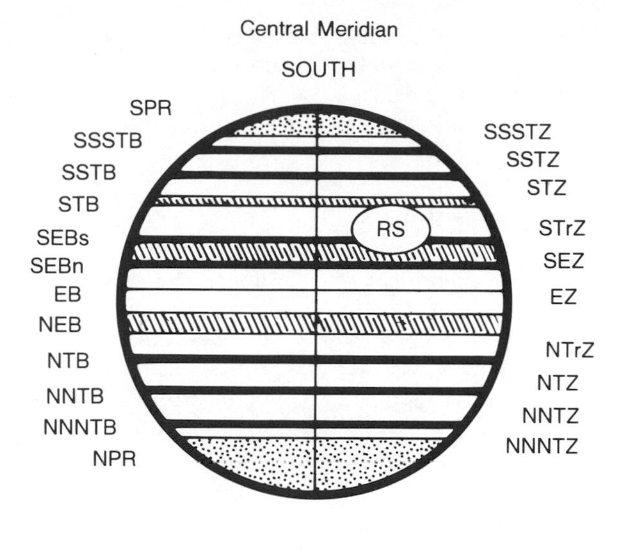

Direction of Rotation

INDEX TO ABBREVIATIONS:

ZONES		BELTS	
EZ	Equatorial Zone	EB	Equatorial Belt
SEZ	So. Equatorial Zone	SEBn	So. Equatorial Belt North
STrZ	So. Tropical Zone	SEBs	So. Equatorial Belt South
STZ	So. Temperate Zone	STB	South Temperate Belt
SSTZ	So. So. Temperate Zone	SSTB	So. So. Temperate Belt
SSSTZ	So. So. So. Temp. Zone	SSSTB	So. So. So. Temperate Belt
NTrZ	North Tropical Zone	NEB	North Equatorial Belt
NTZ	North Temperate Zone	NTB	North Temperate Belt
NNTZ	No. No. Temperate Zone	NNTB	No. No. Temperate Belt
NNNTZ	No. No. No. Temp. Zone	NNNTB	No. No. No. Temp. Belt

OTHER	
RS	Great Red Spot
SPR	South Polar Region
NPR	North Polar Region

NOTE: The Central Meridian (CM) is an imaginary line running from the north pole through the equator to the south pole, as shown in the figure above. This line is used for timing the passage of features so that the correct longitude may be determined, as discussed in the text.

NOTE: During many past apparitions, the South Equatorial Belt (SEB) has not shown two components, but only one. When such is the case, much care must be taken not to confuse the more southerly STB with the SEBs. Likewise, care should be exercised when determining the ZONES associated with these two belts.

NOTE: In several past apparitions, the Red Spot has been absent, but the area where the spot SHOULD be has been well seen. This large "gap" surrounding the missing RS is known as the "Red Spot Hollow" (RSH).

NOTE: It should be emphasized that not all of the belts and zones listed above will be visible during EVERY apparition; many times, atmospheric disturbances prohibit observations of many such features.

FIGURE 9-1. The belts and zones of Jupiter.

The Great Red Spot (RS) appears to be subject to the rotational characteristics of System II although the RS drifts considerably and independently of this system, as do many other transient features.

AMATEUR STUDIES OF JUPITER

The amateur astronomer can provide scientifically valid information that otherwise would be unavailable to professional astronomers. Each area of amateur endeavor in the following list is discussed at length in later pages of this chapter.

- ☆ *observation of transits*—Timing features as they cross the Central Meridian (CM).
- ☆ *determination of longitude*—Computing longitude from the timing of features as they cross the Central Meridian.
- ☆ *determinations of latitude*—Measuring roughly the latitudes of belts and zones with the aid of a micrometer reticle eyepiece, or bifilar micrometer.
- ☆ *observations of Red Spot*—Noting unusual behavior, drift, color changes, intensity, and so on.
- ☆ *estimates of color*—Recording the coloration of Jovian features and the changes of that coloration over a period of time.
- ☆ *patrol observations*—Either visually (through disk drawings) or photographically recording the image of Jupiter in search of the first sign of transient phenomena and time-related changes.
- ☆ *determination of rotational rates*—Directly measuring features as they drift in longitude within particular latitudes.

Irregular phenomena occurring in the clouds of Jupiter may be classified according to the categories in the list below. All such features should be sketched on a disk drawing. Timing of these features across the CM are of the greatest value in determining their positions and rotational velocities.

- ☆ *ovals*—Either gray or white irregular formations. The gray patches are best seen superimposed over the lighter zones; white patches can be observed either on belts or zones since they generally are the brightest of all features. Many such ovals can be found in the South Tropical Zone (STrZ) and the South Temperate Zone (STZ).
- ☆ *white spots*—Smaller than the ovals, usually round and well-defined, the white spots are smaller than shadows cast by Jovian satellites and are commonly found in large belts, such as the NEB and the SEB.
- ☆ *festoons*—A bridge of cloud material, usually quite thin and difficult to observe, that connects two belts across a zone. Festoons are common in the Equatorial Zone (EZ) and both tropical zones.
- ☆ *garlands*—Similar to a festoon, and found in the same areas, but does not bridge completely. Generally, garlands begin as a "bulge" in a belt and extend straight for a short distance into an adjacent zone; the marking then curves, appearing as a "hook", or perhaps curving upward, toward the originating belt to form a closed loop.
- ☆ *rifts*—Segments that appear to bridge one zone to another across a dark belt.
- ☆ *hℓm*—Horizontal Linear Markings are short, belt-like segments seemingly isolated in the belts and zones, with no relation to known belts.
- ☆ *bars*—Short, stubby segments more obvious than HLM and found generally in belts rather than zones. Many times (particularly during disturbances) these bars appear reddish or orange in color.
- ☆ *knots*—A segment within an otherwise thin belt that appears greatly thickened or darker.

JUPITER, 1977

EPHEMERIS FOR PHYSICAL OBSERVATIONS

LONGITUDE OF CENTRAL MERIDIAN OF ILLUMINATED DISK

SYSTEM I

Day (0ʰ U.T.)	JAN.	FEB.	MAR.	APR.	MAY	JUNE	JULY	AUG.	SEPT.	OCT.	NOV.	DEC.
	°	°	°	°	°	°	°	°	°	°	°	°
1	332.8	185.1	281.3	129.1	178.5	25.8	76.2	285.8	137.4	193.4	49.6	109.9
2	130.7	342.8	79.0	286.7	336.2	183.4	233.9	83.6	295.3	351.3	207.5	268.0
3	288.5	140.6	236.7	84.4	133.8	341.1	31.6	241.4	93.1	149.2	5.5	66.0
4	86.4	298.3	34.4	242.0	291.5	138.7	189.3	39.1	250.9	307.1	163.5	224.1
5	244.3	96.1	192.1	39.7	89.1	296.4	347.0	196.9	48.8	105.0	321.5	22.1
6	42.1	253.8	349.8	197.3	246.8	94.1	144.7	354.7	206.6	262.9	119.5	180.1
7	200.0	51.6	147.4	355.0	44.4	251.7	302.4	152.4	4.4	60.8	277.5	338.2
8	357.8	209.3	305.1	152.6	202.0	49.4	100.1	310.2	162.3	218.7	75.5	136.2
9	155.7	7.1	102.8	310.3	359.7	207.1	257.9	108.0	320.1	16.7	233.5	294.3
10	313.5	164.8	260.5	107.9	157.3	4.8	55.6	265.8	118.0	174.6	31.5	92.3
11	111.4	322.5	58.1	265.6	315.0	162.4	213.3	63.5	275.8	332.5	189.5	250.3
12	269.2	120.3	215.8	63.2	112.6	320.1	11.0	221.3	73.7	130.4	347.5	48.4
13	67.0	278.0	13.5	220.9	270.3	117.8	168.7	19.1	231.5	288.4	145.5	206.4
14	224.8	75.7	171.2	18.5	67.9	275.5	326.5	176.9	29.4	86.3	303.5	4.5
15	22.7	233.4	328.8	176.2	225.6	73.1	124.2	334.7	187.2	244.2	101.5	162.5
16	180.5	31.2	126.5	333.8	23.2	230.8	281.9	132.5	345.1	42.2	259.5	320.5
17	338.3	188.9	284.2	131.5	180.9	28.5	79.7	290.3	143.0	200.1	57.6	118.6
18	136.1	346.6	81.8	289.1	338.6	186.2	237.4	88.1	300.8	358.1	215.6	276.6
19	293.9	144.3	239.5	86.8	136.2	343.9	35.1	245.9	98.7	156.0	13.6	74.7
20	91.7	302.0	37.2	244.4	293.9	141.5	192.9	43.7	256.6	314.0	171.6	232.7
21	249.5	99.7	194.8	42.1	91.5	299.2	350.6	201.5	54.5	111.9	329.6	30.7
22	47.3	257.4	352.5	199.7	249.2	96.9	148.3	359.3	212.3	269.9	127.7	188.8
23	205.1	55.1	150.2	357.3	46.8	254.6	306.1	157.1	10.2	67.8	285.7	346.8
24	2.9	212.8	307.8	155.0	204.5	52.3	103.8	314.9	168.1	225.8	83.7	144.8
25	160.7	10.5	105.5	312.6	2.1	210.0	261.6	112.7	326.0	23.8	241.7	302.9
26	318.5	168.2	263.1	110.3	159.8	7.7	59.3	270.5	123.9	181.7	39.8	100.9
27	116.2	325.9	60.8	267.9	317.5	165.4	217.1	68.3	281.8	339.7	197.8	258.9
28	274.0	123.6	218.4	65.6	115.1	323.1	14.8	226.1	79.7	137.7	355.8	56.9
29	71.8		16.1	223.2	272.8	120.8	172.6	24.0	237.6	295.6	153.0	215.0
30	229.6		173.8	20.9	70.4	278.5	330.3	181.8	35.5	93.6	311.9	13.0
31	27.3		331.4		228.1		128.1	339.6		251.6		171.0

MOTION OF THE CENTRAL MERIDIAN

	0ʰ	1ʰ	2ʰ	3ʰ	4ʰ	5ʰ	6ʰ	7ʰ	8ʰ	9ʰ	10ʰ	11ʰ
m	°	°	°	°	°	°	°	°	°	°	°	°
0	0.0	36.6	73.2	109.7	146.3	182.9	219.5	256.1	292.7	329.2	5.8	42.4
5	3.0	39.6	76.2	112.8	149.4	186.0	222.5	259.1	295.7	332.3	8.9	45.4
10	6.1	42.7	79.3	115.8	152.4	189.0	225.6	262.2	298.7	335.3	11.9	48.5
15	9.1	45.7	82.3	118.9	155.5	192.1	228.6	265.2	301.8	338.4	15.0	51.5
20	12.2	48.8	85.4	121.9	158.5	195.1	231.7	268.3	304.8	341.4	18.0	54.6
25	15.2	51.8	88.4	125.0	161.6	198.1	234.7	271.3	307.9	344.5	21.1	57.6
30	18.3	54.9	91.5	128.0	164.6	201.2	237.8	274.4	310.9	347.5	24.1	60.7
35	21.3	57.9	94.5	131.1	167.7	204.2	240.8	277.4	314.0	350.6	27.2	63.7
40	24.4	61.0	97.6	134.1	170.7	207.3	243.9	280.5	317.0	353.6	30.2	66.8
45	27.4	64.0	100.6	137.2	173.8	210.3	246.9	283.5	320.1	356.7	33.2	69.8
50	30.5	67.1	103.6	140.2	176.8	213.4	250.0	286.6	323.1	359.7	36.3	72.9
55	33.5	70.1	106.7	143.3	179.9	216.4	253.0	289.6	326.2	2.8	39.3	75.9
00	30.6	73.2	109.7	146.3	182.9	219.5	256.1	292.7	329.2	5.8	42.4	79.0

(a)

JUPITER, 1977

EPHEMERIS FOR PHYSICAL OBSERVATIONS

LONGITUDE OF CENTRAL MERIDIAN OF ILLUMINATED DISK

SYSTEM II

Day (0ʰ U.T.)	JAN.	FEB.	MAR.	APR.	MAY	JUNE	JULY	AUG.	SEPT.	OCT.	NOV.	DEC.
	°	°	°	°	°	°	°	°	°	°	°	°
1	325.0	300.8	183.4	154.7	335.2	306.0	127.5	100.6	75.6	262.6	242.3	73.8
2	115.3	91.0	333.5	304.7	125.3	96.0	277.6	250.7	225.8	52.9	32.7	224.2
3	265.5	241.1	123.6	94.7	275.3	246.0	67.6	40.9	16.0	203.2	183.0	14.6
4	55.8	31.2	273.6	244.7	65.3	36.1	217.7	191.0	166.2	353.5	333.4	165.0
5	206.0	181.3	63.7	34.8	215.3	186.1	7.8	341.1	316.4	143.8	123.7	315.4
6	356.2	331.4	213.7	184.8	5.3	336.1	157.9	131.3	106.7	294.0	274.1	105.8
7	146.5	121.6	3.8	334.8	155.3	126.2	307.9	281.4	256.9	84.3	64.5	256.2
8	296.7	271.7	153.8	124.8	305.4	276.2	98.0	71.6	47.1	234.6	214.8	46.6
9	86.9	61.8	303.9	274.9	95.4	66.2	248.1	221.7	197.3	24.9	5.2	197.0
10	237.1	211.9	93.9	64.9	245.4	216.3	38.2	11.8	347.5	175.2	155.6	347.4
11	27.3	2.0	244.0	214.9	35.4	6.3	188.3	162.0	137.7	325.5	305.9	137.9
12	177.5	152.1	34.0	4.9	185.4	156.4	338.4	312.1	287.9	115.8	96.3	288.3
13	327.7	302.2	184.1	154.9	335.5	306.4	128.5	102.3	78.2	266.1	246.7	78.7
14	117.9	92.3	334.1	304.9	125.5	96.5	278.6	252.5	228.4	56.4	37.1	229.1
15	268.1	242.4	124.1	95.0	275.5	246.5	68.7	42.6	18.6	206.7	187.4	19.5
16	58.3	32.5	274.2	245.0	65.5	36.6	218.8	192.8	168.9	357.0	337.8	169.9
17	208.5	182.6	64.2	35.0	215.5	186.6	8.9	342.9	319.1	147.3	128.2	320.3
18	358.7	332.6	214.3	185.0	5.6	336.7	159.0	133.1	109.3	297.6	278.6	110.7
19	148.8	122.7	4.3	335.0	155.6	126.7	309.1	283.3	259.6	88.0	69.0	261.1
20	299.0	272.8	154.3	125.1	305.6	276.8	99.2	73.4	49.8	238.3	219.4	51.5
21	89.2	62.9	304.4	275.1	95.6	66.8	249.3	223.6	200.1	28.6	9.8	202.0
22	239.4	213.0	94.4	65.1	245.7	216.9	39.4	13.8	350.3	178.9	160.2	352.4
23	29.5	3.0	244.4	215.1	35.7	6.9	189.5	164.0	140.6	329.2	310.6	142.8
24	179.7	153.1	34.5	5.1	185.7	157.0	339.6	314.1	290.8	119.6	101.0	293.2
25	329.8	303.2	184.5	155.1	335.7	307.1	129.7	104.3	81.1	269.9	251.4	83.6
26	120.0	93.2	334.5	305.2	125.8	97.1	279.9	254.5	231.3	60.2	41.8	234.0
27	270.1	243.3	124.5	95.2	275.8	247.2	70.0	44.7	21.6	210.6	192.2	24.4
28	60.3	33.4	274.6	245.2	65.8	37.3	220.1	194.9	171.8	0.9	342.6	174.8
20	210.4		64.6	35.2	215.9	187.3	10.2	345.1	322.1	151.3	133.0	325.1
30	0.6		214.6	185.2	5.9	337.4	160.3	135.2	112.4	301.6	283.4	115.5
31	150.7		4.7		155.9		310.5	285.4		92.0		265.9

MOTION OF THE CENTRAL MERIDIAN

	0ʰ	1ʰ	2ʰ	3ʰ	4ʰ	5ʰ	6ʰ	7ʰ	8ʰ	9ʰ	10ʰ	11ʰ
m	°	°	°	°	°	°	°	°	°	°	°	°
0	0.0	36.3	72.5	108.8	145.1	181.3	217.6	253.8	290.1	326.4	2.6	38.9
5	3.0	39.3	75.5	111.8	148.1	184.3	220.6	256.9	293.1	329.4	5.7	41.9
10	6.0	42.3	78.6	114.8	151.1	187.4	223.6	259.9	296.1	332.4	8.7	44.9
15	9.1	45.3	81.6	117.9	154.1	190.4	226.6	262.9	299.2	335.4	11.7	48.0
20	12.1	48.4	84.6	120.9	157.1	193.4	229.7	265.9	302.2	338.5	14.7	51.0
25	15.1	51.4	87.6	123.9	160.2	196.4	232.7	268.9	305.2	341.5	17.7	54.0
30	18.1	54.4	90.7	126.9	163.2	199.4	235.7	272.0	308.2	344.5	20.8	57.0
35	21.2	57.4	93.7	129.9	166.2	202.5	238.7	275.0	311.3	347.5	23.8	60.0
40	24.2	60.4	96.7	133.0	169.2	205.5	241.8	278.0	314.3	350.5	26.8	63.1
45	27.2	63.5	99.7	136.0	172.2	208.5	244.8	281.0	317.3	353.6	29.8	66.1
50	30.2	66.5	102.7	139.0	175.3	211.5	247.8	284.1	320.3	356.6	32.8	69.1
55	33.2	69.5	105.8	142.0	178.3	214.6	250.8	287.1	323.3	359.6	35.9	72.1
60	36.3	72.5	108.8	145.1	181.3	217.6	253.8	290.1	326.4	2.6	38.9	75.1

(b)

Explanation of example cited in text:

1. Determine longtitude for date at 0 hours UT—October 25, 1977 — **269.9°**

2. Determine motion of CM for time of observation 03:09 UT — **114.8°**

3. Determine total longitude of Central Meridian for October 25, 1977 at 03:09 UT by adding the two values in Steps 1 and 2.

   269.9°
   +114.8°
   ———
   384.7°

4. Because there are only 360° in total longitude, we must subtract that amount from the longitude computed in Step 3, which was greater than 360°:

5. 24.7° is the correct longitude of the CM (as well as any object on it) for October 25, 1977 at 03:09

   384.7°
   −360.0°
   ———
   24.7°

FIGURE 9–2. Tables for converting Universal Time to longitude of the central meridian. (From *The American Ephemeris and Nautical Almanac 1977*, Washington, D.C., Government Printing Office)

Observation of Transits

The telescopic observation of transits of Jovian features provides either direct or indirect determination of the following:

- ☆ Placement of a feature in longitude to an accuracy of ±1°.2.
- ☆ Direct determination of the longitudinal expanse (size) of a feature.
- ☆ Accurate determinations of rotational rates and changes in normal rotational patterns.

However, you must first have access to a copy of the *American Ephemeris and Nautical Almanac* (AENA) for the current year. The times you log can be converted directly to longitude through use of tables provided in the AENA. The conversion is direct and simple but can be obtained only through the charts shown as Figure 9-2. One chart (Figure 9-2a) corresponds to longitude in System I; the other (Figure 9-2b) shows longitude in System 2.

Always use Universal Time for observations because it corresponds with the data in the AENA. Use a watch or clock set to within a minute's accuracy for transit times, or a shortwave radio set to WWV time-signal frequencies (5, 10, or 15 MHz).

A *transit* is simply the time at which an object (such as the Red Spot shown on the diagram in Figure 9-3) is centered on the face of Jupiter as seen from earth. To facilitate the exact time of transit, first affix a mental image of a line extending from the north pole to the south pole of Jupiter. The process of "imagining" this line is quite simple because it divides the globe into two equal sections (east and west), and the line is virtually perpendicular to the obvious belts and zones. This imaginary line is known as the Central Meridian (CM).

Because Jupiter and its features rotate from west to east, the transit of a feature takes three forms, as follows:

1. *Preceding End* (Figure 9-3a). The time (UT) at which the leading (or eastern) edge of a feature crosses the CM. When you report transits of the preceding end, simply refer to "pr."
2. *Central Transit* (Figure 9-3b). The time (UT) at which the feature you are timing is centered exactly on the globe. At this point, our imaginary CM would cut the feature in

FIGURE 9-3. Aspects of transit timing.

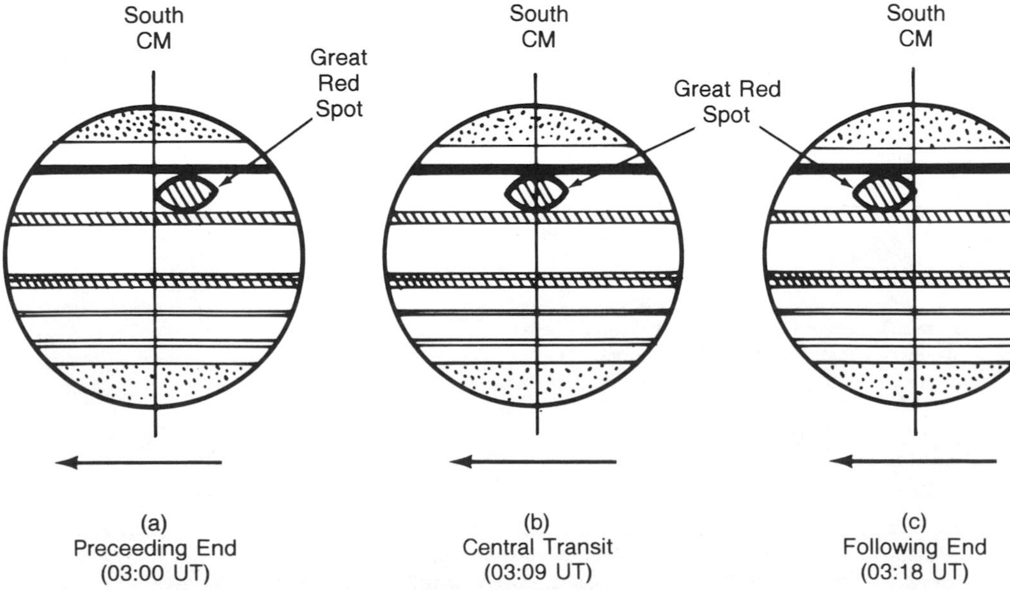

South CM
Great Red Spot

(a) Preceeding End (03:00 UT)

(b) Central Transit (03:09 UT)

(c) Following End (03:18 UT)

half. In your reports, refer to central transit simply as "C".

3. *Following End* (Figure 9-3c). The time (UT) at which the western edge (or "tail end") of the feature reaches the CM. This will be the last timing made for a particular feature, since it is moving from west to east. The following end of a feature is simply referred to as "f" when you report data.

If you can make only one timing, it is essential to determine No. 2–Central Transit. Many features appear to be less than 2° expanse in longitude. Try to get only central timing on these because their preceding and following ends cannot accurately be determined with such tiny features.

Placement in Longitude

To place a Jovian feature in longitude, simply record the time (UT) at which the marking is *centered* (Figure 9-3b). To determine longitude, there is no need to record the preceding or following ends of transit.

The observation in Figure 9-3 was made October 25, 1977 at 03:00 to 03:18 UT. We are interested only in *central transit* for longitude determination, therefore we use the time 03:09 (Figure 9-3b). Knowing that the feature is in System II because of its high latitude, we use Figure 9-3b to extrapolate the correct latitude at 03:09 UT. We find the correct longitude for October 25 by using Fig. 9-2b. This longitude corresponds to 00h 00m UT.

The longitude in System II on that date (October 25) is determined to be 269.9° (see Figure 9-2). However, 3 hours and 9 minutes have elapsed from 0 hours UT to the time of our observation, and a factor for this elapsed time must be applied. We take the factor from the lower chart on Figure 9-2b. Moving across to 3h on the horizontal scale and down 10m on the vertical scale, we find our motion for 3h 09m to be 114.8°.

These two values (the first for the *date* and the second for the *time*) are then added together, giving a longitude value of 384.7°. However, if the value is greater than 360°, as it is in our example, the amount by which it is greater must be subtracted from our determined longitude, because there are only 360° in any longitude system. Subtracting, we get a value of 24.7° as the longitude of the marking in Figure 9-3b.

In 1 minute, features in System II move about 0.6° in longitude. Hence, an accuracy of about 2 minutes for a timing will place a feature within ±1°.2 of its true position. With experience, this is the best attainable accuracy you will be able to attain through visual means.

Determination of Longitudinal Expanse And Size of Jovian Features

Using Figure 9-3 as a guide, we shall determine the longitudinal expanse and size of the feature shown (the Great Red Spot).

The interval between *preceding end transit* (03:00) and *following end transit* (03:18) is 18 minutes. Remembering that 1 minute of time corresponds to 0.6° of longitude on Jupiter, we can derive the simple equation:

$$E = I^m \times 0.6^\circ$$

where E is the total longitudinal expanse (angular size) to be determined, I^m is the interval from beginning to end (in our example, 18 minutes), and 0.6° is the movement of longitude per minute in System II. Thus, the Great Red Spot on that date was:

$$E = 18 \times 0.6^\circ$$

or

$$E = 10.8^\circ$$

which is about 11° total breadth in longitude.

Once having determined the longitudinal expanse (angular size) of this feature, it is a simple matter to extrapolate degrees longitude into actual size in kilometers (or miles), using the formula

$$x^k = E \times 1187 \text{ km}$$

where x^k is the actual size in kilometers to be

determined, E is the longitudinal expanse determined in the first equation above (10.8°), and 1187 km is the number of kilometers per degree longitude of any Jovian feature. Thus, the actual size (in kilometers) of the Red Spot on that date was:

$x^k = 10.8° \times 1187$ km

or

12,819.6 km

Determinations of Rotational Periods Using Longitudinal Drift

If the same Jovian feature can be observed on two dates 20 to 30 days or more apart, it is possible to determine accurately the true rotational period of that feature by noting the amount of *longitudinal drift*, which is simply the *difference* in longitude of a feature between two dates. The procedure for determinations of longitude has been previously discussed.

Table 9-2 gives the correction factors for varying amounts of drift in a 30-day period. It is essential that all intervals be converted to 30 days to correspond to the data in Table 9-2. For example, if a drift of 3° is noted in 20 days, a simple proportion will convert it to a 30-day value, which can be used with Table 9-2:

$$\frac{3°}{20} = \frac{x}{30}$$

or

$20x = 90°$

Therefore,

$x = 4.5°$

or 4.5° drift in 30 days.

TABLE 9-2. Table for the conversion of measured drift to rotational period.

Change in Longitude in 30 Days[a]	*System I (9h 50m 30.003s)*[b]	*System II (9h 55m 40.632s)*[b]
0.1	0.1345	0.1369
0.2	0.2690	0.2738
0.3	0.4036	0.4107
0.4	0.5381	0.5476
0.5	0.6726	0.6845
0.6	0.8071	0.8214
0.7	0.9417	0.9583
0.8	1.0762	1.0952
0.9	1.2107	1.2321
1.0	1.3452	1.3689
2.0	2.6905	2.7379
3.0	4.0357	4.1068
4.0	5.3810	5.4758
5.0	6.7262	6.8447
6.0	8.0714	8.2137
7.0	9.4167	9.5827
8.0	10.7620	10.9516
9.0	12.1072	12.3206
10.0	13.4525	13.6895
20.0	26.9050	27.3790

[a] In degrees.
[b] In seconds.

If the feature drifts *westward* (increasing longitude), the values in Table 9-2 should be *added* to the prime value; if the feature drifts *eastward* (decreasing longitude), the value should be *subtracted* from the prime value. Notice that the prime value (no drift in either System I or in System II in a 30-day period) is the mean rotational rate for each system. The prime values for each system are as follows:

System I	9h 50m 30.003s
System II	9h 55m 40.632s

The values found under each system heading in Table 9-2 should be either added to or subtracted from the *seconds* value of each prime value. If the amount of drift is not found in Table 9-2, it may be computed simply by adding values together. For example, to determine the conversion for 4.5°, simply add the values found in Table 9-2 for 4.0° and 0.5°.

As an example of determining rotational rates from longitudinal drift, suppose that on the night of October 25, 1977, you observe the center of the Red Spot to transit over the CM at 03:09 (Figure 9-3b), corresponding to System II longitude 24.7° (as has been previously determined). Thirty days later, on November 23, the Spot transits at 02:10, which can be computed to 29.2° longitude. Thus, we see that the spot has drifted 4.5° in longitude *westward*.

Notice that the Spot's longitude has *increased*, which is always indicative of a westward drift. Thus, the correction value from Table 9-2 must be *added* to the prime value, 9h 55m 40.632s, because the spot is in System II.

Using Table 9-2, we see that the drift of 4.5° corresponds to a value of 6.1603 sec:

$$\begin{array}{r} 4.0^\circ = 5.4758\text{s} \\ \underline{+0.5^\circ = 0.6845\text{s}} \\ 4.5^\circ = 6.1603\text{s} \end{array}$$

This value is then added to the prime value for System II:

$$\begin{array}{r} 9\text{h } 55\text{m } 40.6320\text{s} \\ \underline{6.1603\text{s}} \\ 9\text{h } 55\text{m } 46.7923\text{s} \end{array}$$

Thus, from October 25 to November 23, 1977 (30 days), the rotational rate of the Great Red Spot has been determined to be *9h 55m 46.7923s*, slightly greater in duration than the mean value of System II, indicating a deceleration of the Spot during this interval.

Determine the rotational values for every marking you can observe on two occasions at least 20 days apart. These values are the single most important determination that you as an amateur astronomer can make in studies of Jupiter.

Determinations of Latitude

It is possible for amateurs to determine the latitudes of Jovian belts and zones, and you should make efforts to do so. This normally is an exacting and difficult task, best done with a filar micrometer. However, micrometers are scarce and the mechanics of their operation are not well known. Two alternate procedures may be used but with substantially less accuracy. Using the table on page 169, convert all values to zenographic latitude.

Reticle Eyepiece Determination

Figure 9-4 demonstrates a simple method of latitude determination using an etched-glass reticle inserted into an eyepiece. Such reticles are commonly available for little investment, but take care in selecting them to make sure they do not have overly large division markings that might obscure faint belts.

Using a reticle is much like using a ruler, where one is measuring on a predetermined scale. The horizontal axis of the reticle must always be placed to run *along the equator* of Jupiter, perfectly parallel to conspicuous belts and zones. One obvious inefficiency of this method is that the determination of the exact equator (0°) is difficult. In some years an obvious equatorial belt is evident, yet it consistently lies south of the true equator. Then, various marks along the vertical axis can be assigned a fraction that corresponds to a fractional part of the distance from the equator (0°) to the pole (90°). Remember that, as on earth, the southern hemisphere is represented by *negative* latitudes, with the equator being 0° and the south pole being -90°.

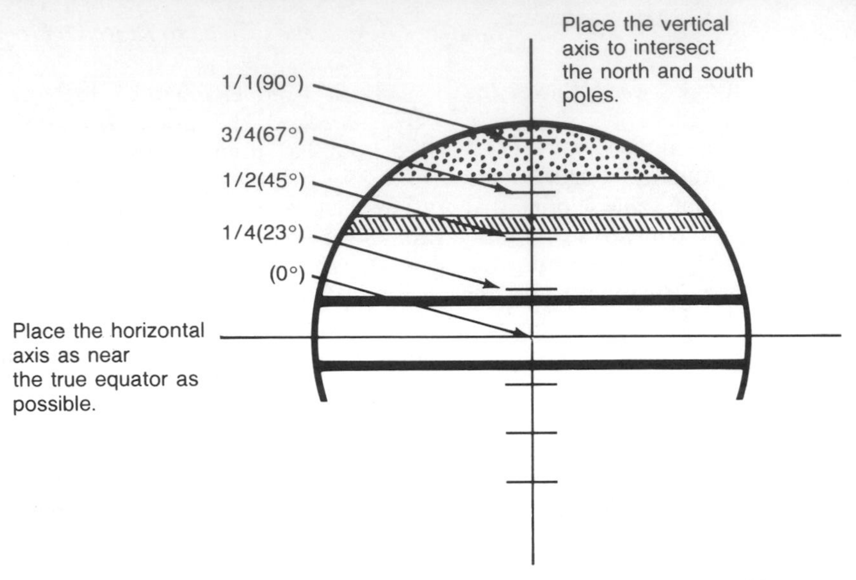

FIGURE 9-4. Determination of latitude using a micrometer reticle eyepiece with quarter subdivisions.

Jovian latitude estimates are quite simple to determine by using the equator (0°) and the poles (either +90° or -90°) as guideposts. For example, if a belt is observed on the reticle as being halfway from the equator toward the north pole, its latitude is half the total distance, or 45°. If another belt is measured a third of the distance from the equator to the northern pole, its latitude is 30°. A faint belt lying exactly halfway between those just determined has a latitude of about 37.5°, and so forth (see page 169).

Photographic Determination of Latitude

Perhaps a bit more efficient than reticle determination, measurement of photographic latitude allows you to measure in comfort and leisure by using a photograph that has enough quality to show all major belts and zones. Although not an accurate zenographic determination (map projection), such a measurement is considered to be better than no measurement at all.

The procedure is similar to that of reticle determination, but you measure, using a millimeter scale, the distance from the equator (0°) toward the pole (90°) and convert millimeter measurement to appropriate angular measurements. In Figure 9-5, measure the distance from the equator to the pole; that value is 25 mm, corresponding to 90° of latitude. For an image of the scale of Figure 9-5, that value (25mm/90°) becomes your standard.

Now measure belt A. You will find that the distance of this belt from the equator is about 12 mm. If 25 mm equals 90° (your standard for this scale image), then 12 mm corresponds to approximately 43.2° latitude. This can be simply computed as shown below:

$$\frac{25\text{ mm}}{90^\circ} = \frac{12\text{ mm}}{x}$$

or

$$25x = 1{,}080^\circ$$

$$x = 43.2^\circ \text{ (Jovicentric)}$$

or -43.2°, since belt is south of the equator.

Now measure belt B and determine its angular latitude. Conversion from these values (Jovicentric) to zenographic latitude is facilitated by using the table on page 169.

Latitude Determinations With a Filar Micrometer

A filar micrometer enables you to determine accurately the placement in latitude of Jovian

Conversion Equivalents for Jovicentric and Zenographic Latitudes of Jupiter.

										a less than 10°	
a°	*b*°	*a*°	*b*°	*a*°	*b*°	*a*°	*b*°	*a*°	*b*°	*a*°	*b*°
11.248	6.55	23.902	15.21	36.908	24.11	49.562	32.77	62.216	41.44	1.758	0.06
11.600	6.79	24.254	15.45	37.259	24.35	49.913	33.01	62.567	41.68	2.109	0.30
11.951	7.03	24.605	15.69	37.611	24.59	50.265	33.26	62.919	41.92	2.461	0.54
12.303	7.27	24.957	15.93	37.962	24.84	50.616	33.50	63.270	42.16	2.812	0.78
12.654	7.51	25.308	16.17	38.314	25.08	50.968	33.74	63.622	42.40	3.164	1.02
13.006	7.75	25.660	16.42	38.665	25.32	51.319	33.98	63.973	42.64	3.515	1.26
13.357	7.99	26.011	16.66	39.017	25.56	51.671	34.22	64.325	42.88	3.867	1.50
13.709	8.24	26.363	16.90	39.368	25.80	52.022	34.46	64.676	43.12	4.218	1.74
14.060	8.48	26.714	17.14	39.720	26.04	52.374	34.70	65.028	43.36	4.570	1.98
14.412	8.72	27.066	17.38	40.071	26.28	52.725	34.94	65.379	43.60	4.921	2.22
14.763	8.96	27.769	17.86	40.423	26.52	53.077	35.18	65.731	43.84	5.273	2.46
15.115	9.20	28.120	18.10	40.774	26.76	53.428	35.42	66.082	44.08	5.624	2.70
15.466	9.44	28.472	18.34	41.126	27.00	53.780	35.66	66.434	44.32	5.975	2.94
15.818	9.68	28.823	18.58	41.477	27.24	54.131	35.90	66.785	44.56	6.327	3.18
16.169	9.92	29.175	18.82	41.829	27.48	54.483	36.14	67.137	44.80	6.679	3.42
16.521	10.16	29.526	19.06	42.180	27.04	54.834	36.38	67.488	45.04	7.030	3.66
16.872	10.40	29.878	19.30	42.532	27.96	55.186	36.62	67.840	45.28	7.382	3.91
17.224	10.64	30.229	19.54	42.883	28.20	55.537	36.86	68.191	45.53	7.733	4.15
17.575	10.88	30.581	19.78	43.235	28.44	55.889	37.11	68.543	45.77	8.085	4.39
17.927	11.12	30.932	20.02	43.586	28.68	56.240	37.35	68.894	46.01	8.436	4.63
18.278	11.36	31.284	20.26	43.938	28.93	56.592	37.59	69.246	46.25	8.788	4.87
18.630	11.60	31.635	20.50	44.290	29.17	56.943	37.83	69.597	46.49	9.139	5.11
18.981	11.84	31.987	20.75	44.641	29.41	57.295	38.07	69.949	46.73	9.491	5.35
19.333	12.08	32.338	20.99	44.992	29.65	57.646	38.31	70.300	46.97	9.842	5.59
19.684	12.33	32.690	21.23	45.342	29.89	57.998	38.55			10.194	5.83
20.036	12.57	33.041	21.47	45.695	30.13	58.349	38.79			10.545	6.07
20.387	12.81	33.393	21.71	46.047	30.37	58.701	39.03			10.896	6.31
20.739	13.05	33.747	21.95	46.398	30.61	59.052	39.27				
21.090	13.29	34.096	22.19	46.750	30.85	59.404	39.51				
21.442	13.53	34.447	22.43	47.101	31.09	59.755	39.75				
21.793	13.77	34.799	22.67	47.453	31.33	60.107	39.99				
22.145	14.01	35.150	22.91	47.804	31.57	60.458	40.23				
22.496	14.25	35.502	23.15	48.156	31.81	60.810	40.47				
22.848	14.49	35.853	23.39	48.507	32.05	61.161	40.71				
23.199	14.73	36.205	23.63	48.859	32.29	61.513	40.95				
23.551	14.97	36.556	23.87	49.210	32.53	61.864	41.19				

NOTES: *a*° denotes Jovicentric (measured) micrometric latitude
b° is corresponding zenographic latitude converted for disk curvature and projection

Adapted by Brian Sherrod and Clay Sherrod

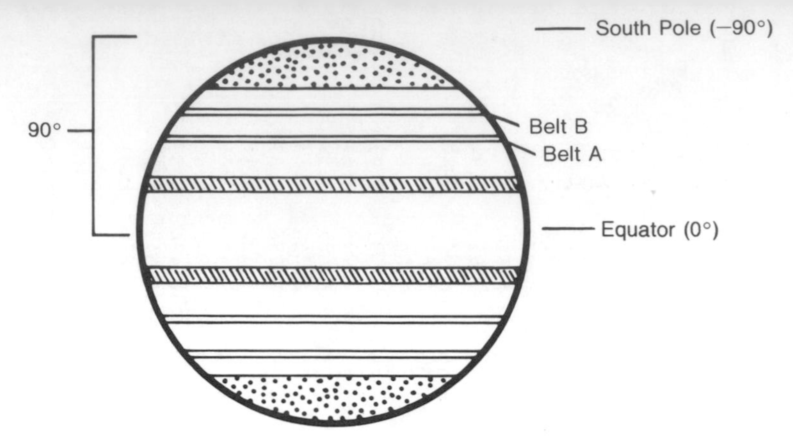

FIGURE 9-5. Measurement of latitude using a photograph and a millimeter rule. (Conversion to zenographic latitudes necessary; see page 169.)

features and to determine precisely the changes in extent of prominent areas (e.g., the size of the Great Red Spot). The filar micrometer is now available from commercial sources. The method of use of each instrument is dependent on the method by which it was made. Most measurements with such an instrument are made in relation to the total disk size. The *American Ephemeris and Nautical Almanac* gives for each date a precise angular size of the Jovian disk (in seconds of arc). Measure the feature in proportion to the disk measurement for that date. Because of the oblate sphere of Jupiter, make sure that the total disk measurement is from south to north and not east to west.

Estimates of Color

The colors of Jupiter and the changes in them are spectacular when you follow them in an organized program. For best results, use reflector telescopes because they do not suffer from the color aberration inherent in refractors. However, the more obvious tones can be accurately determined with *any* telescope, provided that the optics of refractors are properly corrected for secondary color.

When determining color changes, be particularly cautious to account for the planet's altitude in our sky. When it is high in the sky (about 45°, or higher, above either horizon), color estimates can be quite reliable. At lower altitudes, however, atmospheric dispersion will cause the introduction of false colors.

At every opportunity notice the color of the zones, the belts, the Great Red Spot, and all other notable features. Table 9-4 gives examples of the colors that I have recorded in recent years of major features of Jupiter. However, your estimates should not be restricted to the terms I have used. Whatever color combinations that best describe what you see are the ones that you should use.

The colors given in Table 9-4 are average estimates of color from many observations throughout each year cited, and many features that were observed are not listed. As a general rule, *any* feature exhibiting color should be described and reported at every opportunity.

TABLE 9-4. Colors of major features of Jupiter.[a]

Feature	*Colors Observed*
Great Red Spot	reddish-orange (1972); orange-pink (1974) very dim pink (1976)
North Equatorial Belt	very dark brown (1972); deep-brown (1974) blackish-brown (1976)
Equatorial Zone	bright yellow (1972); dusky yellow-brown (1976)
South Tropical Zone	very bright yellow (1976)

[a] As observed by the author.

There is some evidence of a periodicity of color changes in the Jovian clouds, with an average interval of about 11 years. The periodicity seems to be associated with a general reddening of the belts, rather than being an all-inclusive color change. There is currently no suitable explanation to account for the colors and their dramatic changes. A combination of cloud altitudes, chemical constituency, and temperature is perhaps the best—and safest—beginning of any such explanation.

Patrol Surveillance of Jupiter

The most regular systematic study of Jupiter by the amateur should be *patrol studies*, made either visually by drawing the disk of Jupiter on an appropriate form, or photographically by recording Jovian detail. There are distinct advantages in each method.

Regardless of the method used, each patrol observation should include the following data:

1. Name of observer.
2. Date and time in Universal Time.
3. Longitudes of central meridian of both System I and System II at the beginning of each drawing, or at the time of each photograph.

Make patrol surveillance every clear night that seeing conditions permit, so that you can maintain continuity. Examine the following features for any changes:

☆ *The Great Red Spot* Look for changes as described in the following section.

☆ *Polar Regions* Be alert for any appearance of thin belts, gray or white ovals, and sudden appearance of zones in both the north and south polar regions. Compare north vs. south polar regions.

☆ *Equatorial Zone* Note carefully the festoons in this zone, bright spots (quite common), color and brightness changes, and any appearance of the equatorial belt (quite rare). Time any conspicuous features as they cross the central meridian.

☆ *North Equatorial Belt* The north equatorial belt is an area of extremely rapid change. The NEB generally is very wide and dark, containing many knots and spots. Many of the festoons seen in the EZ originate from the NEB. Give attention also to any festoons originating from and moving northward into the NTrZ. Such festoons are very thin and difficult to discern, but they may indicate the beginning of disturbances in the NEB and NTrZ. Many times, particularly when Jupiter is active, a very thin north component of the NEB can be seen in larger telescopes. Give particular attention to the development of NEB "Red Spots," which appear during time of ongoing activity. These spots have the color and intensity of the Red Spot itself, but they are rather short-lived. Comparisons of the rotational periods of the NEB red spots to the Great Red Spot are of the greatest importance.

☆ *South Equatorial Belt* This belt is an area of great activity. Many disturbances originate from this belt, which generally appears to be double. The north component of the SEB (SEBn) is generally the darkest and widest, and it seems to be connected to the south component (SEBs) by fine festoons and, often, series of tiny white spots. The zone separating the two belts (the SEBZ) is the area of origin of many disturbances of major and minor significance. The disturbances originate from fine festoons emanating from the SEBs and crossing southward into the STrZ. Monitor any such festoon at every opportunity because their appearance almost certainly heralds a coming disturbance. Interpret carefully whether the festoon originates from the SEBs or from the activity within the SEBZ. Many tiny white spots and dark bars can be seen rotating within this region, somewhat independently of System II. Be sure to time all CM transits of such features. In addition, make careful observations of the association of the SEBs to the Great Red Spot, noting the degree of interaction of the spot as it pushes the SEBs northward.

☆ *Tropical Zones* As mentioned, you should monitor both the NTrZ and the STrZ for festoons originating from either the NEB or the SEBs, respectively. In addition, make searches to reveal faint, thin belts (HLM), which sometimes appear in these zones. Usually, however, they are only a few degrees in length. Vague, but large, white ovals of irregular shape can often be observed in these zones. These are short-lived phenomena, so observe them as often as possible to determine changes in size, shape, and rotational period.

☆ *North Temperate Belt* This belt has two aspects: (1) it is singular, dark, and rather wide, and (2) it is double, with two thin components extremely close together. There is an increase in activity of this belt when it is single and a decrease when it is double, as well as a marked change in the intensity and color between the two aspects.

☆ *South Temperate Belt* A very curious feature, the STB can be the darkest marking of Jupiter during some apparitions. Contrasting to the somewhat stable NTB, this belt is quite irregular, with large swells, knots, curves, and breaks occurring along its longitudinal expanse. It is generally wide, sometimes separating for short distances into two thin, but dark, components.

The long-enduring white ovals associated jointly with this belt and the STZ just to its south appear to have a definite effect on the activity, width, and intensity of the STB. Primarily, areas of the belt just north of a white oval are darker, wider, and more intense as a rule than isolated areas of the belt. The large ovals often appear to be actually superimposed over the STB and the STZ, resulting in some notable physical interaction between the features. Monitor the ovals constantly for rotation rates, intensity, and the effects of ovals on other features of this region.

In addition to these features listed above, monitor all latitudes north and south of them for any changes. The next section describes the proper procedure for either drawing or photographing Jupiter for patrol purposes.

Disk Drawings of Jupiter

Appendix X shows a proper standardized form for full-disk drawings of Jupiter. Notice carefully the orientation of the form, which corresponds to the image seen in astronomical telescopes. South is at the top, north at bottom, west at left, and east at right. These directions correspond to sky directions in the *astronomical* sense; for example, *west* corresponds to the *direction* west as seen in the earth's sky.

It is extremely important that you maintain the orientation for your drawings. If you use a refractor, use it *without* any right-angle prism at the eyepiece. A Newtonian will conform to this orientation with no alteration. A Cassegrain (or Schmidt-Cassegrain) must also be used *without* the right-angle prism.

Features will be seen as rotating rapidly from *right* to *left* when properly oriented as in the form. Because Jupiter's period is less than 10 hours, a feature will move from the center to the left edge in only 2½ hours. Hence, begin your drawings with features at the *left edge* ("west" on the form) and proceed toward the right ("east" on the form). Because of rapid changes on the planet, as well as its rapid rotation, you must complete your drawings in less than 45 minutes to maintain accuracy.

Ideally, for patrol purposes, make a drawing every 2½ hours. This allows you to construct a zenographic map from each night's observing that covers all 360° longitude. Make a note of the Universal Time of the beginning and end of each drawing. The central meridian is determined for both System I and System II from the *beginning* time [on this form, the central meridians are recorded beside "CM(I)" and "CM(II)"] in the list below the blank disks. Use drawing pencils of at least three grades (hard, medium, and soft), and do not attempt drawings using only one standard No. 2 pencil because shadings are difficult to render accurately with only one type of lead.

Before going outdoors, it is helpful to lightly

designate across the disk the approximate positions (latitudes) of the major belts, but only if you have already determined them on previous nights' observations, or from a recent photograph, or the ALPO's published lists of various Jovian latitudes. Since the belts do not change latitudes significantly in a year's time, this practice is certainly permissible, and will aid you in the rapid rendering of your drawings.

Keep all your original drawings in chronological order, and reference them for rapid retrieval. Make a copy of each drawing (being careful to copy the original exactly), and forward the copies to either the ALPO or the Director of the Planetary (Jupiter) section of your local society.

Photographic Jupiter Patrols

Although not as much fine detail on the planets can be reported photographically as visually, there are distinct advantages to a photographic patrol:

- ☆ Photographs are permanent proof of suspected changes.
- ☆ The photos can be compared at leisure for transient phenomena.
- ☆ Direct measurements can be made from photographic prints.
- ☆ Any changes in the intensity and size of features can be more accurately determined.

The prime limitation of photographic patrols is the long focal ratios necessary for appropriate image scales and the lengthy exposure (1 to 8 sec) needed because of these long focal ratios. If seeing conditions are less than perfect, results will be disappointing and of little value, because air turbulence will cause finer features to blend into invisibility.

Your choice of film is somewhat dependent on your preference. However, some guidelines must be adhered to. The film must meet the following requirements.

1. Have extremely low grain.
2. Have panchromatic capabilities.
3. Be fairly "fast" (high ASA rating).
4. Be fairly inexpensive.

Most people must be aware of costs because at least five photos should be taken at one session to be sure of getting at least *one* clear image. In order of recommendation, some good films for patrol use are the following:

SO 115 Photomicrography (Kodak) (now known as "2415")	ASA 125
High Contrast Copy (Kodak)	ASA 32
Plus-X Pan (Kodak)	ASA 125
Tri-X Pan (Kodak)	ASA 400

These are black and white films. For patrol purposes, color film is not recommended primarily because of its low sensitivity and secondarily because of its high cost. The SO 115 film is excellent for planetary photographs, exhibiting virtually no grain but requiring lengthy exposures. On the other hand, Tri-X Pan is very grainy, but it requires substantially less exposure time. Tri-X is an excellent film for the beginner. Both SO 115 and Tri-X can be "pushed" four times faster using Ethol UFG developer at full strength. With such developer, SO 115 might yield ASA 380, whereas Tri-X can effectively reach ASA 1600 with little increase in grain. Such high speeds can increase the efficiency of a planetary photograph by reducing the length of exposure, yet maintaining high EFL's (see Chapter 14). Developing either Tri-X or SO 115 in full-strength UFG, at 68°F for 8 min will provide this high ASA rating.

The left part of Figure 9-6 is a graphic aid in determining exposure times with differing focal ratios when photographing Jupiter. The focal ratio changes in a telescope according to the length of the eyepiece. You can determine the necessary focal length by referring to Figure 9-6 (see also Chapter 14). The ASA (speed) of the film is given at the top of each column in the left part of the illustration.

To process the films listed, follow the manufacturer's specifications, but alter them to provide fine-grained and high-contrast negatives. Many developers of film negatives are available for just that purpose; use them if possible. When printing the photographs, use only high-contrast paper, such as Kodak Kodabromide, Hard or Ultra-Hard surface, to accentuate fine detail.

Colored filters aid tremendously in capturing fine details otherwise not seen in white light. Some filters and the features best observed with them are the following:

- ☆ Medium Blue—White ovals, spots, equatorial and temperate belts, detail in the Great Red Spot.
- ☆ Green—Highly recommended for all belts, festoons, the Great Red Spot, white ovals, and spots.
- ☆ Yellow—Not recommended.
- ☆ Orange—Some white spots, some belts, detail within the zones.
- ☆ Red—Not recommended.

When you use filters in your photography work, remember that filtered light requires longer exposures than does white light, and this must be considered when using the data in Figure 9-6. For those filters listed above, a factor of 3.5 (3.5 times the time cited) should be included.

A Program of Observation For the Great Red Spot

One of the most intriguing features of the solar system is the famous Great Red Spot of Jupiter. At maximum size, this large circulating storm can be 40,000 km in length and 8000 km in breadth. Its origin and its physical makeup are uncertain, although theory indicates that it originated as a thermal "cyclonic" feature. Recent observations indicate that the reddish color of this spot may be due to crystals of phosphine (hydrogen phosphide).

FIGURE 9-6. Determining exposure time. (From *Graphic Determinations of Photographic Exposures,* J.E. Henry. Midsouth Astronomical Society Publications, Little Rock, Ark., 1974)

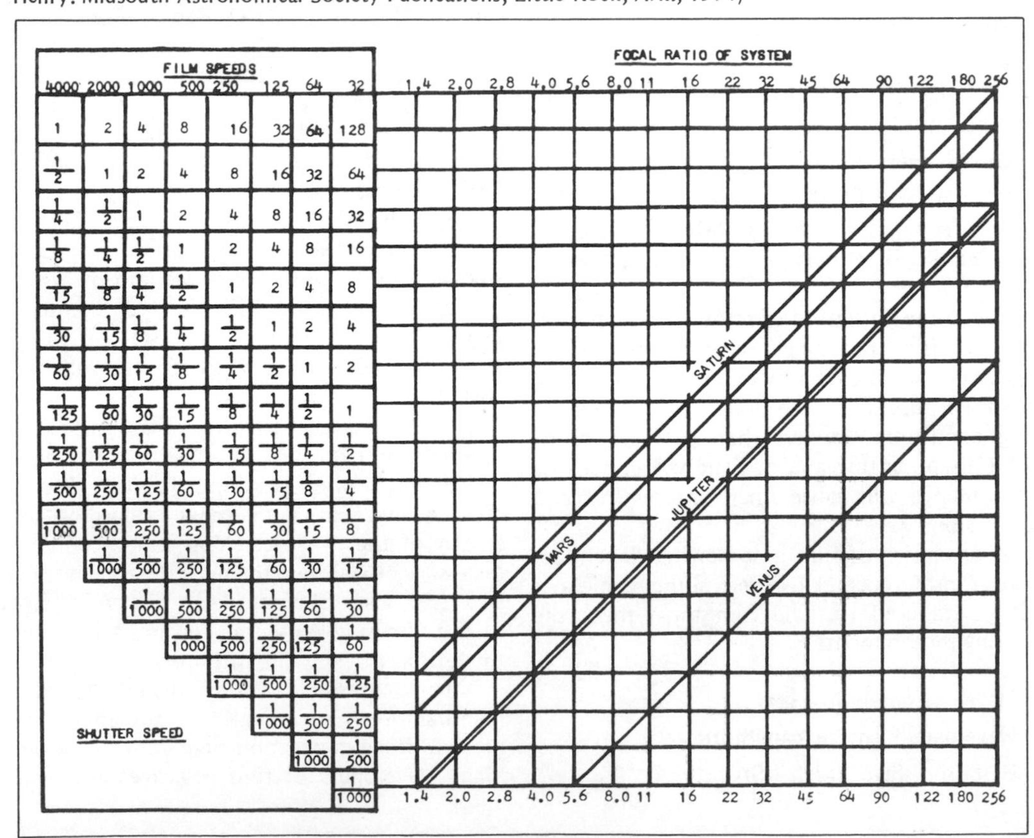

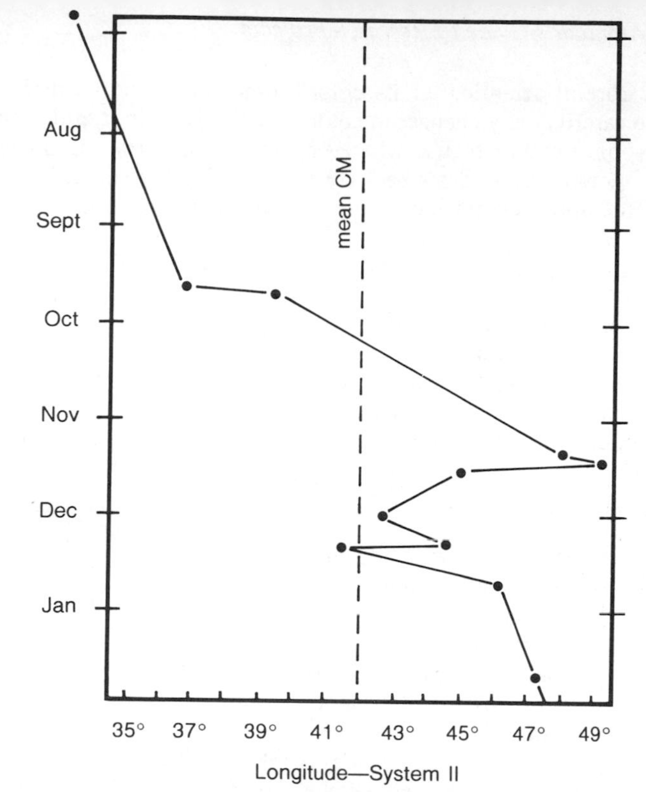

FIGURE 9-7. The drift of the Great Red Spot, based on the author's 1975-1976 observations. (Courtesy, Midsouth Astronomical Research Society, Inc.)

The Spot is commonly associated with System II, although its motion is very erratic and largely independent of the System II current. The mean rotational rate of the RS since 1910 has been 9h 55m 38s. Figure 9-7 demonstrates the erratic drift of the Red Spot in 1975-1976 and is indicative of normal motion of the Spot during turbulent activity in its vicinity.

A complete observing program for the Red Spot should include the following:

1. Estimates of color.
2. Determinations of longitude.
3. Determinations of size.
4. Determinations of drift.
5. Computations of rotational period.
6. Strip sketches of RS vicinity.
7. Observations of interactions of the RS with other Jovian features.

In some years, the Red Spot has been invisible, or absent. In those years, however, a conspicuous hollow or "bay" in the position where the spot *should* be can readily be followed. During such years, continue your observations in the normal way, recording data on the hollow, rather than on the Spot itself.

Estimates of Color

As indicated previously, the color of the Spot is at least in part determined by the chemicals of which it is comprised. In addition, there is some evidence that the vortex size of the RS in the Jovian clouds may affect the color and the intensity as the Spot seen from earth. Hence, your observations of the Spot's color are of great importance as part of your observing program. The Spot can vary from a tannish-orange to a salmon pink, and it sometimes has even been gray. During times of great cloud disturbances near the RS, give careful attention to its color changes. Also note carefully any changes in color tone within the spot. Be sure to use whatever descriptive color phrase best describes your impression of the Red Spot's coloration.

Determinations of Longitude

Determination of the longitude of Jovian features has been described on pages 164–165. To place the Red Spot in longitude, use the time of central transit (Figure 9-2b), which yields the longitude of the center of the Red Spot.

Determinations of Size

By recording the three times (preceding end, central transit, and following end) of the central meridian transit of the Red Spot, its size in longitude (length) can be determined as described on page 165. It is important to monitor the size (both *angular* and *actual*) because the size varies rapidly and to a great extent.

During some apparitions, the RS assumes a football shape, with slightly pointed ends. When it has that shape, take care during both the preceding and following end transits because the time at which the point is actually at the CM is rather difficult to determine.

Determination of Drift And Rotational Period

Because the Red Spot varies greatly in drift and rotational rate (Figure 9-7), it is critical that both values be evaluated as often as possible as described on pages 166–167. Many factors affect the drift patterns of the RS, including interactions with SEB disturbances, STZ white ovals, and undefined factors (see Figure 9-8). Correlations between such factors and RS drift is an important aspect of amateur research.

Observations of Red Spot Interactions

It is important when observing Jupiter that you watch for possible interactions of the RS with features in its vicinity before the interactions actually occur, particularly with the STZ white ovals and SEB disturbances. Such features many times have normal drift patterns opposite those of the Red Spot and, consequently, the countering drifts carry the objects toward one another, resulting in some physical disturbance or change. During interactions, record the following data:

1. Date and time of interaction.
2. Approximate difference in latitude of the Red Spot and the other feature.
3. Rotation period of the Red Spot *before* interaction.
4. Rotation period of other feature *before* interaction.
5. Rotation periods of both objects *during* interaction.
6. Change in rotation period of each feature *after* interaction.
7. Strip sketches in series throughout the interaction.

Plot the drift patterns of both features graphically, as shown in Figure 9-7, covering the time interval before, during, and after the interaction. Such plotting yields visual information regarding the drift changes and their relationships to one another. An example of data that can be obtained from recording the drift patterns during RS interactions with other features is shown as Figure 9-8. The graph shown as Figure 9-7 corresponds to the data of Figure 9-8. Notice carefully the evaluation of possible *causes* for this drift phenomenon.

Strip Sketches of the Vicinity Of the Red Spot

To render accurate information concerning the appearance of the RS, make *strip sketches*, rather than disk drawings. Strip sketches isolate a narrow range of latitude and generally are drawn to larger scale than full disk drawings in order to exhibit smaller detail. Aspects you should look for and draw, if you see them, are the following:

☆ Borders—Examine for a dark or light distinct border surrounding the Red Spot.

DATE from	to	DRIFT	ROTATIONAL PERIOD system II	SPOT CENTER from	to
June 26	Sept 19	− 4°.3	9h 55′38″.989	33°.3	37°.6
Sept 19	Nov 15	−11°.9	9h 55′31″.735	37°.6	49°.5
Nov 15	Nov 17	+ 4°.9	9h 57′23″.200	49°.5	44°.6
Nov 18	Dec 2	+ 2°.5	9h 55′47″.476	44°.9	42°.4
Dec 3	Dec 10	− 2°.0	9h 55′31″.597	42°.4	44°.4
Dec 10	Dec 12	+ 3°.0	9h 56′42″.235	44°.4	41°.4
Dec 12	Dec 22	− 4°.8	9h 55′20″.919	41°.4	46°.2
Dec 22	Jan 24	− 1°.6	9h 55′38″.716	46°.2	47°.8

Possible Explanations for Drift Phenomena

DATE	PHENOMENON	R.S. CENT.	POSSIBLE EXPLANATION
Sept 19	Acceleration (rapid)	36.6°	Begin passage of Preceding End of South Tropical Zone Disturbance (noted by P. Budine) over Red Spot—duration 58 days.
Nov 15	Deceleration (very rapid)	49.5°	Conjunction of Preceding End of Red Spot with prominent festoon from South Equatorial belt s. component, leading into major #1 SEB Disturbance.
Nov 16 17 18	Rapid Change in Deceleration	45.0° 44.7° 44.9°	Conjunction with very dark STB streak and major section of SEB disturbance #1. Also conjunction with STrZ oval BC?
Dec 3	Acceleration	42.4°	Near normal drift rate—appears to be attempting to stabilize eastward drift?
Dec 10	Deceleration (rapid)	44.4°	Begin physical change in appearance of Spot. Preceding end beginning to disassociate, tiny white spots recorded on N. edge of RS by Sherrod, possible remnant of SEB disturbance #1. Vertical gas flows under RS?
Dec 12	Acceleration (very rapid)	41.3°	Predicted date of conjunction of following end of STrZ oval FA with preceding end of RS. Combined eastward drift could accelerate rapidly as observed (prediction by Sherrod).
Dec 22	End Acceleration	46.3°	Resume almost normal eastward drift of 9h 55′38″.716 (Sherrod). Probably end of passage of STrZ oval FA across Red Spot. Spot loses all identity of shape, size, color.

FIGURE 9-8. Drift data and possible explanations for the Red Spot in 1975. (From *Observations of the Great Red Spot and Jupiter, 1975,* P.C. Sherrod. Project Publication 003. Midsouth Astronomical Research Society, Inc., Little Rock, Ark., 1975. Courtesy, the Society)

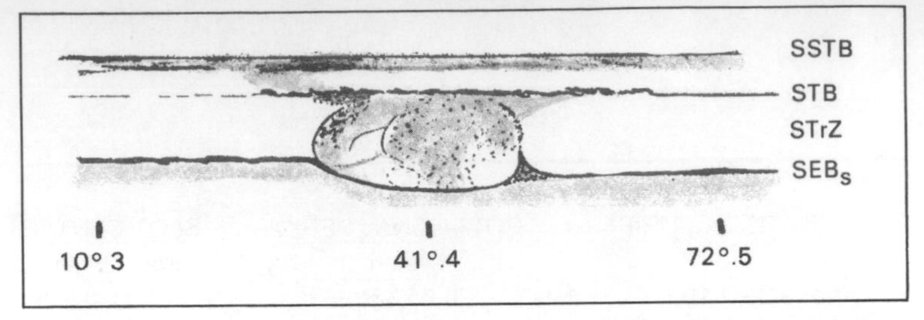

FIGURE 9-9. A strip sketch of the vicinity of the Great Red Spot, made on December 12, 1975. (Courtesy, Midsouth Astronomical Research Society, Inc.)

- ☆ Shape—Carefully draw the shape of the Spot exactly as it appears.
- ☆ Interactions—Look for interactions with belts and zones in the vicinity of the Spot.
- ☆ Interior—Examine the interior of the Spot for bright spots, color changes, central dark areas, and the like.

When making strip sketches, make a note of the longitude at various intervals along the drawing and label (using abbreviations) all belts and zones included in that sketch. Be sure to include the date and time (UT). Figure 9-9 is an example of a strip sketch of the RS vicinity made by me on December 12, 1975 at 02:20 UT.

VISIBILITY OF JUPITER

Each year the astronomer is afforded a view of Jupiter. The great cloud layers vary so rapidly that it is essential that each night be used to study the Jovian clouds and their changes. Oppositions in summer show Jupiter positioned quite low in the southern sky from northern latitudes, but the steadying potential of atmospheric inversion layers often offset the low altitude of the planet, affording excellently steady views.

Because of its great brightness, Jupiter can

TABLE 9-4. Oppositions of Jupiter, 1982-2000.

Date	*Apparent Diameter*	*Magnitude*
April 25, 1982	44.4	-2.0
May 27, 1983	45.5	-2.1
June 29, 1984	46.8	-2.2
August 4, 1985	48.5	-2.3
September 10, 1986	49.6	-2.4
October 18, 1987	49.8	-2.5
November 23, 1988	48.7	-2.4
December 27, 1989	47.2	-2.3
January 28, 1991	45.7	-2.1
February 28, 1992	44.2	-2.0
March 30, 1993	44.5	-2.0
June 1, 1995	45.6	-2.1
July 4, 1996	47.0	-2.2
August 9, 1997	48.6	-2.4
September 16, 1998	49.7	-2.5
October 23, 1999	49.8	-2.5
November 28, 2000	48.5	-2.4

[a]In seconds of arc.

be seen telescopically in morning twilight only 13 days after conjunction with the sun. Climbing higher each night as the earth-Jupiter distance decreases, the planet begins its retrograde motion 60 days before opposition and continues this reversal until 60 days after opposition. After retrograde ends, the planet remains visible in the evening sky for about 125 days.

BIBLIOGRAPHY

American Ephemeris and Nautical Almanac. Washington, DC: U.S. Government Printing Office. Published for each year.

Gehrels, Tom, *Jupiter*. Tucson: University of Arizona Press, 1976. Paperback.

Hartmann, William K., *Moons and Planets*. New York: Bogden and Quigly, 1972.

Kuiper, G.P., "Lunar and Planetary Laboratory Studies of Jupiter—I, II," *Sky & Telescope*, Jan.-Feb., 1972.

Kuiper, G.P., and B.M. Middlehurst, *Planets and Satellites*. Chicago: University of Chicago Press, 1961.

Michaux, C.M. et al., *Handbook of the Physical Properties of the Planet Jupiter*, NASA SP-3031. Washington, DC: U.S. Government Printing Office, 1967. Paperback.

National Aeronautics and Space Administration. *Pioneer Odyssey—Encounter with a Giant.* Washington, DC: U.S. Government Printing Office, 1974.

Peek, Bertrand M., *The Planet Jupiter*. London: Faber and Faber, 1958.

Roth, Gunter D., *Astronomy: A Handbook*, pp. 373-381. Cambridge, MA: Sky Publishing Corp., 1975. Paperback.

Sherrod, P. Clay, *Analysis of the 1978-79 Apparition of Jupiter*. Project Publ. 005 of the Midsouth Astronomical Research Society. Little Rock, 1979.

Sherrod, P. Clay, *Analysis of the 1979-80 Apparition of Jupiter*. Project Publ. 006 of the Midsouth Astronomical Research Society. Little Rock, 1980.

Sherrod, P. Clay, *Observations of the Great Red Spot and Jupiter 1975*. Project Publ. 003 of the Midsouth Astronomical Research Society. Little Rock, 1975.

10

OBSERVATIONS OF SATURN

Saturn is the most impressive object in the solar system and surely one of the most beautiful in the universe seen by man. No longer is it unique in having a striking ring system; we have discovered the Jovian and Uranian rings as well. Saturn, however, still remains unique as the only ringed planet visible in the amateur's telescope. On a clear, steady night, nothing rivals the sharp divisions and contrast seen in the ring system, nor the delicate shadings of the globe. Every observer surely remembers the circumstances of that first glance at the incredibly haloed world of Saturn.

In this chapter, I present data needed by you as an amateur to interpret your study of Saturn, and I describe meaningful endeavors that you can undertake while viewing the planet. To the novice or the unconcerned amateur, Saturn seems to be merely a static and uneventful place. The discovery of transient phenomena is of great interest and scientific importance, and the trained observer can participate in studies vital to our knowledge of the physical nature of this planet.

TABLE 10-1. Physical data and orbital statistics of Saturn.

Distance from sun	1428 million km
Least distance from earth	8.00 AU
Greatest distance from earth	11.07 AU
Sidereal period	10,759.2 days
Equatorial diameter	120,800 km
Polar compression	1:10
Surface area (earth = 1)	84
Volume (earth = 1)	762
Mass (earth = 1)	95.2
Density	0.69 gr/cm^3
Rotation period	10h 14m
Maximum inclination as seen from earth	26° 45′
Albedo	0.42
Maximum apparent magnitude	-0.6
Maximum apparent diameter	21″
Minimum apparent diameter	15″

THE USE OF TELESCOPES IN OBSERVING SATURN

At a distance of 1,428 million km from the sun, Saturn is a much more difficult object for scrutiny than Jupiter, its nearest planetary neighbor (see Chapter 9). Therefore, higher powers must be used to obtain scientifically valid information.

At maximum, Saturn's disk is 21 sec (″) of arc, less than half the apparent diameter of Jupiter. This appearance of size is caused primarily by distance, and not by actual size. Saturn's disk measures 120,800 km equatorially, only 22,000 km less than Jupiter's.

FIGURE 10-1. Saturn displaying its magnificent system of rings tilted their maximum toward earth. (Photograph by the author)

As in most areas of amateur study, a telescope of large aperture (20 to 30 cm) will reveal a great deal of fine detail invisible in smaller instruments. The finest view of Saturn I have witnessed was through an 8-cm (3-inch) refractor. However, the effect was fine aesthetically rather than scientifically. The refractor generally affords a steadier view, with Saturn contrasted sharply against the background sky. Any type of telescope will suffice, but generally, when conditions permit, use the one with the largest aperture available.

The magnification you use will depend on the steadiness of the atmosphere and, primarily, on the size of the telescope. Table 10-2 suggests magnifications for various apertures on an average night.

Of course, the values given in Table 10-2 can vary. On nights of perfect steadiness, you can safely double the values, but such nights are rare. Above all, don't sacrifice a clear image for the sake of a large one. Many trained observers use powers lower than those given in the table to obtain the finest detail and sharpest contrast possible. A little practice will soon result in your knowing how best to use these lower powers. Generally, you will need to use much higher magnifications when viewing the ring system than when viewing the globe in order to discern the fine gaps separating the main ring components.

It is essential that your telescope be equipped with an equatorial mounting, preferably with a motor drive, when the higher magnifications are to be used.

Have among your equipment pencils, paper, and predrawn blank forms of the outline of Saturn *for the year of observation*. The angle of tilt of the ring system varies from year to year so you must have current drawings. Such drawings ensure accuracy, and they eliminate wasted time at the telescope when you are trying to render accurately what you see in the eyepiece.

TABLE 10-2. Telescope magnifications for observations of Saturn.

Aperture	*Magnification*
2.4 in (6 cm)	125x
4.0 in (10 cm)	200x
6.0 in (15 cm)	250–300x
8.0 in (20 cm)	300x
10.0 in (25 cm)	300–400x
12.5 in (32 cm)	400x+

OBSERVATIONS OF THE GLOBE

The body of Saturn consists of gases increasingly compressed toward the center of the planet. The similarity in physical and chemical makeup between Saturn and all the other three Jovian planets is remarkable. You will not see the actual surface of Saturn, as with Jupiter. Rather, you will view the tops of the cloud layers and changes occurring at that level. Because of differential rotation (the equator rotates almost 1 hour faster than belts and zones in higher or lower latitudes), the cloud layers are divided into belts and zones in a pattern identical to that of Jupiter (see Figure 9-1, Chapter 9). Anyone who has ever compared Jupiter and Saturn realizes immediately that the globe of Saturn is not as remarkable in detail and color as the globe of Jupiter. Basically, there are two explanations for this absence of color and contrast in Saturn. First, Saturn is nearly twice the distance from us as is Jupiter and thus is seen as a much smaller disk, making

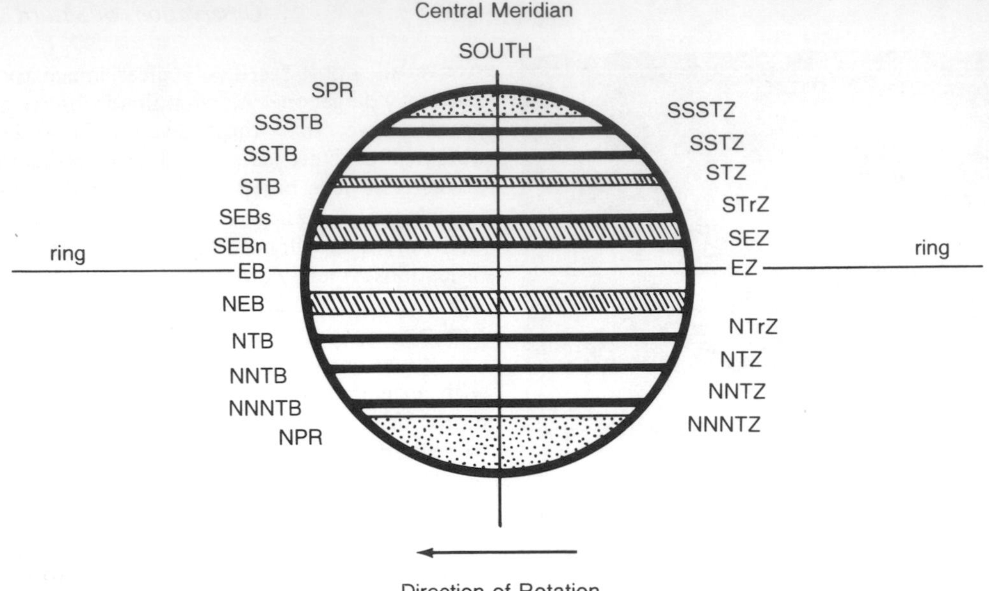

INDEX TO ABBREVIATIONS:

ZONES		BELTS	
EZ	Equatorial Zone	EB	Equatorial Belt
SEZ	So. Equatorial Zone	SEBn	So. Equatorial Belt North
STrZ	So. Tropical Zone	SEBs	So. Equatorial Belt South
STZ	So. Temperate Zone	STB	South Temperate Belt
SSTZ	So. So. Temperate Zone	SSTB	So. So. Temperate Belt
SSSTZ	So. So. So. Temp. Zone	SSSTB	So. So. So. Temperate Belt
NTrZ	North Tropical Zone	NEB	North Equatorial Belt
NTZ	North Temperate Zone	NTB	North Temperate Belt
NNTZ	No. No. Temperate Zone	NNTB	No. No. Temperate Belt
NNNTZ	No. No. No. Temp. Zone	NNNTB	No. No. No. Temp. Belt

FIGURE 10-2. Diagram of Saturn's zones and belts.

finer detail difficult or impossible to discern. Second, the thermal output of Saturn is much less than that of Jupiter, making for considerably less convective action in the cloud layers of Saturn than in those of Jupiter. Much of the rapid activity and color changes observed on Jupiter are a result of this thermal convection.

However, there are occasional minor outbreaks on the globe of Saturn that warrant study and surveillance by the amateur. There is very little activity change in the higher and lower latitudes but frequently, in the equatorial and tropical regions, you can see large white spots or ovals similar to those described in Chapter 9 as occurring on Jupiter. It is interesting that most of these rare spots have been discovered by amateur astronomers.

The Equatorial White Spots

The rotational spin of Saturn is quite rapid, exceeded only by that of Jupiter. Every 10 hours 14 minutes Saturn shows the observer on earth the same face. Consequently, any feature spinning within the equator of Saturn can be timed to such accuracy that revisions of the equatorial rotational rate can be made. However, the rarity of visible rotating features prohibits one from making regular refinements on this length of Saturn's day. When white ovals or spots are discovered, professional and amateur astronomers alike give their attention to revising rotational patterns.

System I assumes a sidereal rotation rate of 844.0° per day (period = 10h 14m 13.08s), intended for use with features in the NEB, EZ, and SEB. System II, intended for the remainder of the ball, has a sidereal rotation rate of 812°.00 per day (period = 10h 38m 25.42s). These rates are approximations only, because latitude-dependent rotation rates for Saturn are more uncertain than, say, for Jupiter. However, longitudes calculated from these data should give conveniently small drift rates for most features. Observers of Saturn are urged to make central meridian timings, combined with latitude measures (or at least estimates), whenever possible so that these rotation rates, and any future CM tables, can be made more accurate.

If you have only a small telescope, you can still undertake regular surveillance of the equatorial and tropical regions of Saturn in an effort to discover the white ovals. Generally, these spots are very vague, but there's just enough contrast to enable you to discern the spots against the bright yellow equatorial zone (EZ). On discovery, take the following steps:

1. Record in UT the date and time of discovery.
2. If possible, have someone else verify what you see. We often see what we are looking for if we are *expecting* to see it.
3. Draw the approximate location of the spot on a disk that represents the globe of Saturn.
4. Try to obtain times when the spot crosses the central meridian. If possible, try to time to the nearest minute the passage of the preceding end, the center, and the following end of the spot. This affords a rough estimate as to the true size of the spot, and provides the valuable data necessary for refinements of the rotational period.
5. Send copies of all your observations to:

   IAU Telegrams
   Smithsonian Observatory
   60 Garden Street
   Cambridge, MA 02138

   Lunar & Planetary Laboratory
   University of Arizona
   Tucson, AZ 85721

6. Estimating that the spot will cross the meridian every 10 hours 14 minutes from the first timing, estimate the next opportunity to view the transit of that spot.
7. Go out on that date and time and record all the above information obtained in Steps 1 to 5 once more, noting any apparent drift from the predicted time of transit.

Discovery of the white spots is rare, but the scarcity of these features justifies the search. Besides the determination of changes in the rotational rate of the equatorial current, the spots give you an opportunity to witness outbreaks of a thermal nature that eventually might result in our knowing the true nature of the interior processes of the Jovian planets.

Color on the Globe

Occasionally the otherwise dull-yellow globe of Saturn exhibits some unusual color changes. The most striking color contrast normally seen on the globe is that of the EZ, which appears bright against the higher and lower latitudes. Record any color changes you observe, and send the details of your observations to the ALPO for evaluation. Make every effort to patrol the globe nightly for color changes. It is generally the case that observers who are looking for color are the ones who find the white ovals.

Belts and Zones

The belts and zones are less clearly defined on Saturn than on Jupiter. In most amateur instruments, you may expect to see the equatorial belts, and maybe one belt either north or south of them. Noting the intensity (or darkness) of these belts is important because they change unpredictably. Also, the sudden appearance of a belt hitherto not noted in your observations warrants being recorded and, if possible, being drawn on the blank form to show the approximate latitude on the globe.

The zones on Saturn are almost invisible, probably because the contrast between them is extremely low. Also there are no belts on either side of them to delineate their borders. The most prominent zone, which can be seen in almost all telescopes, is the Equatorial Zone, followed by the North and South Tropical Zones, which are seen only occasionally in amateur instruments.

Festoons and Related Activity

Although not as common or as spectacular as the festoon activity seen on Jupiter, some thin festoons do occur on Saturn and are visible in amateur instruments. Most such activity is concentrated in the equatorial zone, seemingly bridging the North and South Equatorial Belts. Festoons can often be seen emanating from either the North or South Equatorial Belt into higher latitudes—for example, from the South Equatorial Belt into the South Tropical Zone. Large festoons that possibly may be related to white spot activity often appear. You should record them as accurately as possible. Reporting the exact time (in UT) of the passing of any feature across the central meridian of Saturn is of the greatest importance because of the rarity of pronounced markings of long duration.

OBSERVATIONS OF THE RING SYSTEM

The rings encircling Jupiter, Saturn, and Uranus are thought to be the result of the gravitational disruption of a solid body orbiting close in. Another possibility is that the rings are many small bodies that never did coalesce into one large satellite. If the body is far enough away from the planet, the latter's gravitational pull on the near side of the body is only slightly more than on the far side. This is simply because of the inverse-square law. But if the body is too close, then (if it is of astronomical size), its near side will experience much more pull by the planet than will the far side. This gives rise to enormous strain which rends the body apart. Once the pieces are small enough so that the difference in gravitational pulls experienced by their near and far sides is too small to disrupt them, the pieces can still collide in their orbits around the planet and so millstone down to a quite small size. In the Saturnian rings, the pieces now are known—thanks to Voyager 1—to range in size from dust motes to car-size boulders.

FIGURE 10-3. Resonance between orbital periods (forming Kirkwood gaps). (After drawing by Paul. T. Covington, personal correspondence, 1981)

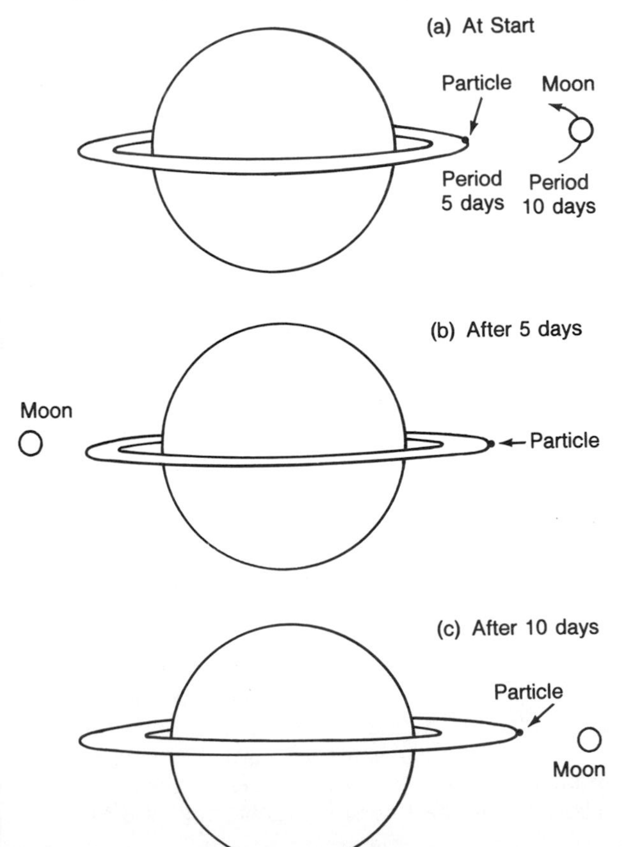

Roche's Limit—2.44 times the radius of the planet—is the distance from a planet at which this disruption will take place. This varies according to the mass and size of the planet and the density and size of the small body. Strictly, that value holds only for a fluid satellite with the density of water. The stronger the material, the closer the satellite could be before disruption.

If there are satellites orbiting outside the rings, Kirkwood gaps will appear in the rings. This is because particles at a certain distance from the planet will have a period of revolution just half that of some one moon. Hence, every time they make two complete orbits, they again experience the maximum possible pull from that moon. That is, every so often the particles will get pulled outward especially strongly and be swept out of their original path into one higher above the planet that is by the moon (see Figure 10-3).

The planet's mass does affect the situation in that it determines the orbital period of every ring particle and every moon at whatever distances they happen to be. What periods occur restrict what Kirkwood gaps can occur, because their distances from the planet are at points where an orbiting particle would have a period just one-half, one-third, one-fourth, two-thirds, three-fifths, or some such integral ratio of some more remote moon. If there are two or more moons, then there will be gaps caused by each, conditioned by the period of each. Also, the more massive the moon, the better it is at sweeping out the gaps by the means just discussed. A similar pattern, is caused by Jupiter, is found in the asteroid belt.

The newly discovered "sheepdog" satellites of Saturn orbit more closely to Saturn than do even the innermost rings and act to keep particles in the latter by a mechanism that shall not be detailed here. Were these "sheepdogs" massive enough, they too would cause noticeable Kirkwood gaps. That is, a satellite can be closer to its planet than the rings and still cause such gaps. Additionally, the outermost ring of Saturn is now known to be braided, like a woman's hair.

There is much more than first meets the eye that you as an amateur can pursue in your studies of the ringed planet, Saturn. If you can set up a regular schedule of observing, useful studies of the ring system that you can undertake are the following:

TABLE 10-3. Data for rings and Kirkwood gaps visible in amateur telescopes.

Description	*Distance from Globe*		*Width*[c]
Outer edge of Ring A	20.1[a]	138,600[b]	–
Encke's Division	19.0	129,500	0.35
Inner edge of Ring A	17.5	120,400	–
Center of Cassini Division	17.2	118,700	0.53
Outer edge of Ring B	16.9	117,000	–
Division IV	15.6	107,000	0.18
Division III	13.5	93,500	0.65
Inner edge of Ring B	12.9	88,100	–
Division V	12.7	87,500	0.20
Inner edge of Ring C	10.7	70,900	–
Equatorial radius (Saturn)		60,500	–

[a]In seconds.
[b]In kilometers.
[c]In seconds of arc.

☆ *periodicity in visibility*—The rings appear to be inconsistent in their visibility. If so, it may be because of the amount of reflectivity from various parts of the ring disk. Therefore, it appears that we can assume that the rings are variable in brightness, and some studies to determine if this variability is periodic should be made.

☆ *relationship of rings to other components*—If the rings do change their visibility it must be determined if all the ring components change simultaneously and in proportion to other components. If so, this would lead one to believe the changes result from periodic gravitational events caused by the influence of the orbiting satellites.

☆ *ring gaps*—It must be determined if the ring gaps (Kirkwood Gaps) are completely uniform in their widths or if the gaps appear broader at some points than at others. It should also be determined if the gaps have uniform intensities (i.e., equal darkness). If any inequalities are found, efforts to correlate these with the positions of the satellites and/or ring brightenings should be made.

☆ *positions of the ring gaps*—On the nights of best seeing, draw as carefully as possible the positions within the ring system of the gaps that can be seen.

☆ *the crepe ring*—This ring definitely changes in intensity, sometimes being quite obvious, sometimes invisible. An important study would be the relationship between the visibility of the crepe ring and brightening of other ring components.

☆ *color estimates*—All components of the rings that are visible on any given night should be defined according to their colors. Care must be given to those times when Saturn is observed either rising or setting, at which time atmospheric refraction will influence any color estimates.

☆ *bright spots*—Occasionally there appear on the rings what are known as *Terby White Spots*, areas of scintillating brightness easily distinguishable from the seemingly dull ring on which the spot is superimposed. These spots are quite short-lived and should be monitored when they are known to exist and searched for when none are suspected. Observations of any white spots over a one-week period, noting carefully the exact minute at which the spot is on the central meridian of the ring as seen from earth, will provide accurate determinations of the rotational values of the ring components, data that are badly needed. A Wratten No. 12 yellow filter will accentuate any such brightening and aid in their discovery.

☆ *edge-on rings*—This valuable study is discussed completely in a following section of this chapter.

☆ *the bicolored aspect*—Certain filters reveal that one portion of the ring system may occasionally appear brighter with those filters than without them. This valuable study is discussed completely in a following section of this chapter.

Most of these aspects are discussed in detail in the following pages. To be successful in observations of the ring system, attempt to observe at every opportunity, and keep detailed records of all your observations. A standard ALPO observing form for Saturn is shown in Appendix XI.

Studies of Ring Divisions

An interesting aspect of the ring divisions is that they can be easily seen on some dates, yet not even the largest telescopes will show them on others, even if observing conditions are equal. Careful observations and drawings that can place the positions of these divisions are badly needed, particularly since Voyager I revealed a plenitude of material within the gaps. It is possible that the density of particles varies, visible as brightening or dimming of these regions.

The two most prominent divisions in the ring system are the *Cassini division* and the *Encke division*. The Cassini division divides the outer *A* ring from the bright middle *B* ring. Encke's division separates the outer *A* ring into two components and sometimes is very difficult to find.

Saturn's rings appear to close up into a line and then open again as the earth passes through the plane of Saturn's orbit. The visibility of all

FIGURE 10–4. A drawing of Saturn made by an amateur in 1973 through a 10-inch (30-cm) Newtonian reflector. Encke's division (outer) and Cassini's division (dark middle line) are clearly visible. The crepe ring is particularly obvious. South is at the top in this drawing.

the rings changes according to the angle at which we view the rings. When the rings are open (i.e., tilted about 26° to our line of sight), the gaps are much easier to see than when the rings are closed (i.e., seen edge-on from our line of sight). Cassini's division is visible until just before this edge-on aspect occurs, and it does not seem to be as subject to variation at other times as the other gaps are.

In your studies of the ring divisions, check the following aspects of Saturn's rings over a period of time:

☆ If a gap changes in visibility more than once, check you observational records for any sign of periodicity.

☆ If one gap suddenly becomes much easier to see, examine carefully the other areas of the ring (particularly within the bright *B* ring) to see if other gaps that before were invisible have simultaneously become visible. These observations are critical.

☆ Determine if minor gaps can be seen uniformly in both ansae (the points of greatest curvature of the rings, located on each side of the globe opposite each other). If not, note if a gap seen in one ansa is at all visible in the opposite ansa. There usually is an unexplainable difference in the two.

☆ Note the intensity of such gaps, relative to the nearly black intensity of Cassini's division.

☆ Draw the positions of all gaps as seen in the telescope, beginning with Cassini's division.

Details concerning the positions, intensities, and so on of suspected gaps in the ring system are given in Table 10-3. Familiarize yourself with the locations and possible visibility of each separation. Be aware that many ring gaps may exist other than those noted.

Studies of the Ring Components

Three main rings circling Saturn are telescopically visible—the *A ring*, or outer ring, the *B ring*, which is the middle and brightest ring, and the *C ring*, also known as the crepe ring, which is quite dim and transparent. In addition, there is evidence of a ring known as the *D* ring which appears in width up to the globe of Saturn inside the *C* ring. There is also a *D′* ring, which circles Saturn outside of the *A* ring, as well as many other rings not visible in earth-based telescopes, but discovered by Voyager I in 1980.

Because the ring system is composed of particles, each a tiny satellite of Saturn, it is natural to suppose that dynamic changes might occur throughout these particles. Some areas of study for the amateur astronomer are the following:

☆ Visibility of the crepe ring to note the rapidity and degree of changes over periods of time. In addition, note the appearance of a gap separating the crepe ring from the bright *B* ring.

☆ Determine roughly the width of the crepe ring relative to the width of the *B* ring.

☆ Estimate the colors of each ring component compared to the others.

☆ Intensity variations, or "ripples," might occur within the ring system. These probably are caused by similar forces but not as intense as those producing ring gaps. Note how long the ripple is visible, and draw its position on the form in relation to other gaps (preferably Cassini's).

☆ Examine each night for Terby Spots, bright, large, oval-shaped patches sometimes seen in the *B* ring.

☆ When the ring system is edge-on (every 14 years), several special observing programs should be planned, among them the following:

Intensity Clumpings. These are areas that can distinctly be seen when the rings are edge-on although other portions of the rings are invisible. Be careful not to confuse them with inner satellites.

Visibility. Note when the entire extent of the rings can last be seen, and the exact day on which they again become visible. Also note how far from the globe the rings can be seen and aperture telescope that is required to see them.

☆ The phenomenon known as the *bicolored aspect* is discussed at length on the following pages.

Physical data concerning the ring system of Saturn are in Table 10–3. It is best to be quite familiar with every aspect of the physical nature of the rings so that immediate interpretations can be made at the telescope and pursued if necessary.

It is quite important to record all aspects of observing conditions. Both air turbulence and transparency affect the accuracy of detailed observations and should be noted on the observing sheet.

Edge-on Rings Studies

Throughout Saturn's sidereal revolution period of 29.5 years, the intersection of the orbit of the earth and the plane of the ring system takes place only twice at intervals of 13.75 and 15.75 years. Astronomically speaking, such events are considered quite rare and particularly noteworthy. The two periods, incidentally, are not equal, owing to the ellipticity of Saturn's orbit. During the 13.75-year period, the south face of the rings and the southern hemisphere of the globe of the planet are inclined toward the earth; Saturn passes through perihelion during this time. In the longer 15.75-year interval, Saturn passes through the aphelion point in its orbit, and the north face of the rings and northern hemisphere of the globe are exposed to observers on the earth.

An interesting and worthwhile project that provides better understanding of the thicknesses of various ring components is to monitor the dates of theoretical edge-on presentation. This project is suitable for observers with telescopes of various apertures, particularly those in the 12-inch (30-cm) to 14-inch (35-cm) sizes. This study leads to knowledge of the differences in particle clumping on one edge of the ring as compared to the other. If the sunlit side of the ring is angled toward the observer on earth, it is possible for observers with instruments of the type mentioned above to see the rings within hours of exact edge-on orientation. However, if the dark side of the ring is angled toward earth (i.e., the face of the ring not illuminated directly by sunlight), the rings should become invisible days or even weeks prior to the edge-wise presentation. If the rings are equal in both angle and thickness, then it stands to reason that the amount of sunlight reflected from each ansa should also be equal. However, this is rarely the case.

Visual observers can measure the effective brightness of any given portion of the edge-on ring by using a numerical scale from 0 to 10. An area totally invisible, such as a dark shadow, is denoted by 0, whereas the number 10 refers to areas more brilliant than any portion of the globe of Saturn. It is important that you record not only the intensity of bright spots appearing in the rings but also their positions relative to Saturn. This can be done in two ways. In the first method, measure the radius of Saturn. Then estimate the spot's distance from Saturn's limb in radii, being particularly careful to orient properly the directions east and west. Second, and more precise, if you have a filar micrometer, is to actually measure in seconds of arc from the center of Saturn's globe the distance that the bright spots appear.

It is very easy, particularly when the dark spaces of Saturn's rings are angled toward earth, to mistake many of the inner satellites for bright spots in the rings, because these appear to pass through the linear expanse of the plane of the rings during edge-on presentation. Data concerning the positions for every major Saturnian satellite for any given date are in the *American Ephemeris and Nautical Almanac*. If you notice bright spots, consult the reference to make sure you are not confused as to the nature of the bright area.

The study of bright clumping in the rings can be better initiated by using a photoelectric photometer and the magnitudes of Saturn satellites as comparisons. The procedure for this comparison is discussed in detail in Chapter 13.

Observations of Saturn's Satellites

Observe the satellites of Saturn particularly when the rings are presented edge on. At that time the angle of the earth–sun–Saturn system is such that accurate timings of the satellites across the globe of Saturn can be determined, thus facilitating considerable refinements in the nature of their orbits. Because such edge-on apparitions occur only once every 13.75 or 15.75 years, it is very important that all observers with telescopes larger than 3 inches (8 cm) monitor these events. The satellites Titan, Mimas, and Hyperion are those that you should study most during transit events. Events you should give attention to include the following:

- ☆ Occultations of the satellite by the globe of Saturn.
- ☆ Eclipses, timing the shadow as it enters and exits the apparent disk of Saturn.
- ☆ Transits of the satellite as it enters and exits the apparent disk of Saturn.
- ☆ Mutual phenomena—those events during which two satellites occult or eclipse each other.
- ☆ Satellite/ring encounters.

It is important that you time any event to the nearest minute (UT) as determined by WWV time standard broadcasts. If you can obtain at least 5 sec accuracy, your observations become even more valuable.

During any apparition of Saturn, observations of the satellite Iapetus are of utmost importance because this satellite varies in brightness almost one full magnitude from about magnitude 10.8 to almost magnitude 12. You can make comparisons of the brightness of this satellite by comparing it with other satellites of known brightnesses.

Iapetus, first noted for its variability by G.D. Cassini in 1672, is considerably brighter when it is on the western side of Saturn than when it is on the eastern side. It is possible that the variation depends on three elements as follows:

1. Iapetus is an irregularly shaped satellite in a tide-locked orbit about Saturn, so that, at any given point in its orbit, the same features are presented to earth. One side is always facing Saturn.
2. One hemisphere of the satellite is more reflective than the other, causing variations in the light we receive.
3. The amount of sunlight reflected from Saturn varies in such a way that more light is reflected off its western limb than the eastern.

In any case, you can quickly appreciate the importance of both magnitude and colorimetry estimates that are made on a regular basis. For brightness comparisons, you might choose one of the other major Saturnian satellites, because they will be in proximity and thus subject to minimal extinction. A few of the major satellites of Saturn within the range of Iapetus are given in Table 10-4.

TABLE 10-4. Some major satellites of Saturn.

Satellite	*Period*	*Magnitude*
Titan	14d 23h	8.3
Rhea	4d 12h	9.8
Tethys	1d 21h	10.3
Dione	2d 17h	10.4
Enceladus	1d 8h	11.9

When at its faintest (less than magnitude 11), Iapetus is beyond most amateur telescopes for photoelectric photometry. A 12-inch (30-cm) or 14-inch (35-cm) telescope usually provides enough brilliance for some deflection to be seen in the V mode.

The Bicolored Aspect of Saturn's Rings

According to definition, the *bicolored aspect* is an apparent brightening of one ansa (i.e., the extreme curve of the rings as seen from earth) over the other. The bicolored aspect can be seen with some filters but not with others. The effect has long been attributed to extinction by the atmosphere of certain wavelengths of light, which indeed does seem to cause some of the observed brightening. However, studies made by me in 1974 and 1976 for ALPO reveal that the effect is not entirely owing to extinction by our atmosphere but might be attributable to a physical change within the rings themselves.

To observe the bicolored aspect, two steps are necessary. First, decide if any suspected brightening of one ansa over the other is possibly caused by refraction of light through our atmosphere. Second, if it is not, make a series of filter observations, recording the change in brightness throughout the filter set. You can often easily notice a refractive brightness when Saturn is low on the horizon if you use a red filter, but you will not be able to observe it if you use a blue filter.

Filters recommended for use in the bicolored study are red (No. 25A), orange (No. 21A), yellow (No. 12A), green (No. 58A), blue (No. 80A), and violet (No. 47A). It is best to begin with the violet and progress through the filters until all are used. *Always use the filters in this same sequence*. Keep careful records of each observing session, making notes of the following data:

1. Your name, location, and telescope used.
2. Date and UT.
3. Altitude of Saturn above horizon (in degrees).
4. Which horizon (east or west).
5. Results of filter series.

To record the results of filter observations, it is best to list all filters, beginning with the violet, and record next to each which ansa (if any) is the brightest as seen in that filter. A quick way to record these is as follows:

UV	-W+	(west ansa is brighter in UV)
BLUE	-O	(neither appears brighter)
GREEN	-E+	(east ansa is brighter in green)

There is no quick way to differentiate between a true physical brightening of the ring ansa and the bicolored aspect of atmospheric refraction. The best method to differentiate is to observe Saturn—on one night—when it is low in the *east*, then again when it is nearly overhead, and then if possible when it is setting in the *western* sky, and compare results of this one night. If the same effects are seen at all altitudes, then the effect is most likely attributable to physical changes in the ring system. However, effects seen when Saturn is low on the horizon that disappear at higher altitudes are most probably related to the atmosphere.

If several instances of brightening caused by changes are seen, it is important to observe on successive nights until the effect is no longer noticeable. It is possible to photograph the physical brightenings, and you should make every attempt to do so. On a regular basis send all observations of the bicolored aspect to ALPO at the address below:

ALPO
Box 3AZ
University Park, New Mexico 88003

FIGURE 10-5. Description of the term *ansa* as it is applied to Saturn's rings. (Photograph by the author)

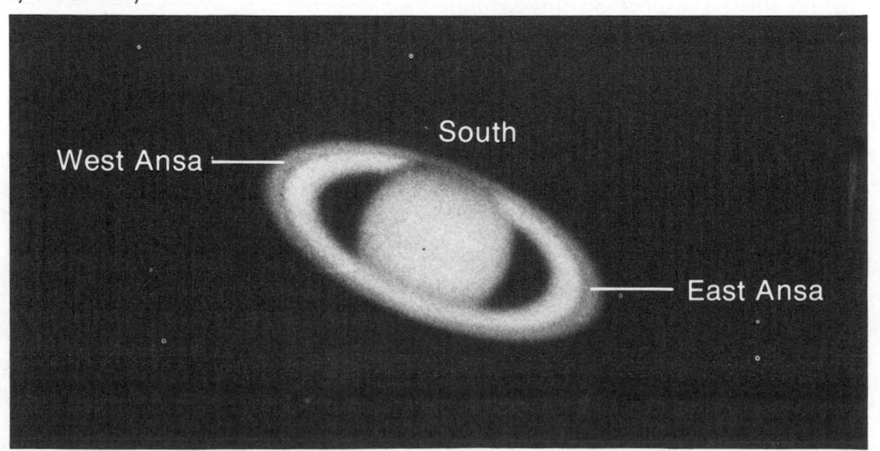

VISIBILITY OF SATURN

Anyone who has ever looked at Saturn through a telescope will not forget that first view—it is truly an outstanding sight principally because of the magnificent system of rings. However, Saturn's ring system is not always visible to us, and when viewed edge-on, Saturn may be quite disappointing to the observer. Because Saturn requires 29.5 years to orbit the sun, its rings disappear after 13 years, slowly open up again, and then gradually disappear again after 16 years. It is the ring system that reflects most of the light from this great distance. When the rings are not visible to us on earth, therefore, Saturn appears much dimmer than when angled at its maximum tilt of 26°.

After being obscured by the glare of the sun, Saturn appears in the early morning sky about 18 days after conjunction. The planet reaches opposition 189 days after conjunction, and remains in the sky for another additional 171 days. Sixty-four days prior to opposition, Saturn begins its backward retrograde motion, which continues for 128 days, a longer period than Jupiter's.

TABLE 10-5. Oppositions of Saturn from 1982 to 2000.

Date	*Magnitude*
April 8, 1982	+0.5
April 21, 1983	0.4
May 3, 1984	0.3
May 15, 1985	0.2
May 27, 1986	0.2
June 9, 1987	0.2
June 20, 1988	0.2
July 2, 1989	0.2
July 14, 1990	0.3
July 26, 1991	0.3
August 7, 1992	0.4
August 19, 1993	0.5
September 1, 1994	0.7
September 14, 1995	0.8
September 26, 1996	0.7
October 10, 1997	0.4
October 23, 1998	0.2
November 6, 1999	0.0
November 19, 2000	-0.1

BIBLIOGRAPHY

Benton, J.L., "Observational Notes Regarding the 1979-80 Apparition of Saturn: The Edgewise Presentation of the Rings," *Journal A.L.P.O.*, Vol. 28, 1-2 (1979), 1-5.

Benton, J.L., *The Saturn Handbook*. (ALPO) Savannah: Review Publishing, 1975. Paperback.

Alexander, A.F. O'D., *The Planet Saturn*. London: Faber and Faber, 1962.

Evans, E.C., *The Remote Sensing of Saturn's Rings: I—The Magnetic Alignment of the Ring Particles*, NASA TN-D-4775. Washington, DC: Government Printing Office, 1973. Paperback.

Goodman, J.W., "The Edgewise Presentation of Saturn's Rings," *Sky & Telescope*, Vol. 30, No. 3 (1965), pp. 128-131.

Hartmann, William K., *Moons and Planets*. New York, Bogden and Quigly, 1972.

Kuiper, G.P., and B.M. Middlehurst, *Planets and Satellites*. Chicago: University of Chicago Press, 1961.

National Aeronautics and Space Administration, *Rings of Saturn*, SP-343. Washington, D.C.: Government Printing Office, 1974. Paperback.

Roth, G.D., *Astronomy: A Handbook*, pp. 381-383. Cambridge, MA: Sky Publishing Corp., 1975. Paperback.

——, *Handbook for Planet Observers.* London: Faber and Faber, 1970.

11

VISUAL PHOTOMETRY OF THE MINOR PLANETS

Although appearing just as tiny points of light in amateur telescopes, the minor planets—or asteroids—offer much ground for modest research by the nonprofessional astronomer. Even if you have only a small refractor, you can contribute significantly to the study of the minor planets.

Many asteroids are irregularly shaped and, because each spins on its axis of rotation, the total light from the sun reflected from their surfaces varies from asteroid to asteroid. Of the many known to be variable in reflectance, only a handful are within the scope of study of amateur astronomers. These are listed in Table 11-1, along with pertinent observational data.

For adequate study, use at least a good 8-inch telescope for consistent asteroid photometry. However, even a 2-inch refractor can be used effectively to estimate the brightness changes of 433 Eros when it passes near the earth and sun because Eros attains a brightness only one magnitude below the naked-eye limit.

Photometry (i.e., estimates of magnitude or light intensity) of asteroids is a highly subjective study whether it is done visually, photographi-

TABLE 11-1. Asteroids that vary in brightness.

Asteroid	*Magnitude*[a]	*Amplitude*	*Estimated Period*	*Minimum Telescope Required*	
				(in.)	(cm)
15 Eunomia	8.7	0.4–0.5	6h 05m	2.4	6
43 Ariadne	10.4	0.5	11h 28m	6	15
44 Nysa	9.9	0.2–0.4	6h 25m	4	10
216 Kleopatra	10.0	0.4–1.6	5h 36m	6	15
321 Florentina	14.4	0.3	2h 52m	14	35
433 Eros	10.0	0.0–1.5	5h 16m	3.1	8
1580 Betulia	12.0	0.5	6h 08m	12.5	32

[a]At average periodical opposition; brighter magnitude at favorable opposition.

cally, or photoelectrically. Many factors can cause deceptive "light changes." Atmospheric refraction, haze, and even your mental and physical condition when you make the estimates are three of those factors. Because of the possibility of uncertainty, I recommend that when you initiate a visual photometric study of the minor planets you find an "observing buddy" and work as a team for the effort. Such an arrangement helps to make cold nights a little less hostile and adds excitement when the results begin to trickle in. It also helps add credibility to the findings because of the dual confirmation.

PURPOSE OF ASTEROID PHOTOMETRY

The purposes of conducting photometric observations of the minor planets are extensive. Some of them are as follows:

- ☆ From the light curve (page 197), the rotational rate of the asteroid can be accurately derived.
- ☆ The light curve also indicates the *amplitude* of the asteroid as it spins; this value is simply the magnitude range for a given date. Such information, coupled with various characteristics of the light curve, gives rise to *relative* shape and dimensions of the minor planet, as well as its axial angle presented to earth.
- ☆ Any light variation is indicative of irregularities in the asteroid's shape or peculiarities in the reflectance of the surface. Studies of the shapes, sizes, rotational rates, and peculiar surface features provide valuable information as to the origin and history of minor planets.
- ☆ Visual photometry of irregular asteroids is a seriously neglected area of professional astronomy, and observations are quite valuable because of this scarcity of effort.

INSTRUMENTATION FOR VISUAL PHOTOMETRY

As mentioned previously, a telescope with an 8-inch aperture will put all the minor planets listed in Table 11-1 into adequate reach for study during favorable apparitions. The telescope should definitely be equatorially mounted and preferably motor driven, because high magnifications are sometimes used for fainter asteroids. Setting circles are a necessary luxury to aid in rapid acquisition of the field in which the asteroid is located. Many amateurs have telescopes with setting circles but never use them, assuming that they are difficult and time-consuming. However, it is best to spend time *observing* and not finding. An observer experienced in the use of setting circles can move to a field of the correct location within 30 sec; I have seen amateurs "star hop" to a location in 20 min while the nice, engraved setting circles on their mount collect dust.

Because most of the asteroids change brightness very rapidly, it is essential that you have either a WWV time signal broadcast at the telescope, or a watch accurately set by WWV prior to observing. If a watch is used, check and reset it if necessary immediately before each observing session.

You may wish to use a portable tape recorder as a temporary record, rather than bother with paper, pencil, and flashlight. This is quite convenient when used in combination with the short-wave WWV signals. The recorder not only logs your voice signal of magnitude change but also the exact time at which the change occurred, accurate to better than 1 sec. This method cuts out a lot of the "busy work" at the telescope and allows you to reduce data quickly during daylight hours or on cloudy nights.

STAR CHARTS

For successful asteroid photometry, select a good star atlas that has reference catalogs listing the magnitudes of stars to be used for comparison. The best currently available such atlas and catalog for amateur studies is the *Smithsonian Astrophysical Observatory Star Catalog* and the *AAVSO Star Atlas.* This work contains stars to about magnitude 9.3. Actual visual brightness to 1/10 magnitude accuracy is given for every star in the atlas. This atlas will suffice in locating and estimating asteroids to magnitude 9.5, such as Eros and Eunomia. However, to locate minor planets fainter than this requires an atlas of fainter stars and larger scale. The Vehrenberg *Photographic Star Atlas* is an excellent choice for faint asteroids (to magnitude 13.5), but it has no catalog giving magnitudes of comparison stars. Hence, photometry for faint asteroids (e.g., Betulia) must be relative and not actual. (The procedure for determining relative magnitude change is given on page 196).

Plotting the Minor Planet's Path

Do not begin locating and observing the asteroid until you have plotted the path of the object on the star chart. Many minor planets move rapidly through the star field, so it is advisable to determine the position for each hour of the observing period. This quickens your initial search each night (which can be quite fatiguing), and gives you an opportunity to locate comparison stars within the predicted path before beginning your observing session.

Ephemerides (positions, motions, distances, magnitudes, etc.) are usually printed at least a month prior to the apparition in the journal *Sky and Telescope*. A more comprehensive ephemeris can be found in the circulars obtainable as a subscription from:

Minor Planet Center
Smithsonian Observatory
60 Garden Street
Cambridge, MA 02138

Plot the positions for each date on the charts. Show the asteroid's path from day to day by lightly connecting those dates by pencil. If the asteroid has a rapid daily motion, it may be necessary to determine the *hourly* positions to facilitate location of the asteroid because ephemeride positions are given for 00 hours Universal Time.

To determine hourly positions, you may use the following method. It is intended only to predict *approximate* positions for visual observers and is not adequate for precise professional work, because the curved path of a close asteroid allows for differing amounts of space to be covered in equal amounts of time as seen from our position on earth.

1. Suppose we must derive the hourly positions of 1580 Betulia for December 3, UT. The ephemeride gives the position for this asteroid on December 3 as right ascension (RA) 01h 26m, Declination (Dec.) +06° 30′. The asteroid will be moving from the position on December 3 to that given for December 4, so we also note the position of December 4 as RA 01h 16m, Dec. +01° 27′, at 00 hours UT.
2. Using the appropriate star atlas, plot *both* positions (for December 3 and December 4) and connect them with a fine pencil line.
3. Now measure the distance between the two points with a rule having a millimeter scale. In Figure 11-1 we find that distance to be 49 mm.
4. If the "geometric" motion is 49 mm in one day (24 hr), then the motion during *one* of those hours will be roughly 1/24 of 49 mm, or

$$\frac{49 \text{ millimeters}}{24 \text{ hours}}$$

or,

2.0 millimeters per hour

Of course, the use of millimeters for sky

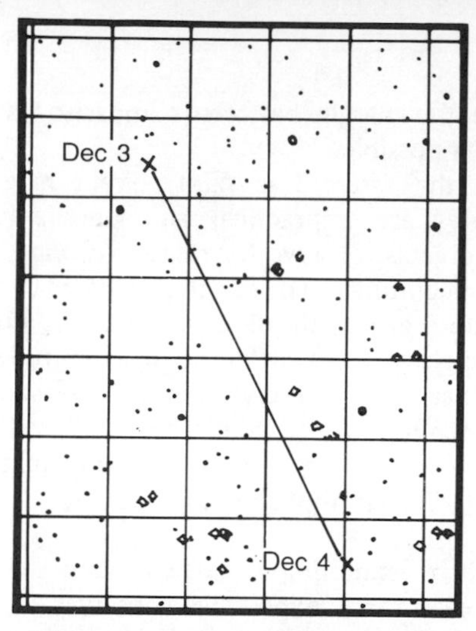

FIGURE 11-1. Measurement of the distance of one day's motion.

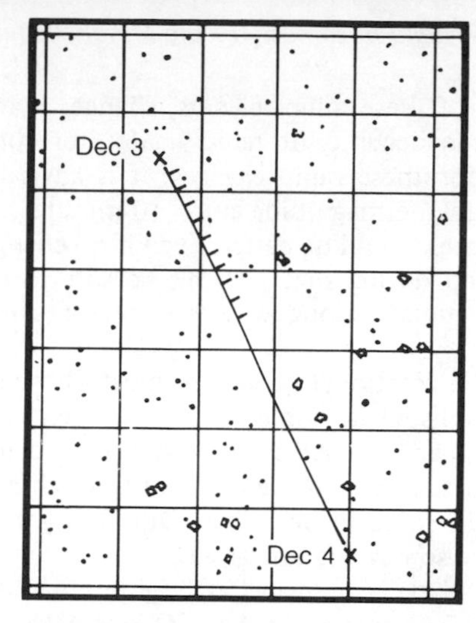

FIGURE 11-2. Hourly position marks inserted at 2-mm spacing.

measurement seems a bit awkward, and it should be remembered that such measurements apply *only* to our star charts, affording a rough—but very convenient—way to determine hourly positions.

5. Starting with the position for December 3, we now measure in 2-mm increments (determined in Step 4 on page 194) along the pencil line toward December 4. Make a mark every 2 mm across the pencil line to indicate the position from hour to hour, as shown in Figure 11-2.

Because the apparent *daily* motion varies considerably from day to day during a close apparition of any minor planet, you must compute each day separately for hourly positions. Always begin your computations anew, using the positions for the current day and those of the following day.

Because this is an indirect, graphic determination of hourly position, you must be aware that the millimeter marks may be slightly off from the actual observed position. Nonetheless, remembering that our objective here is *quick acquisition* of the asteroid, and not astrometric accuracy, the method is sound and serves the purpose as rapidly and easily as possible.

OBSERVING PROCEDURES AND METHODS

Two conditions affect the brightness of minor planets:

1. Irregularities in the shape of the body.
2. Albedo variations caused by areas of low and high reflectivity on the asteroid. These variations can be the result of mineral deposits, crystal orientation, or other surface irregularity.

Both conditions can be monitored by amateurs, after the construction of the light curve (page 196), you can visually determine the difference between rotation (regular) variations and albedo (irregular, but small and recurring) variations.

To derive a light curve that is scientifically valid, it is necessary that you observe on as many dates as possible during the apparition of a minor planet. Because of jagged, pointed, or broken shapes of these minor planets, changes in magnitude can occur quite rapidly. Monitor continuously for these rapid changes; it is advisable to estimate the magnitude every 10 minutes—every 5 minutes would be better. Even if no change has occurred during that interval, be sure to record your estimate along with the correct Universal Time.

The rotational periods of most of the minor planets listed in Table 11-1 have been roughly determined. A dedicated observer of asteroids must attempt to brave the cold through *at least* one full rotation per night. Otherwise, the data become somewhat ambiguous.

If the minor planet is brighter than magnitude 9.5 (average), the *AAVSO Star Atlas* should be used. Before the observing session, record the data concerning the object on the correct chart. For the estimated period of observation, select comparison stars very close to this path. Label them and assign them their correct magnitudes as given in the companion catalog.

At the telescope, make comparisons between those selected stars and the asteroid, finding a star that matches (or closely matches) the magnitude of the asteroid. It may be necessary to estimate *between* two stars, one slightly brighter and one slightly dimmer than the object. Try to reduce the margin between comparison stars as much as possible.

If the asteroid is faint, direct magnitude estimates are impractical and unreliable. For these objects, observe for relative *change*, rather than magnitude. To do this, locate two stars very close in magnitude to the asteroid but each very slightly (0.1) different in magnitude from the other. Then merely estimate the asteroid as "brightening 0.2" or "dimming 0.1," and so on. On the observing form, this information is indicated with the date as +0.2, -0.1, and so forth.

Such estimating is difficult and subject to error, even for an experienced observer. However, if you maintain *consistency* in the estimates and instrumentation, the purposes of the program will be maintained—that is, to determine the times of *brightest intensity* and those of *least intensity*. If you can determine these, regardless of what happens in the interval, you can construct a light curve and obtain valuable data.

It is a good idea to prepare before you start observing some sort of form to facilitate easy recording of data, since many observations are made at one session. (Eros, with a period of about 5h 16m requires over 30 entries for each rotation.)

REDUCTION OF DATA

The Light Curve

Construct the light curve as shown in Figure 11-3, with separating each hour of time by 3 cm, and each magnitude by at least 1 cm. For an asteroid with an amplitude of 0.5 or less, space each magnitude at a distance of 2 cm. Express the time (in UT) along the horizontal axis, running from left to right. Place the magnitude along the vertical axis, with the brightest at the top and the dimmest at the bottom. Once the data have been carefully plotted on the graph, the amplitude and the period of rotation can be interpreted.

The Amplitude

An important aspect of minor planet research, studies of amplitude yield data on polar alignment and polar shifting of the planet, and help astronomers to understand better the nature of the light variations. Amplitude varies somewhat for successive apparitions of a minor planet. If the object is oriented such that we observe in the plane of its equator, the light variations will be relatively large. However, if it is oriented so that its pole is pointing toward us, the variations will be absent, or quite small (see Figure 11-4).

Determination of amplitude is quite simple and should be recorded on every date. The value is simply the *range* of variation, from maximum intensity to minimum intensity, and is expressed in magnitude, or tenths of magnitude. The value "A" represents the amplitude of this light curve, the distance on the vertical scale from midpoints of maximum to minimum, shown in Figure 11-5. If a minor planet varies between magnitude 7.2

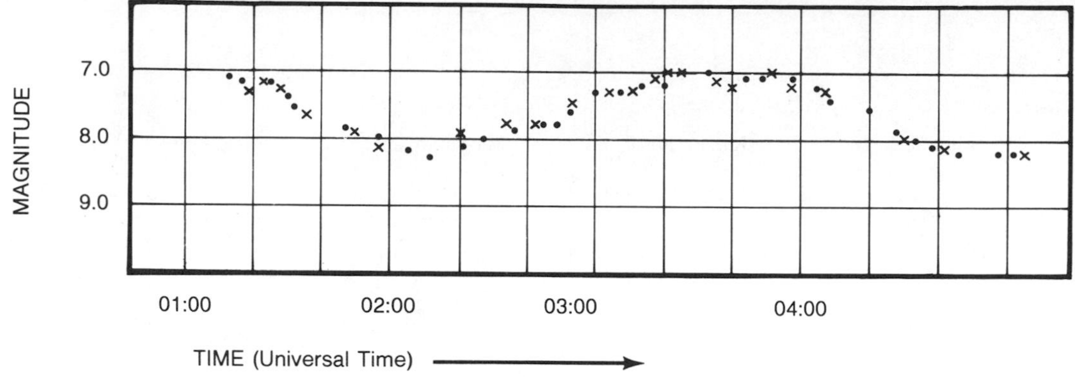

FIGURE 11-3. Plotting the magnitude of 433 Eros.

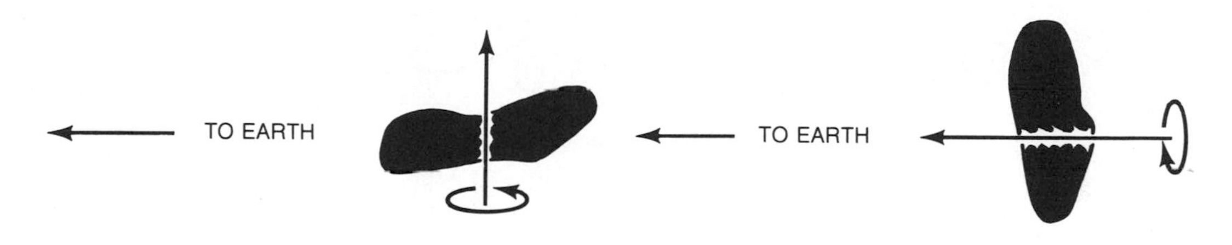

FIGURE 11-4. Demonstrating maximum and minimum amplitudes.

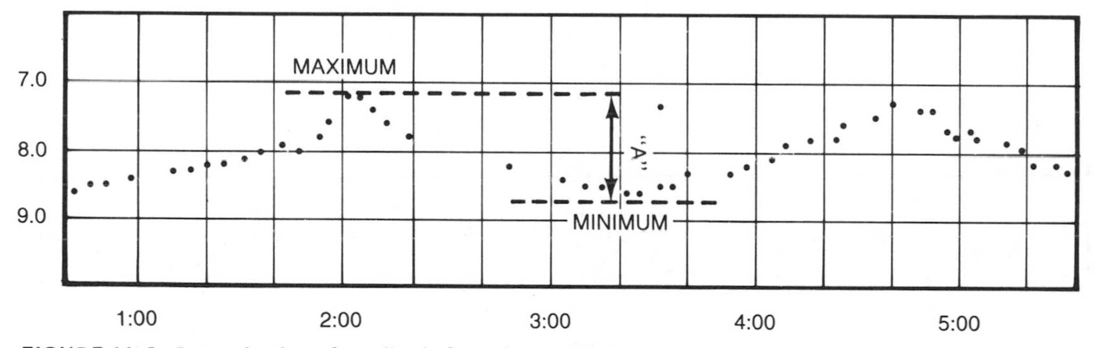

FIGURE 11-5. Determination of amplitude from the graphic light curve.

and 8.6, as in Figure 11-5, the amplitude is simply the difference in the two values, or 1.4. Strive to observe on as many dates as possible for amplitude changes, and send all results to the address given on page 200.

Period of Rotation

Although the rotational values have been roughly determined for most minor planets, amateurs can give valuable assistance in refining these determinations. It may be possible that some minor planets undergo sudden unexpected changes in the observed value, and amateurs should monitor for these.

The period of rotation of a planet is defined as one complete 360° turn of that planet about its axis. For an irregularly shaped (i.e., elongated) object, the light curve of one rotation will include two maxima and two minima, as can be seen in Figure 11-6.

With this in mind, we now refer to the light

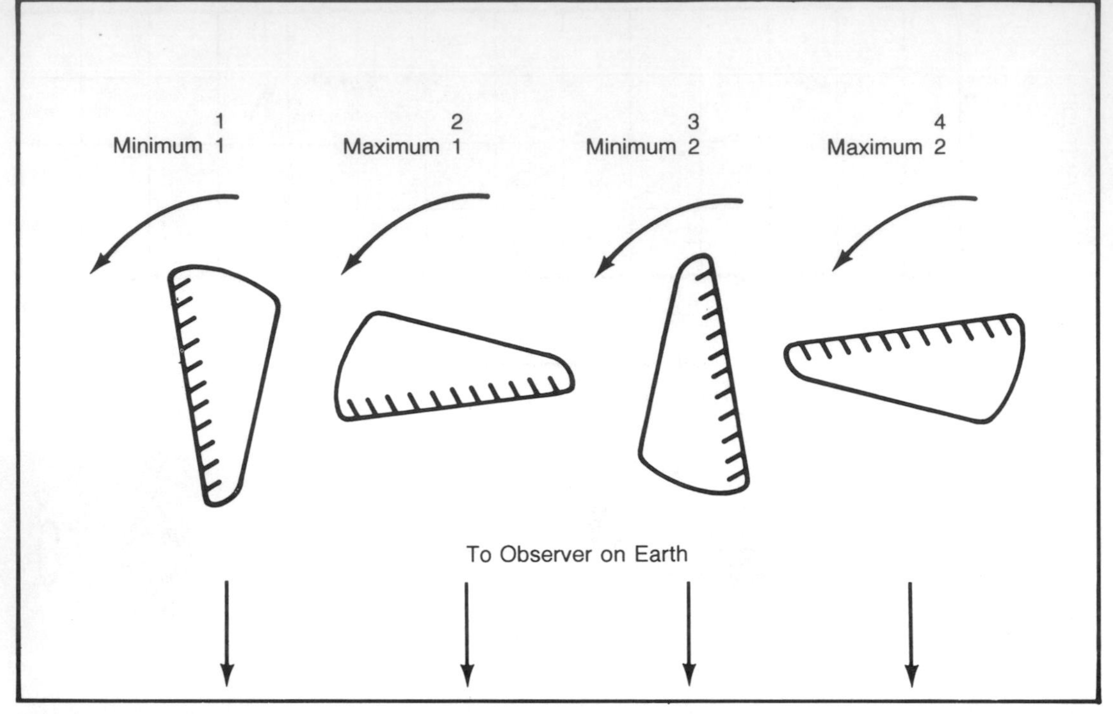

FIGURE 11-6. The appearance of variations as seen from earth.

FIGURE 11-7. Determining the period of rotation from a light curve. The position of the asteroid (facing earth toward bottom) is shown to aid in interpretation.

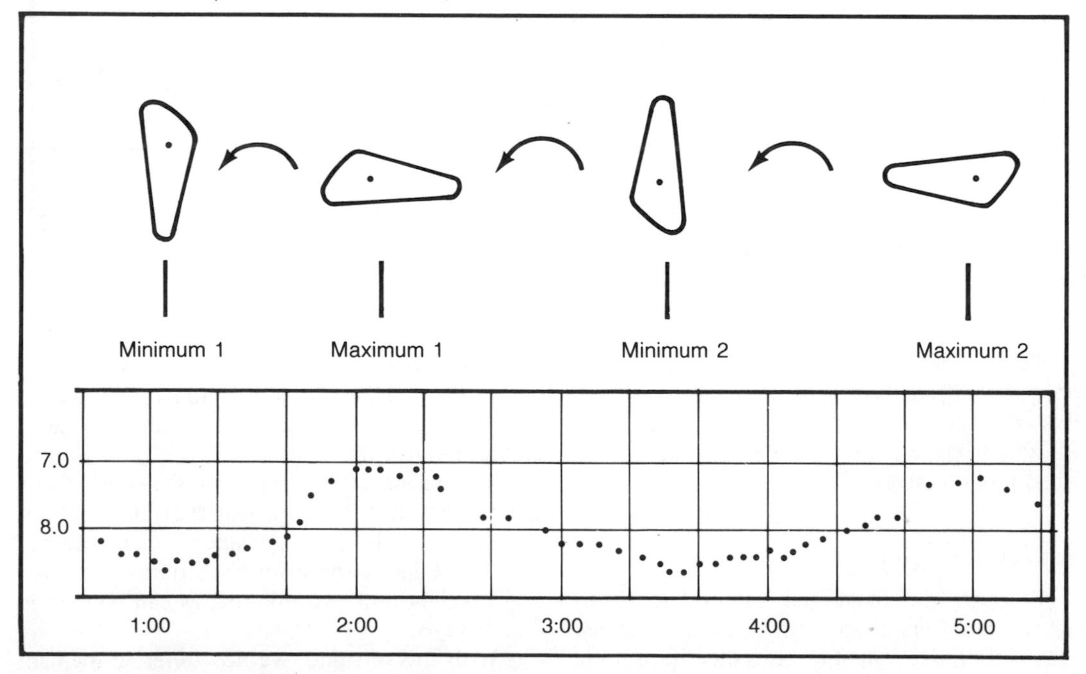

curve to determine the period of our full rotation. Measure from Minimum 1 to Minimum 3 to obtain one complete rotation. (Although minimum 3 is not shown in Figure 11-7, we will assume it is, and we can measure it.) Likewise, the interval from Maximum 1 to Maximum 3 (also not shown) comprises one complete rotation. The duration from Maximum 1 to Maximum 2 is 236m. However, because the rotation will probably include 2 maxima, measure to maximum 3 (during Maxima 1 and 3 the same face of the planet is toward earth). This interval is 5h 18m, a close value to the rotational rate of Eros.

To minimize error, make a second measurement from Minimum 1 to Minimum 3, which is 5h 16m. Taking an average of these two determinations gives a rotational rate for that night of 5h 17m.

There is *much* less chance for error if you derive the rotational rate from the light curve rather than by making a mathematical determination from estimated times.

PHOTOGRAPHIC PHOTOMETRY

The camera can be a useful too in the photometry of asteroids. However, because most asteroids are quite faint, it is necessary to take asteroid photographs at the prime focus of a moderate-sized (20-cm) telescope. If the exposure must be longer than 10 min, as would be the case for an asteroid fainter than magnitude 8, no results can be evaluated. In 10 min, the asteroid can brighten or dim considerably, rendering a photograph useless.

Many rapid black-and-white films are available (e.g., Kodak 2475) that can be processed at ASA 1000 or greater. Such films are recommended for a study of this type. With an f/5 prime-focus system it is possible to reach a magnitude fainter than 11 in a 5-min exposure.

Take photographs every 10 min, maintaining careful records for every shot. Likewise, when processing, take care to number every print to ensure correct times of exposure. You can study the photographs to determine periods of maxima and minima. By using a star chart for reference, you can obtain amazing accuracy when you compile a light curve from the photograph.

PHOTOELECTRIC ASTEROID PHOTOMETRY

With the advent of photoelectric measuring devices available at a modest cost, the amateur astronomer can contribute a wealth of valuable information through the use of these devices for asteroid photometry. The substitution of the electrometer for your eyepiece can provide accuracies and verification of data approaching 0.01 magnitude.

You should monitor the asteroid consistently in V filtration, which will more closely match those estimates made with the human eye. The method for obtaining the estimates is identical to that for visual observers described in the preceding section. Even if you use electronic measurements, however, it is of the utmost importance to make estimates every 5 to 10 minutes.

Unlike the visual method, though, two determinations are necessary when using photoelectric equipment. Because magnitude accuracy approaching 0.01 is attainable, many factors can affect the intensity of the light as recorded by the photoelectric photometer. The most noticeable of the effects are atmospheric turbulence and scintillation. To minimize these effects choose a comparison star quite near the asteroid, preferably one whose magnitude can be obtained to 0.01 accuracy from published values in common star catalogs. As quickly as a reading is taken from the comparison star, move immediately to the asteroid and determine its value.

The process can be greatly speeded up if two persons participate in the program. One person centers the comparison star and the asteroid in the photometer viewer while the other takes readings of deflection as the photometer is activated. After all data has been accumulated for one evening's observing it is reduced as has been described in the previous pages.

Colorimetry Determinations of Asteroids

Using the standard color index system of astronomical photometry, the photoelectric photometer can be a valuable tool in the determinations of the reflectivity, constituency, and the thermal absorption characteristics of the tiny planet that is an asteroid. Photometric colorimetry is discussed in Chapter 13.

As in stellar photometry, narrow pass-band filters are used that comprise a U, B, V, R, and I sequence. For this system to be effective, however, you must know beforehand the accurate color indexes of the star being used for comparison. Otherwise, you can determine only arbitrary color indexes for the asteroid in relation to your equipment, because all filters and photometry vary in sensitivity throughout various wavelengths of light.

REPORTING DATA

Full records of the following data should be reported for your every observation of asteroids:

- ☆ Date.
- ☆ Beginning UT.
- ☆ Magnitude every 5 to 10 min.
- ☆ Amplitude.
- ☆ Estimated rotation period.
- ☆ Peculiarities.

Send copies of your observations to the following:

Association of Lunar and Planetary Observers
Box 3AZ
University Park, New Mexico 88001

Minor Planet Center
Smithsonian Astrophysical Observatory
60 Garden Street
Cambridge, Massachusetts 02138

CONCLUSION

The observation of light changes by the minor planets is an exciting realm of study for the amateur astronomer, and provides the professional with data otherwise unavailable. There are simply too many irregularly shaped minor planets visible throughout the year for the observatories to monitor. In recent years more and more minor planets have been suspected of variability. As this is written, astronomers have announced the potential variability of the asteroid 201 Penelope by as much as 0.5 magnitude. The observations are badly needed, and time is the main requisite. As an asteroid's angle of axial orientation changes with respect to earth, the degree of variability will probably also change. And there may be other minor planets that have not yet been determined to be variable simply because time has not allowed detailed study of the light curves of each and every one.

BIBLIOGRAPHY

Chapman, Clark R., "The Nature of Asteroids," *Scientific American*, Vol. 232, No. 1 (1975), pp. 24–33.

Harvard College Observatory, Annals of. Vol 72, Part 5: *Observations of Eros and Other Asteroids*. Cambridge, MA, 1903.

Harvard College Observatory, Annals of Vol. 72, Part 8: *The Light Curve of Eros in 1914*. Cambridge, MA.

Moore, Patrick, *Astronomy Facts and Feats*, pp. 80–97. London: Guinness Superlatives, 1979.

National Aeronautics and Space Administration, *Physical Studies of Minor Planets*, SP-267. Washington, DC: Government Printing Office, 1971. Paperback.

Pilcher, F., and J. Meeus, *Table of Minor Planets*. Private Printing, 1973. Paperback.

Roth, G.D., *The System of Minor Planets*. London: Faber and Faber, 1963.

Sherrod, P. Clay, *433 Eros*. Proj. Publication 002 of the Midsouth Astronomical Research Society. Little Rock, AR, Feb., 1975.

Vehrenberg, Hans. *Photographic Star Atlas*. Düsseldorf: 1962. Privately published.

Wallentinsen, Derek, "The 1976 Apparition of 1580 Betulia," *Journal ALPO*, Vol. 27, Nos. 7–8 (1978), pp. 130–132.

12

STUDIES OF VARIABLE STARS

The scientific study of variable stars is one well suited for amateur astronomers, for two reasons: (1) the suitability of the amateur's equipment, and the speed at which the amateur can make estimates; and (2) the availability of time in which the study is done, a factor not available to the professional astronomer. Consequently, amateur astronomers contribute, either to the American Association of Variable Star Observers or the Variable Star Section of the British Astronomical Association, hundreds of thousands of observations of variable stars. The data amassed by these two organizations are carefully evaluated and stored for use by professional and amateur astronomers. This is information that would otherwise be lost if not for the efforts of the amateur and his or her modest equipment.

INSTRUMENTS OF THE VARIABLE STAR OBSERVER

One need not avoid the study of variable stars for lack of appropriate equipment. Indeed, even a pair of eyes is a useful tool for studying some of the brightest variables. In the following pages I detail some specifics for observing using apertures ranging from the eye through telescopes of 32cm (12.5 inches). Table 12-1 gives recommended instrumentation for various stellar magnitudes.

Notice that for telescopes of larger aperture the recommended brightest star in the magnitude range becomes increasingly fainter (e.g., 9.0 for a 10-inch scope). A 12-inch telescope is capable of showing stars *brighter* than magnitude 9.5, but it is not capable of accurate magnitude estimates of these brighter stars. The concentrated brilliance of such stars affects the reliability of your eye for differentiating between subtle intensity differences and slight color interference. It would be best to stay within the recommended range for a given telescope. Of course, if your eye and telescope are capable of observing stars fainter than those recommended for a given aperture, be sure to include these in your observing schedule. Estimates of stars near the threshold limit (i.e., faintest magnitude visible in a particular instrument) are far more reliable than those brighter than the optimum recommended magnitude.

The use of telescope setting circles will greatly increase the proficiency of your variable star program. Not only can you locate and estimate a greater number of stars in a given time, but the estimates are indirectly more accurate. Because patience is a key factor in any serious astronomical observing program, it is essential that locating and identifying star fields be as simple and painless as possible. Star-hopping through the sky in search of a variable star field is slow and tedious, and you will tire more rapidly than if you use setting circles. Many experienced observers are now able to estimate one variable per 2 min of time. Such speed requires memorization of the field and comparison stars as well as proficiency with the setting circles. Nonetheless, it is something that with practice you will master in time.

TABLE 12-1.

Instrument	*Recommended Magnitude Range*
Naked eye	1.0– 4.5
Binoculars	4.0– 8.5
2 in. telescope	4.0– 8.5
4 in. telescope	7.0–12.5
6 in. telescope	8.0–12.5
8 in. telescope	8.5–13.0
10 in. telescope	9.0–14.0
12 in. telescope	9.5–15.0

Naked-Eye Variables

Many stars visible to the naked eye are variable, but observations of these stars are difficult for the reasons discussed in the following paragraphs.

Amplitude

In variable star observing the *amplitude* of a star is simply the range (in magnitude) of the star's light variations. For the naked-eye variables, the amplitudes are usually quite small, as can be seen in Table 12-2. Estimates of bright stars with small variations allow for great discrepancies among observers, owing primarily to atmospheric effects, the observer's mental and physical condition, and largely to the *Purkinje Effect*.

The Purkinje Effect

Some observers' eyes are much more sensitive to red light than others. Such persons are prone to *overestimate* the brightness of reddish stars, and—in some cases—*underestimate* the brightness of stars with blue color, particularly if reddish comparison stars are used.

Comparison Stars

Even with the wide-angle view afforded by the human eye, there is a lack of suitable comparison stars with which to compare the naked-eye variable. The brighter the star, the fewer comparison stars there will be. It is not a good practice (either when using the naked eye or a telescope) to compare with stars that are out of the field of view (i.e., that cannot be viewed simultaneously), because this necessitates memorizing the image of the star when looking away to compare (see Table 12-2).

Variables for Binoculars Or for Telescopes up to 2.5 Inches

Stars to about magnitude 8 can be observed adequately with good binoculars, but finder charts are necessary to locate most of these stars. Such charts (available through the AAVSO—see pages 204–205) also list comparison stars known to be constant in magnitude. Unless suitable charts are used, magnitude estimates can be subjective and erratic.

Because fainter stars can be more suitably studied in binoculars than with the naked eye, the problem of finding suitable comparison stars is somewhat less troublesome. However, you must first make certain to center the variable and then the comparison star in the field to eliminate aberrations encountered near the edges of most optical fields. Another method to achieve equalization of extinction in a telescopic field is to position the variable and the comparison star equidistant from the center, one on each side, the same distance from the center as the other.

Variables for Telescopes with Apertures of 4 to 6 Inches

Stars to magnitude 12 can be studied in telescopes of 4 inches aperture. These are more stars than even the most experienced observer could possibly observe. Remember that only stars

TABLE 12-2. Naked-eye variables.

Star	*RA*	*Dec.*	*Magnitude Range*	*Period*[a]	*Class*
Betelgeuse	5h 53m	7°24′	0.2–1.3	2070	SR
Antares	16h 27m	-26°22′	0.9–1.8	1733	SR
α Cassiopeiae	00h 53m	60°27′	1.4–2.3	Irregular	IRR
β Pegasi	23h 01m	27°49′	2.3–2.8	40	SR
Algol	3h 05m	40°46′	2.2–3.5	2.8	ECL
μ Cephei	22h 27m	58°10′	3.7–4.6	5.7	CEP
η Geminorum	06h 12m	22°31′	3.1–3.9	234/2983[b]	SR
α Herculis	17h 12m	14°27′	3.0–4.0	50	SR
n Aquilae	19h 50m	00°53′	3.7–4.7	7.2	CEP
γ Cassiopeiae	00h 38m	56°16′	2.1–2.4	Irregular	IRR
β Lyrae	18h 48m	33°18′	3.4–4.3	12.9	ECL

[a]In days.
[b]A double variable occurs when a shorter period is imposed on the longer period.

within 5 magnitudes (Table 12-1) of the limiting magnitude of a particular telescope can be suitably observed. With telescopes with apertures in the range of 4 to 6 inches, the possibilities exist for extensive observation of many of the brighter irregular variables and novae through much of their cycles.

Observing Variables with Telescopes Of 8 to 12 Inches of Aperture

In larger amateur telescopes, say from 8 to 12 inches of aperture, the true nature of cyclic variations in stars is within reach. Such instruments make it possible for you to follow a star far into its quiescent phase. In addition, many classes of variable stars expand into greater numbers within reach, particularly the irregular group of variables. In smaller instruments, many of the R Geminorum and Z Camelopardalis stars are visible only during the height of their activity.

More and more large-aperture telescopes are being acquired by amateurs today, and the study of variable stars with these instruments is perhaps the most valuable contribution to be made. Contrary to several published estimates giving somewhat conservative figures, I have consistently been able to follow stars through minima reaching the fifteenth magnitude with a 12.5-inch (32-cm) Newtonian telescope on exceptionally clear nights.

Unlike high-resolution observing, the large-aperture instrument can be used in less-than-perfect times of steadiness for observing variable stars. Many times when the sky is incredibly dark, the seeing conditions are quite poor in telescopes with apertures larger than 8 inches. But these are the times when such scopes can penetrate deep into space, reaching magnitudes far below their theoretical limit. Poor seeing will affect variable star observing only by causing stars at the threshold limit to intermittently "blink away." I have never found this to be a discouraging or troublesome factor.

CHARTS FOR LOCATING VARIABLE STARS

I urge any observer who desires to make a meaningful contribution to the study of variable stars to apply for membership in the American Association of Variable Star Observers (AAVSO). This is a well-organized group consisting of both amateur and professional astronomers. The number of estimates contributed thus far by amateurs is well into the millions, compiling an invaluable storehouse of light curves and data otherwise unavailable. Indeed, even the very physical processes of stellar systems have been determined through estimates provided by amateur astronomers of the AAVSO.

As well as serving as a clearinghouse for variable stars, the AAVSO has developed perhaps the finest collection of variable star comparison charts in the world, all available to participants. Charts for every variable star are available, many in a wide choice of scales. Each scale is designed for use depending on the brightness of the star through its cycle and the field of view in a specific instrument. Five chart scales are available:

Scale a	5′ = 1 mm
Scale b	60″ = 1 mm
Scale c	40″ = 1 mm
Scale d	20″ = 1 mm
Scale e	10″ = 1 mm

Obviously, the limiting magnitude of each successive star chart increases, and the field covered on each chart decreases. Chart a or Chart b might serve well as a finder chart to locate the variable field, whereas Chart c or Chart d might then be used to estimate the magnitude of the faint variable.

In addition to providing quick identification of the variable star field, these charts give precise magnitudes of comparison stars in the field, with an accuracy of 0.1 magnitude. On most charts the values given for these comparison stars are accurate. However, a few apparent discrepancies exist on some charts. I urge you to examine carefully each comparison star relative to other comparison stars before using it for a final estimate.

Proper identification of a variable can be made only when the chart is oriented with respect to the sky directions as seen in the eyepiece. Thus, it is important that you be familiar with such orientation. All comparison charts are printed with *South* at the top, as it appears in a typical inverting telescope. Using a Newtonian reflector, this is a natural orientation. However, most refractors and compound (catadioptic and Cassegrain) telescopes are supplied with right-angle prisms for observing comfort. With such prisms, the field appears upright and reversed left to right, as if seen reflected in a plane mirror. Consequently, these prisms should be removed. Otherwise, field identification is made very difficult and unbelievably frustrating. Without this prism, the field is astronomically correct in these instruments.

For membership in the AAVSO and a list of their available charts, write to the following address. Excellent descriptive literature is provided to new members, detailing the use of these charts and the particulars of variable star observing.

The American Association of Variable Star Observers
187 Concord Avenue
Cambridge, MA 02138

OBSERVING PROCEDURE FOR VARIABLE STARS

Once the variable star has been located in the field, the estimation of brightness is straightforward. Simply stated, make the estimate by finding a star—or several stars, ideally—that matches the brightness of the variable. That is the ideal situation. Normally, no star in the field will exactly match the variable, or if it perfectly matches, it might not have any assigned magnitude on the chart.

In such cases, it is necessary to interpolate the true magnitude between two stars—one brighter and another dimmer than the variable. Most variable star fields have close sequences in 0.1 magnitude increments that give you fairly reliable comparisons. As an oversimplified example, suppose that the variable is found to be slightly brighter than star *A*, at magnitude 8.6, but not as bright as star *B*, at magnitude 8.4.

Obviously, the variable can reasonably be assigned a magnitude of 8.5. When two comparison stars are not so ideally close together in brightness, it is necessary to adopt a somewhat less reliable scheme. Suppose that the variable is brighter than either star *A* or star *B* and those are the only accessible comparison stars. With such a situation, a *relative* estimate is adequate. If the variable is about a third brighter than star *A* is to star *B*, then you can designate its brightness as:

var (1)*A* (3)*B*

If the variable is fainter than either star, and if it appears to be a third fainter than star *B* is to star *A*, then you can designate its brightness as:

A (3) *B*(1)var

Be very careful when choosing comparison stars, particularly when a variable is near maximum. Various colorations—red objects in particular—will appear brighter to some observers than they actually are. This phenomenon is the *Purkinje Effect* (previously discussed), and it will cause overestimation of a reddish variable (such as a nova on the decline) or underestimation of a variable if a red comparison star is used. In addition, care must be exercised by all observers when dealing with reddish stars. Estimates of red variables must be made as quickly as possible, because most eyes will begin to perceive reddish objects as being brighter after some time is spent looking at them.

Many times a star in its cycle will become too faint to be seen in your telescope. It is important to continue observations even if you cannot see the star. Simply record the faintest star visible in the field of view, and record the brightness of the variable as "fainter than ——" (written as "<——") the magnitude of the faintest star.

The most reliable method of making comparisons that I have found involves *defocusing* the eyepiece until all star images become small disks of light, rather than points of light. This method enables you to estimate a uniform brightness spread over a greater area, and it substantially reduces problems introduced by averted vision as well as star "blinking"—a phenomenon of the eye and atmosphere that causes a point source to appear to vary frequently during a short observation. Defocusing is not effective for stars that are near the limiting magnitude of the telescope in use, and it is best that you use precise focus for such stars. A neat trick for very faint stars is to turn the focusing slightly until the faint variable disappears in the field. With some practice, it becomes quite easy to estimate a variable relative to comparison stars, which are very close in brightness and difficult to differentiate. Stars slightly dimmer will disappear first, while the somewhat brighter ones remain faintly visible. One by one, you can eliminate the faint stars until you find one as bright as the variable.

I must emphasize a word of caution before concluding this section. An observer, after monitoring a star through many of its cycles, is greatly prone to *bias*—that is, preconceived ideas about the degree of brightness the variable should have, based on its past performance or on its normal cyclic variations. Although it is difficult to overcome bias, you should strive consciously to eliminate any speculation about the brightness of a variable before observing it. This is particularly true of the long-period variables, and the novae, whose light curves are rather predictable.

RECORD KEEPING AND THE JULIAN DAY

Most observers adopt a method of recording the brightness of variables at the telescope that works best for them. Using some sort of standardized method is not important. What *is* important is that whatever method used is used consistently. Consistency in record keeping prevents error both in the magnitude estimate and in the date/time you made the estimate. At the telescope, any method will suffice—tape recorder, tablet, or prepared form. However, to *report* your observations to an organization (for example, to the AAVSO), you must use standardized forms from that organization. These forms are normally supplied to observers free of charge (see Appendix XIII).

On these reporting forms, it is necessary that the time of the observations be expressed in Julian days and decimals of those days, rather than in Universal Time (in hours, minutes, and seconds), as is conventionally done in most other astronomical work. Most observers find it convenient to record the time of the observation at the telescope using Universal Time, and then convert it to the Julian day later. In 1582 the Julian day was arbitrarily set to zero at noon on January 1, 4713 BC, and every day that has elapsed since then is added to the total. The Julian day for January 1, 1984 (Universal Time) is 2445701.

Each Julian day begins at noon Universal Time (12 hours UT), and thus is half a day behind Universal Time. During the day, observations are expressed in decimal parts of the 24-hour period, rather than in hours, minutes, or seconds. Appendix II gives decimal equivalents for various time intervals throughout the day. Use of the Julian day greatly simplifies the compilation of light curves, allows quicker mathematical use of the data (by insertion into a computer program, for example), and eliminates the problem encountered by having to change dates while an observation is in progress, as is customary with Universal Time.

CONSTRUCTING ASTRONOMICAL LIGHT CURVES

The graphic light curve is an important aspect of astronomical recording. It allows you to interpret visually the changes shown on paper, whether the observations be of a variable star, rotating asteroid, or nova. However, the interpretation of the light curve cannot be made if the graph is not properly organized and constructed. There are two axes (directions) on the light curve, as shown in the diagram in Figure 12-1. The vertical axis is known as the "y" axis; the horizontal axis is called the "x" axis. It is essential that the light curve be constructed as described so that it will conform to other similar curves supplied by other observers.

In the typical astronomical light curve, the "y" axis represents the magnitude, with the brightest magnitude recorded at the top, the faintest at bottom. In some cases, the "y" axis represents magnitude *change* rather than actual magnitude. This is common practice when recording stars that change only fractions of a magnitude, and when the actual visual magnitude is of little importance compared to the *change* in that magnitude.

The "x" axis should always represent a progression of time. In visual studies, the unit of measure of time should always be the day. Actual dates (U.T. or Julian day) should be expressed along this axis and should be clearly indicated. Dates will run from left to right, showing the progression of time in that direction.

The scale of the light curve may be somewhat arbitrary, but it must remain consistent for each observer. Do not use one scale for one star and another for a different star. A recommended scale for variable stars is outlined in the two lists that follow (p. 208).

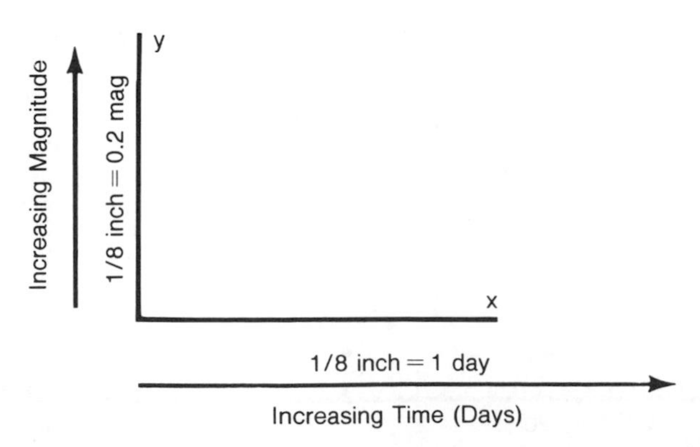

FIGURE 12-1. Axes of light curve.

"y" Axis (Magnitude)

1. If the period is less than 100 days, represent the magnitude as ⅛ inch per 0.2 magnitude, or ⅝ inch per magnitude.
2. If the period is from 100 to 200 days, represent this axis as ¼ inch per 0.2 magnitude, or 1 ¼ inch per magnitude.
3. If the period is over 200 days, maintain the "y" axis scale at ¼ inch per 0.2 magnitude, and change only the scale of the "x" axis as described below.

"x" Axis (Time)

1. If the period is less than 100 days, represent the time as ⅛th inch per day, or eight days per inch.
2. If the period is from 100 days to 200 days, represent the "x" axis as two days per ⅛ inch, or 16 days per inch.
3. If the period is over 200 days, represent the scale as 4 days per ⅛ inch, or 32 days per inch.

Ideally, it would be best to put all variables, regardless of period, on the same scale. The most workable scale, if you wish to assume this consistency, is ⅛ inch per 0.2 magnitude, and ⅛ inch per day. By putting all variables on the same scale, intercomparisons may be made to visually evaluate similarities between different stars. Above all, remember to put the brightest magnitude at the *top* of the "y" axis and the earliest day at the far left of the "x" axis. A sample light curve is constructed in Figure 12-2. Compare the dates given at its left and the placement of the observed magnitudes on the graph.

Observations of a particular object should not cease until one phase of that object is finished. A phase of a variable star would simply be one complete cycle (bright to dim to bright again for a pulsating star, for example), or for an asteroid a phase would simply be one complete rotation.

For objects with irregular light changes, it is essential that you make as many observations close together as possible so that any rapid fluctuations in brightness can be acknowledged on the light curve.

Label all light curves properly when you complete them, giving your name, the object for which you prepared the graph, the beginning and ending dates, the optical equipment you used, and your address. When the curve is complete and properly recorded, you have at your disposal a history of an object's behavior recorded by means of the phenomenon of light. This history is much like a fingerprint, serving to identify that object in a particular stage of its existence. If the observations are not made and the curve not constructed, a possibly vital link in the life history of that object could be lost forever.

The interpretation of the astronomical light curve follows. However, you must first be completely familiar with the fundamentals of the graph before the patterns emerge. I advise that you compile light curves of at least two variable stars before attempting any sort of analytical work that will help you understand what the light curve is telling. The story is there, but you must learn the language with which to read that story.

FIGURE 12-2. Sample light curve.

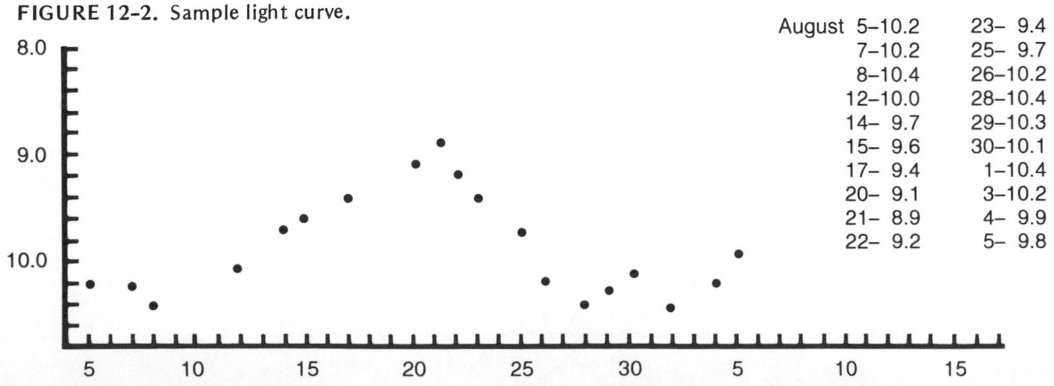

August 5–10.2
7–10.2
8–10.4
12–10.0
14– 9.7
15– 9.6
17– 9.4
20– 9.1
21– 8.9
22– 9.2
23– 9.4
25– 9.7
26–10.2
28–10.4
29–10.3
30–10.1
1–10.4
3–10.2
4– 9.9
5– 9.8

Pogson's method for determining the exact dates of maximum and minimum from a plotted light curve is the most reliable method available to the amateur. Rather than using the peak, or high point, of the variation, this method is representative of the *total event* (i.e., an outburst of a star from beginning to end).

Once you have plotted the light curve, draw several chords at various heights across the rise and decline of the event, as shown in Figure 12-3. Start near the top and draw a line through the midpoints of each chord. This will indicate on the "x" axis the precise date of minimum or maximum.

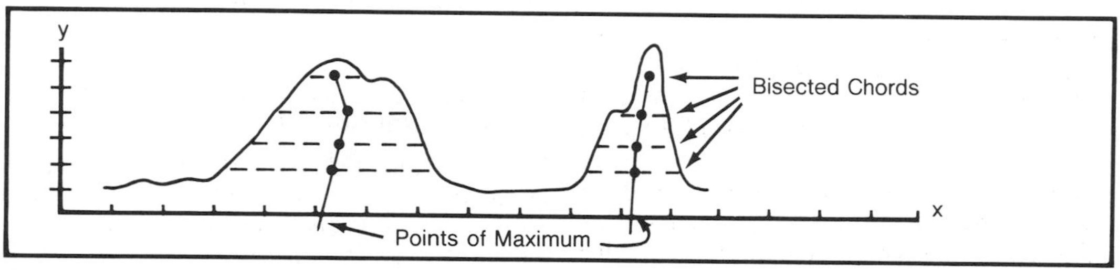

FIGURE 12-3. Determining the time of maximum from intersection of chords.

CLASSES OF VARIABLE STARS

The thousands of variable stars are classified into major categories, depending either on physical causes for their variability or on similarities in their light curves when the cause of variability is not well defined.

Eclipsing Variables

The most observed group of variable stars is the class known as *eclipsing variables*. In these stars, there is usually no intrinsic change responsible for the varying light from the star. Rather, an outside—or extrinsic—cause results in the variability.

These systems are composed of at least two stars which, in their orbits, pass in front of each other. This eclipse will rapidly block (either partially or totally) the light from the occulted star, thus causing a quick drop in the normal brightness as seen from earth. Figure 12-4 and its accompanying light curve demonstrate the process of variation.

Studying the eclipsing variables and their light curves provides valuable information regarding each system. You can easily obtain the shape of the orbit of the stars by noting the

FIGURE 12-4. Processes of variation in an eclipsing binary system. Notice that star B is a fainter star than star A.

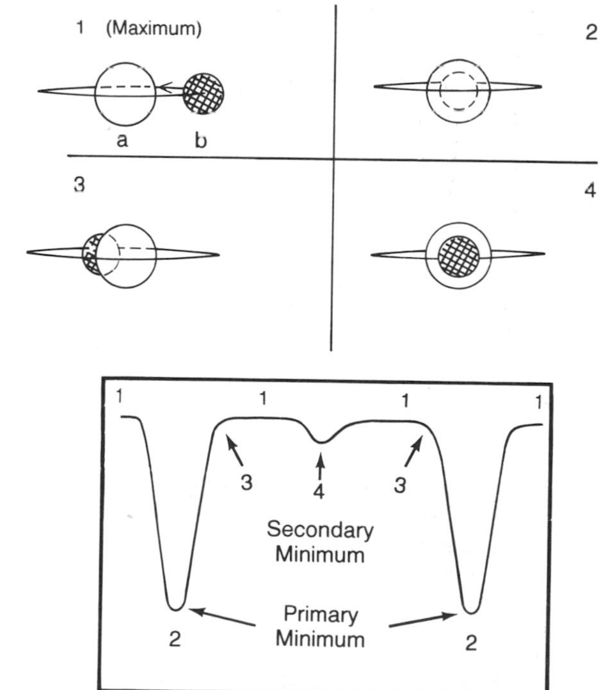

positions of the primary and secondary minimums relative to each other. For example, if the secondary minimum occurs exactly midway between two successive primary minima, as in the light curve shown in Figure 12-4, the orbit is nearly circular. However, if the secondary minimum is displaced greatly toward either primary minimum, the orbit is strongly elliptical.

In addition, the amplitude on each minimum provides an accurate measure of the *relative* luminosities of each star. Two stars of the same brightness will result in two successive minima of equal depth, regardless of which star is in front or in back, assuming of course that the eclipse is a total one.

Another determination that you can make through observations of eclipsing variables is the total mass of the binary system, and thus, the mass of each component. Kepler's laws of mass and motion apply to binary star systems as well as to the entire solar system. To make such a determination, you need only to have information from the light curve (provided usually by the professional astronomer). Hence, the value of visual estimates comes to light in the making of light curves. The spectrum allows astronomers—through the Doppler Effect—to determine the eccentricities of the binary orbit and, consequently, derive information regarding the size of the orbit, the velocities of the stars in the orbit, and even the diameters of each star within the system. Like the information on most categories of variable stars, information on the eclipsing binary stars is provided by the amateur astronomer so that such determinations can be made.

FIGURE 12-5. Sample light curve for eclipsing binary.

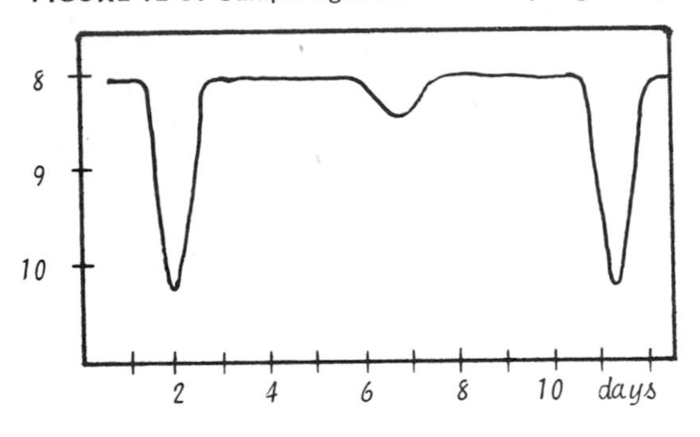

Example: Eclipsing Binary

S Cancri

Maximum magnitude	8.00
Minimum I magnitude	10.22
Minimum II magnitude	8.05
Period (days)	9.48
Spectrum	AO, G5
Right ascension	8h 41m
Declination	19° 13m
AAVSO No.	083819

Pulsating Variable Stars

In 1913 a yardstick was discovered that enabled astronomers to measure fairly accurate distances to star clusters and nearby galaxies. Leavitt found that a relationship existed between the absolute brightness of Cepheid variables and their periods of variation. She discovered that the brighter a Cepheid, the longer its period. Consequently, if we derive the absolute magnitude of these stars, the distance can easily be calculated for any Cepheid variable. No matter where they are found in space—in distant galaxies or in star clusters—the rule still holds true.

In the class of pulsating variable stars, there exist two subclasses, the Cepheids and the RR Lyrae stars, each named after the classic example for the subclass. In both subclasses, the light variation is caused by the actual pulsating of the stars, just as the name implies. Physical studies have shown that maximum light does not occur when the star has grown to its greatest diameter but rather three-fourths its greatest diameter. During the expansion, the temperature of the star drops as the density of the expanding gases decreases. Thus, the brightness as a factor of temperature subsides before the star's diameter reaches maximum.

RR Lyrae Stars

The pulsating RR Lyrids have very rapid periods, all shorter than one day. They are also termed *cluster variables* because of their dominance in globular star clusters. They are extremely old stars, unlike the Cepheid stars, and are part of the class of stars known as Population II.

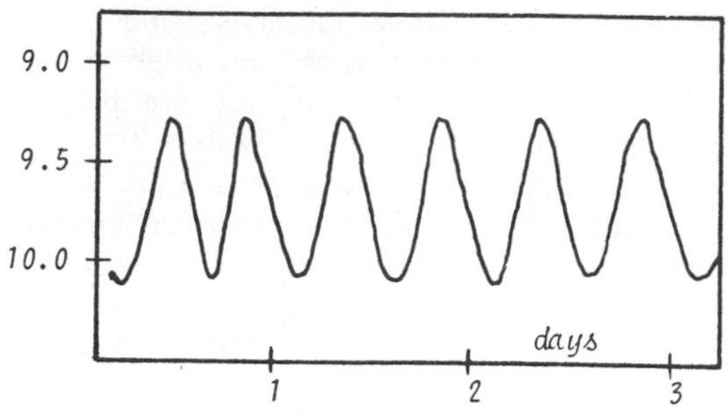

FIGURE 12-6. Sample light curve for RR Lyrae stars.

Example: RR Lyrae Stars

SW Andromedae	
Maximum magnitude	9.3
Minimum magnitude	10.3
Period (days)	0.4
Spectrum	A3 to F8
Right ascension	00h 21m
Declination	29° 07m
AAVSO No.	001828

Cepheids

There are several distinct differences in the nature of the Cepheid stars when compared to RR Lyrae stars. The main difference is that Cepheid variables belong to a younger group of stars—Population I—and are found primarily in the spiral arms of galaxies, rather than in the core, or in clusters. The Cepheids are further divided according to their periods of variation into two subsequent groups: the *classical Cepheids* have periods ranging from 1.5 to 28 days, and the *long-period Cepheids* have periods of 28 days and longer.

FIGURE 12-7. Sample light curve for Cepheid variables.

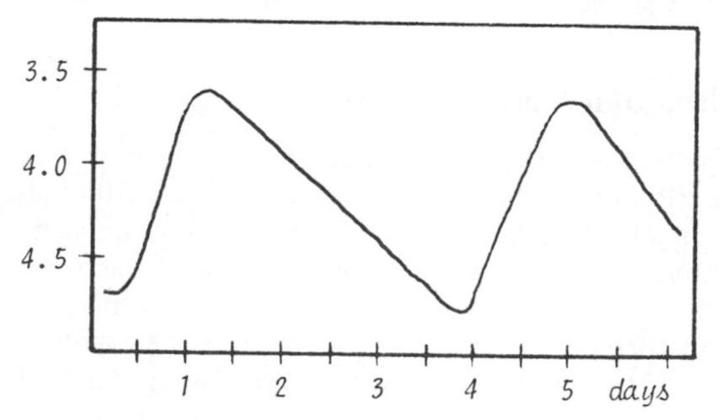

Example: Cepheid Variables

Delta Cephei	
Maximum magnitude	3.8
Minimum Magnitude	4.6
Period (days)	5.36
Spectrum	F5 - G2
Right ascension	22h 27m
Declination	58° 10m
AAVSO No.	222557

Long-Period Variables

The stars classified as long-period variables are like the Cepheids in that the light changes are caused by the pulsing instability of the star. These stars have periods of 100 to 700 days, and they are somewhat regular and predictable in their light changes. All long-period variables are red giant or supergiant stars of spectral class M or later. The pulsations are caused by great shock waves emanating from deep within the star. These stars are very well observed because the

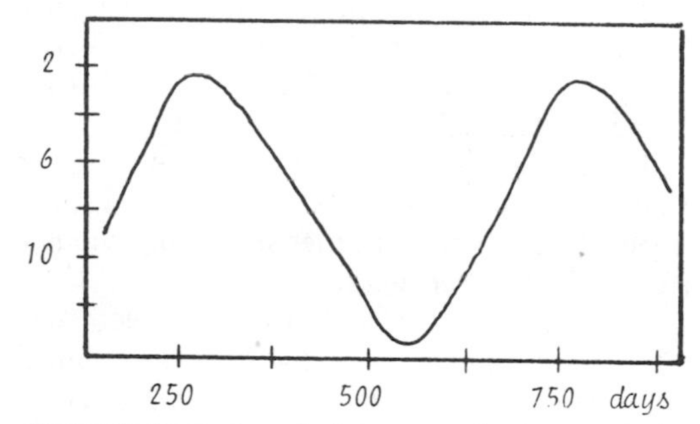

FIGURE 12-8. Sample light curve for long-period variables.

Example: Long Period Variables

χ Cygni	
Maximum magnitude	2.3
Minimum magnitude	14.3
Period	407 days
Spectrum	M
Right ascension	19h 49m
Declination	32° 47m
AAVSO No.	194632

long period allows for a casual observing schedule. Over the course of one period, the light curves of the long-period variables are quite spectacular because the amplitudes usually are quite large.

Semiregular Variables

Very similar to the long-period variable star, the semiregular variable is a red giant star, but it is not nearly so smooth and predictable in its light changes as is the long-period star. In addition, the total change in light is quite small when compared to the spectacular changes of the long-period variables. A secondary light cycle can often be seen superimposed on the major light curve.

Also like the long-period variables, the semiregular variables are in some cases "over-observed" because their light curves are somewhat predictable. This does not mean that you should neglect either of these two classes, however.

Make observations once a week on the long-period variables and once every three days on the semiregulars, unless some unusual rapid activity can be detected in the semiregulars. If such is the case, make observations nightly until this activity ceases. Observing long-period and semiregular stars every night is overkill, and results in inconsistent data that seem to reveal minute changes that may not actually have taken place.

There are several subclasses of semiregular variables because many seem to show distinct changes characteristic of other stars and obviously different from other semiregulars.

SRa. There is a very thin line between stars in this class and long-period variables. Small amplitude constitutes classification as SR. These are giant or supergiant stars.

SRb. Again, these are very similar to long-period variables. Stars in this class undergo infrequent and irregular variations. The changes are very slow at times, and sometimes the light remains constant for long periods of time. An example is U Boötes.

SRc. These semi-regular stars are characterized by extremely small amplitudes, most often being of less than one magnitude total change. Many of the naked-eye stars previously mentioned as being variable belong in the SRc class. Several of these stars show secondary cycles of even lower amplitude superimposed on the primary light curve. Observers find it very difficult to accurately assess the brightness of these stars because they usually are brighter than any comparison stars within the field of view of the naked eye. Estimates are further complicated by the distinct coloration inherent in most bright stars. An example is Betelgeuse.

SRd. Some of the most luminous stars in the sky belong to this class of variables, which have much earlier spectra than in any other of the semiregular subclasses. The light curve is characterized by a somewhat deep minimum followed by a shallow minimum, whereas variation of the height of the maximum peak is quite negligible. An example is R Scuti.

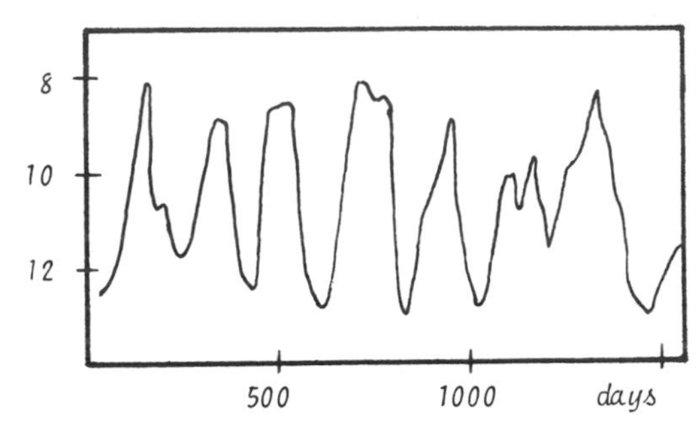

FIGURE 12-9. Sample light curve for semiregular variables.

Example: Semiregular Variables

U Boötes

Maximum magnitude	8.4
Minimum magnitude	13.3
Period (days, mean)	185.8
Spectrum	M4
Right ascension	14h 52m
Declination	17° 54m
AAVSO No.	144918

Irregular Variable Stars

With the irregular variable stars, all traces of periodicity and predictability vanish. Although not all irregular stars vary in light for the same physical reason, it is my opinion that in each star there is some change of recurring nature, an obscure periodicity, or some predictable event.

Stars not falling into some other class of

eruptive variables (as will be discussed) and showing no predictable variations are considered irregular. Therefore, the number of such stars is great. Normally, these stars vary only small amounts, and many remain quiescent for long periods of time, with a very sudden outburst of activity. Since the times of such outbreaks are not yet predictable it is quite important for amateur astronomers to monitor the irregular variables at every opportunity, at least once every clear night. Several irregular variables are thought to be rapidly rotating stars which eject luminous clouds of mass from this great force of motion.

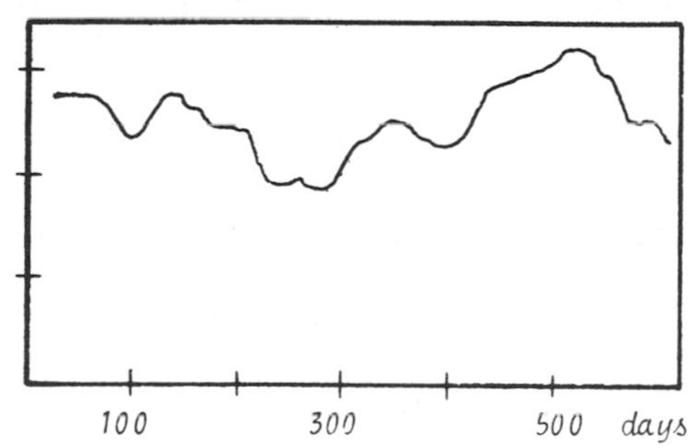

FIGURE 12-10. Sample light curve for irregular variables.

Example: Irregular Variables

	μ Cephei
Maximum magnitude	3.6
Minimum magnitude	5.1
Period (days)	?
Spectrum	cM3e
Right ascension	21h 42m
Declination	58° 33
AAVSO No.	214058

The Nebular Variables

The nebular variables are unique. They differ from the other classes of variable stars in that their variability is caused by *external* factors and is not a physical change or property of the star itself. In most cases the stars themselves are stable and nonvarying. Nebular variables are embedded in either dark or bright nebulosity, which occasionally will obscure or reduce the amount of light seen on earth. Consequently, the changes in brightness are expected to be irregular, and the amplitudes of these stars small. Studies of the nebular variables enable astronomers to deduce the movements and densities of the clouds of nebulae throughout our Milky Way.

Many of the stars embedded in the Orion Nebula have been discovered to be nebular variables, and they afford the amateur astronomer a challenge limited only by the time and the aperture of the telescope available.

As with the semiregular variable stars, the nebular variables have also been assigned subclasses, each according to specific characteristics common to each subclass.

RW Tauri. This class consists of large amplitude (up to magnitude 3). Their light curves are rapid and irregularly changing. The stars are G-type dwarfs, and many are found in the Orion Nebula.

T Orionis. Stars in this class also have large amplitude (magnitude 2 to 3). Their light curves show rapid and irregular changes. These are giant stars, not dwarfs. They are probably new stars, still condensing from the original gas clouds.

T Tauri. These stars generally have low amplitudes with occasional outbursts of greater magnitude. They are probably red dwarf stars surrounded by orbiting disks of gas, which some astronomers believe are planetary systems in the making. Generally, the nebulae surrounding these stars are small and dark. Consequently, the brightness of the nebula itself will vary somewhat from the luminosity of the stars. In the telescope, the T Tauri stars are quite faint and usually appear reddish.

FIGURE 12-11. Sample light curve for nebular variables.

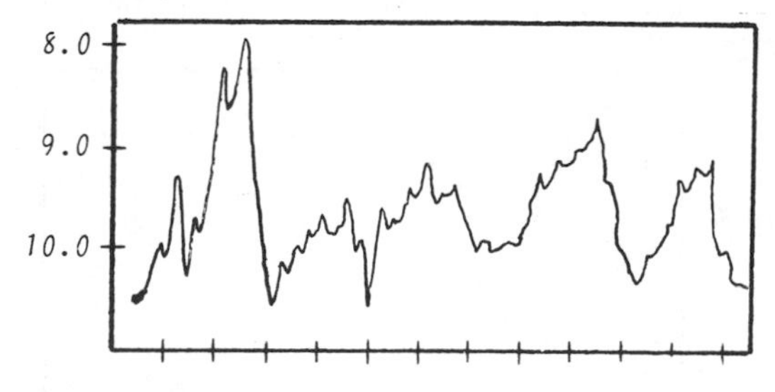

Example: Nebular Variables

UX Orionis	
Maximum magnitude	8.9
Minimum magnitude	10.6
Spectrum	A2e
Right ascension	05h 02m
Declination	−03° 51m
AAVSO No.	053005[a]

[a]Indicates southern declination.

R Coronae Borealis Variables

Perhaps the most unusual of all the variable star classes, the R Coronae Borealis stars are characterized by very long durations at maximum followed by sudden and unpredictable drops in magnitude to minimum. These drops occur at irregular intervals, with the maximum preceding them being smooth with no minor fluctuations. Conversely, the minimum is marked by great fluctuations in light, many of which are very rapid. Maximums have been recorded in these stars that lasted as long as nine years before any change toward minimum took place.

Although the exact cause of the unusual variations in R Coronae Borealis stars is not known, scientists have one hint as to what might be taking place. Changes in the spectrum provide the only clue thus far. From the spectrum of these stars, it appears that tiny particles of carbon (or some other light-absorbing particle) are intermittently obscuring the light we would normally see. This is made evident by the changing spectrum during the light changes. The main difference, as it now appears, between the R Coronae Borealis stars and the nebular variables is the RCB stars somehow eject the carbon that obscures them and that forms a light-absorbing shell about them. This shell eventually settles back into the stellar atmosphere and dissipates in the intense heat of the star.

FIGURE 12-12. Sample light curve for R Coronae Borealis stars.

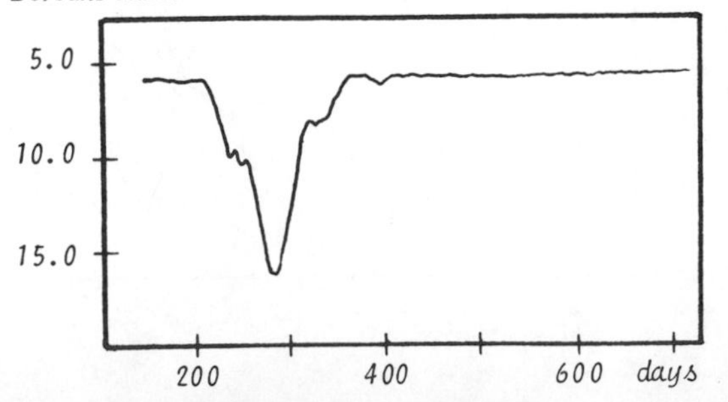

Example: R Coronae Borealis Stars

R Coronae Borealis	
Maximum magnitude	5.8
Minimum magnitude	14.4
Period	?
Spectrum	cG0
Right ascension	15h 47m
Declination	28° 19m
AAVSO No.	154428

The Eruptive Variables

The class of eruptive variables includes the novae and the supernovae, perhaps the most spectacular of celestial events. The class is characterized by stars that are normally quite inconspicuous that rapidly rise by several magnitudes. Several subclasses of eruptive variables exist and are explained on the following pages. Observe these variables at every opportunity and report all unusual activity immediately to the AAVSO.

Dwarf novae. These stars are also known as the *U Geminorum* stars, after the classic example of the subclass. The most popular of these stars is SS Cygni, a spectacular eruptive star. The dwarf novae are characterized by rapid rises to maximum, which occur at erratic intervals, but are not easily predictable. The rise can often appear vertical on the light curve. The maximum is maintained at a somewhat turbulent plateau for a short interval and then drops off suddenly to minimum. The minimum is marked by rapid fluctuations in brightness. During a rise, or at the end of a normal minimum, it is essential that you observe these stars at every opportunity. When these stars are rising or falling near a maximum, you can make observations at 5-minute intervals.

It is believed that all dwarf novae are binary systems, each star being connected to the other by a bridge of flowing gases. The binary consists

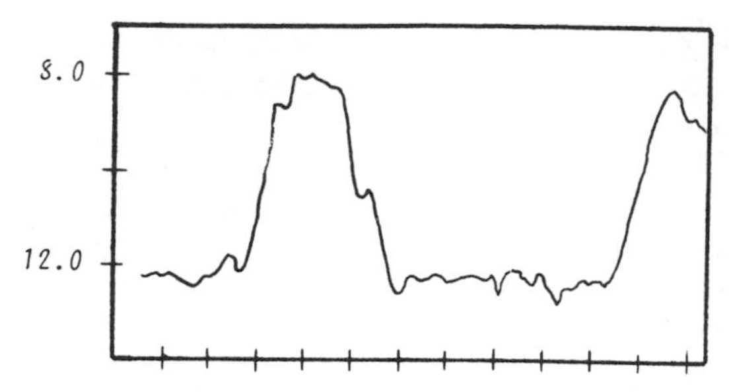

FIGURE 12-13. Sample light curve for dwarf novae.

Example: Dwarf Novae

SS Cygni

Maximum magnitude	8.1
Minimum magnitude	12.1
Period (days)	50.4 (avg.)
Spectrum	A1 + G
Right ascension	21h 41m
Declination	43° 30m
AAVSO No.	213843

of a solar-type star and a white dwarf; the gas flow is from the normal star to the dwarf. Outbursts appear to be caused by an increased gas flow from the solar-type star, which raises the temperature of the dwarf by the heat produced when the "fresh" gases fall into the outer layers of the dwarf. The only difference thus far known between the dwarf novae and the Z Camelopardalis stars (discussed below) is the shape of their orbits. It is not known if there are any different physical processes at work in each star; any that are apparent are not fully understood. The orbit of the dwarf novae binary is typically nearly circular; that of the Z Camelopardalis binary system is typically highly elliptical.

The Z Camelopardalis stars are a poorly understood subgroup of the dwarf novae stars. In many ways they resemble the U Geminorum variables discussed above. Generally, the Z Cam stars are very faint, like the U Geminorum stars, but the periods of variation are much shorter in the Z Cam stars. Many times, after maximum, the star does not fade all the way to a minimum. Rather, it reaches a midpoint in its decline and then it stays on a plateau for a long time before fading any more into minimum. Sometimes the plateau period can last several months, followed by a very gradual decline in magnitude to a "normal" minimum brightness.

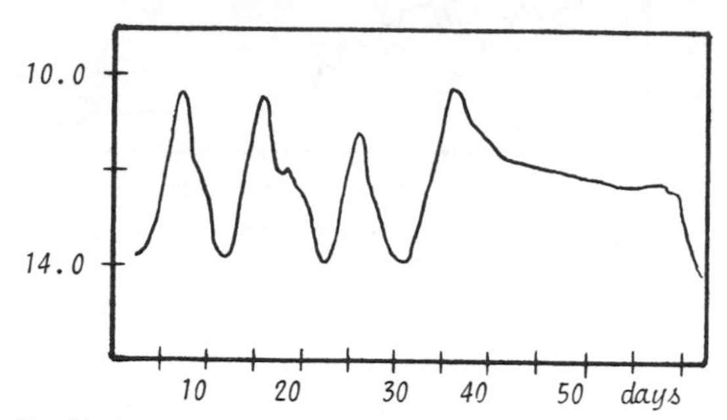

FIGURE 12-14. Sample light curve for Z Camelopardalis variables.

Example: Z Camelopardalis Variables

Z Camelopardalis

Maximum magnitude	10.2
Minimum magnitude	13.4
Period (days)	20.0 (avg.)
Spectrum	Peculiar
Right ascension	8h 19m
Declination	73° 17m
AAVSO No.	081473

The Novae

These spectacular stars are characterized by rapid increases in light from a star generally not recorded previously as having variable activity. Many reach naked-eye visibility, and many have been found by amateur astronomers. An average of 25 novae occur each year, and most are concentrated in the Milky Way. There are three main groups of novae, mostly derived from the shapes of their light curves.

Rapid novae. The rapid novae are characterized by a rapid increase to maximum. The maximum lasts only a short time, followed by a rapid drop, although the drop is not as fast as the initial increase. A transition phase (Figure 12-15) follows the decline and is characterized by a gradual slope with marked and rapid fluctuations. By definition, a rapid nova is one that fades 3 magnitudes from maximum in 100 days or less. The amplitude in the rapid nova can be as great as 13 magnitudes.

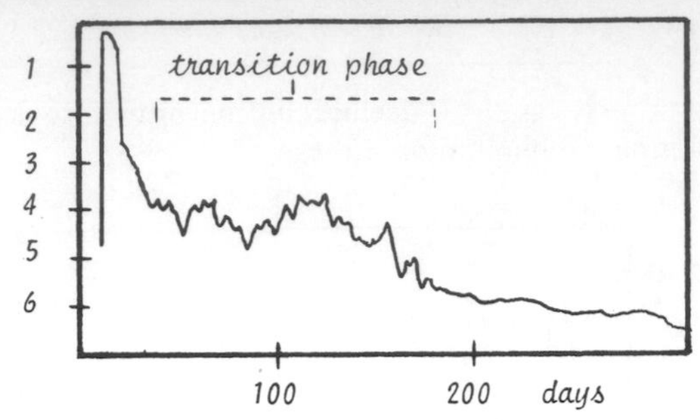

FIGURE 12-15. Sample light curve for a rapid nova.

Example: Rapid Nova

GK Persei 1901

Maximum magnitude	0.2
Minimum magnitude	14.0
Spectrum	Q
Right ascension	3h 24m
Declination	43° 30m
AAVSO No.	032443

Slow novae. By definition, the slow novae require longer than 100 days to decline at least 3 magnitudes from maximum. The maximum of the slow nova is not as sharp as in the rapid nova, and it may last as long as several months. Occasionally during the initial decline there will be a secondary slow rise to maximum, although not as bright as the first. The decline from this secondary maximum is slow. Total amplitude is less than with rapid novae.

FIGURE 12-16. Sample light curve for a slow nova.

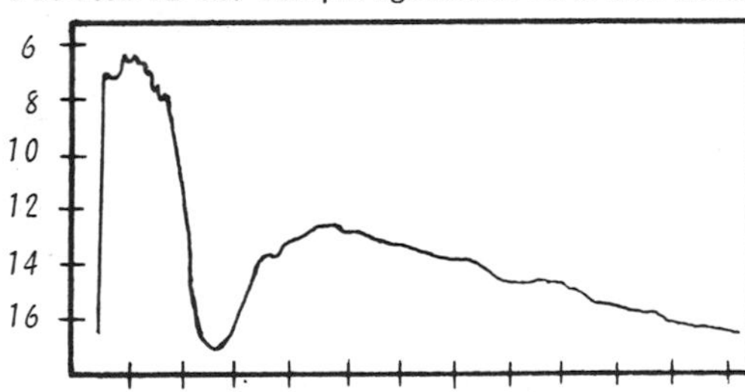

Example: Slow Nova

HR Delphini

Maximum magnitude	3.6
Minimum magnitude	12.4
Spectrum	Q
Right ascension	20h 40m
Declination	18° 45m
AAVSO No.	203718

Ultra slow novae. The ultra slow novae are similar to the slow novae, but years are required for total emergence through the transition phase. The minimum after eruption of these novae will not be as faint as the star was originally, but the light will bottom out at this magnitude. During minimum, there will be few minor oscillations; those that do occur will be slow and infrequent.

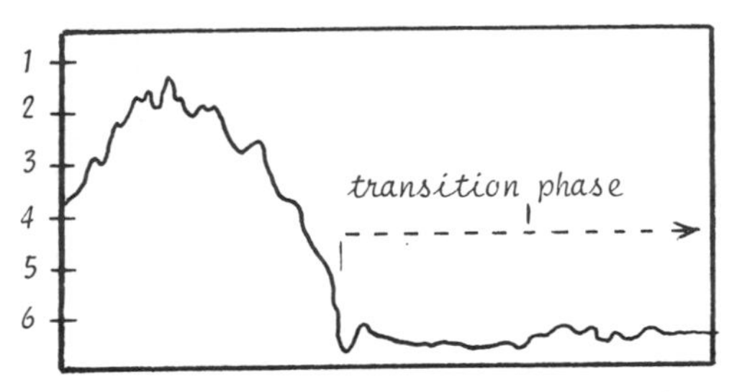

FIGURE 12-17. Sample light curve for an ultra slow nova.

Example: Ultra Slow Nova

FU Orionis

Maximum magnitude	9.9
Minimum magnitude	16.1
Spectrum	cF5
Right ascension	05h 42m
Declination	09° 03m
AAVSO No.	053909

Recurrent novae. The recurrent novae are stars in which outbursts occur somewhat frequently, although not predictably. The amplitudes are relatively small when compared to the

FIGURE 12-18. Sample light curve for a recurrent nova.

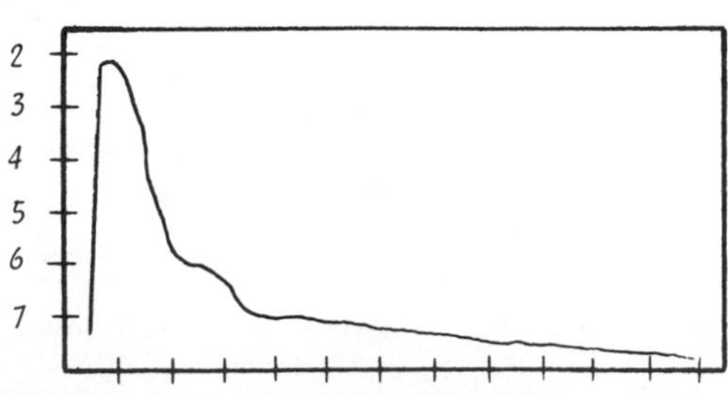

Example: Recurrent Nova

T Coronae Borealis

Maximum magnitude	2.0
Minimum magnitude	11.0
Spectrum	Q + M3
Right ascension	15h 57m
Declination	26° 04m
AAVSO No.	155526

other classes of novae, because the recurrent nova is apparently in a constant state of unrest. The pattern of the light curve during an outbreak is often identical in most respects to each outburst preceding it. During the decline, many rapid and unpredictable variations occur. This important class of eruptive stars deserves your careful attention. Make observations nightly, and more frequently during an outburst.

The Supernovae

The light curve of the supernova is similar to that of the nova, but the energy output and the amplitude of the supernova are many times greater. Even at a distance of 10,000 light years, a supernova would appear to be as bright as Venus; the amplitude could exceed 25 magnitudes. Stars that have used up the nuclear fuel of their cores cannot keep their fires going and so collapse inward, heating their outer layers into fusion reactions that consume a billion years' worth of hydrogen in a minute or so, resulting in the supernova event. All supernovae are distinctly grouped into two types, as follows:

Type I. Type I is characterized by very sharp maximum, followed by a slightly less rapid decline, which becomes progressively slower in time.

Type II. The maximum of Type II is not as sharp as that of Type I. The maximum lasts as long as 80 days, followed by a very rapid decline.

FIGURE 12-19. Type I supernovae.

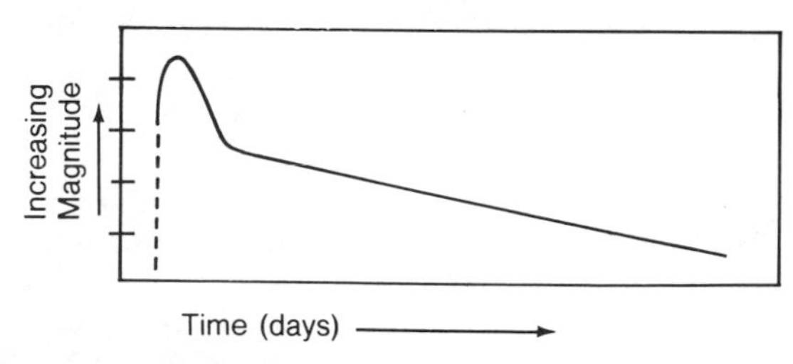

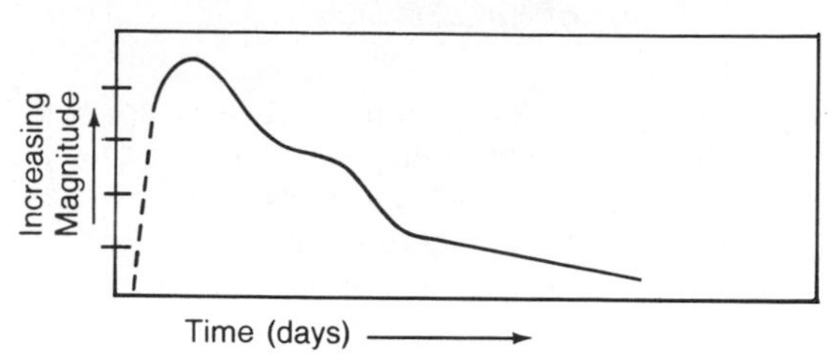

FIGURE 12-20. Type II supernovae.

Flare Stars

This minor group of eruptive variables consists of red dwarfs from which brightness increases are seen that last only a fraction of a day. The outbursts on these stars are apparently caused by actual solar-type flares of extremely high luminosity. The flare stars are extremely active; monitor them as often as possible. The flares occur almost instantaneously and recede just as quickly; make your observations at 5-minute intervals should an outburst occur.

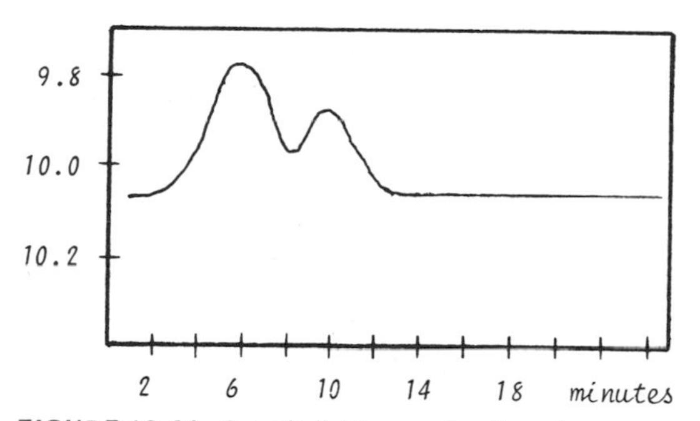

FIGURE 12-21. Sample light curve for flare stars.

Example: Flare Stars

EV Lacertae

Maximum magnitude	8.2
Minimum magnitude	10.2
Spectrum	dM4
Right ascension	22h 45m
Declination	44° 04m
AAVSO No.	224243

Secular Variable Stars

Among the brightest stars in the sky are many that have changed in brightness by at least 2 magnitudes in the course of history. Ancient records of the sky indicate that there are many stars that once were much brighter than they are today, and that just as many have risen to brightness only in modern times.

Other than historical importance, these stars offer no scope for research for either the amateur or the professional. Nonetheless, the information derived from ancient literature is fascinating for those persons pursuing this subject.

A SEARCH FOR NOVAE AND SUPERNOVAE

The thrill of *discovery* in astronomy is second to no other experience, and it is why the comet searcher religiously scans the skies in the early mornings of the year, spending hundreds or thousands of hours peering at the same regions of sky. All the time is well spent, but not forgotten, at the moment a new object is found.

To help in understanding the physics involved in the evolution of stars, no event is as important as the explosion of a star—the nova or supernova. These phenomena can occur instantaneously, and valuable records can be lost forever if the beginning of the nova is not monitored by the professional astronomer. Consequently, many amateur astronomers dedicate themselves to searching methodically through the sky for these new stars, either photographically or visually. If they are motivated by serious purpose, eventually they will find a previously undiscovered nova or supernova giving the professional a jump on a newly erupted star.

Being Ready for the Nova Event

It is likely that no amateur will be fortunate enough to be viewing, at just the right time, a starfield out of which one star will rapidly increase in brilliance by a magnitude of thousands. The rise to maximum light of the nova is quite rapid, requiring only hours to increase perhaps as much as 15 to 20 magnitudes.

For discovery work, you should be concerned only about detecting a new nova as soon after the event takes place as possible. Others may jointly discover and report the new star, but it takes no worth away from your discovery. It is quite important for amateur astronomers to continue to monitor a nova as long as it is visible.

Maximum light of a nova remains for only a short time. If it is a very bright outburst, this maximum light is the signal that something unusual is happening in the nighttime sky, and draws your attention to the area in which the new star is located. However, the outburst soon subsides, and a bright nova can be quickly lost in a field of third and fourth magnitude stars if you are not familiar with the star fields and patterns of the constellations.

Such a bright nova can result in the accidental discovery of the star by perhaps hundreds of independent observers. The competition and chance of being first to report such a discovery is slim, and it is not this type of discovery that we are dealing with in this discussion. Rather, we should concentrate on the planned, scheduled, and methodical searching of specific areas over intervals of time for novae that are at or closer than the naked-eye limit. Of the many novae that occur each year in our galaxy, only a small fraction are ever detected. The brightest ones are discovered usually by happenstance while those closer than the naked-eye limit go largely unnoticed.

Galactic novae, or those occurring in our galaxy, occur more frequently in the dense clouds of the Milky Way. However, the detection of the novae in these regions is very difficult because of the enormous numbers of stars in those regions. Ordinary novae that occur in other galaxies are well beyond the range of ordinary amateur equipment. However, the most violent of all stellar events—the supernova—can be, and often is, detected in earth-based telescopes even though the event occurs in galaxies millions of light years distant.

A Visual Program For the Detection Of Galactic Novae

If you have only small instruments, or even only binoculars, you can still search for novae within our galaxy with some degree of success.

If you have a telescope, each night select an area of the sky and examine it visually at low power and with as wide a field of view as possible. Although some type of star chart is necessary, you might wish to concentrate on three or four general areas. Sketch the field of view as it appears in the telescope. This method has the distinct advantage that all stars seen in the telescope will be on the chart you have drawn. If later you see any that do not appear on the chart, they are candidates for further scrutiny.

Concentrate on the densely packed Milky Way regions of our galaxy; the known novae occur more frequently there than in the sparsely populated areas of sky. If you choose and examine four areas at least every other night, then you are not overtaxed with work. About 15 minutes for each area will be required once you become familiar with the star fields. It is remarkable how well you can memorize a star field once you have scrutinized it a few times.

If you have only binoculars or your naked eye, you can discover bright novae if you examine the sky each night as soon as twilight ends. For such naked-eye searches, concentrate on regions of the Milky Way and learn the constellations and the stars within them to the naked-eye limit. A great help is a good star atlas such as Hans Vehrenberg's *Handbook of the Constellations*, which has each constellation indexed for quick reference and shows stars down to the naked-eye limit.

A Photographic Program For Galactic Novae

A more consistent and more rewarding search for galactic novae can be initiated through photographic searching. The considerable benefits of such a program include the following:

- ☆ No memorization of star fields is necessary.
- ☆ The photograph provides a permanent record of any discovery, from which a magnitude and position may be derived.
- ☆ Examination of the field can be done at leisure.
- ☆ More stars can be recorded on the photograph than can be discerned by the human eye.
- ☆ The chance of discovery increases because the number of stars on the photograph is greater than can be seen by the eye.

Attempting to photograph a star field through an average telescope in search for novae is a perplexing and tedious task. A better method is to photograph as wide a field as possible on a single photograph and reach perhaps magnitude 10 or 11.

The Equipment Necessary

To systematically search for novae using the camera, only two things are required—a camera and some provision for tracking on the stars. A simple clock-driven equatorial mounting provides accurate tracking for standard 50-mm camera lenses for up to 10 min if the polar axis has been accurately aligned to north and set at the proper latitude of your station. However, telephoto lenses are advantageous for this study, and you must have some way to correct the tracking of the instrument. Thus, the *piggyback method* is recommended. Lenses of 100-mm to 200-mm focal length are recommended for the search, and exposure times on fast (ASA 400) black-and-white or color film should be around 10 min. If your observing station is located in a bright suburban area, exposure times should not be greater than about 5 min, although the limiting magnitude achieved will be considerably less than if the photograph were taken in dark skies and exposed longer.

The Method of Search

Search the sky for novae with two successive photographs, taking them about 10 days apart. Any new object in the 10-day interval will appear on one photograph but not on the other, and thus give you initial findings that something did appear in that interval. If so, you will need furthur confirmation using a telescope visually.

This all sound rather simple until you try to compare two photographs, each containing thousands of stars. And remember, the *faint* stars are the ones under surveillance! No one has time for inspection of each star, one by one, so you must make some other provision. The professional astronomer uses a large desk-mounted instrument known as a *blink comparator*, in which two negatives centered on the same star field can be alternately compared through a binocular viewing eyepiece. If the two negatives are precisely lined up in the comparator, the effect is that the field is merely flashing on and off rapidly as the comparator shows first one negative and then the other. However, if a new object is apparent on one film and not the other, that new object will alternately flash on and off as the comparator shows first the negative with the object and then the negative without it. This method is fast and reliable, but the instrument is quite expensive and beyond the budget of most amateur astronomers.

Therefore, as in other cases of amateur research, some makeshift method must be devised by which you can accomplish similar results. Several small portable blink comparators aimed at the amateur astronomer have entered the market in recent years. Most of these are adequate, although somewhat difficult to use.

A simpler method is to devise a light box using a translucent material on top of which you can position the two negatives (or color transparencies). Then precisely aim two viewing eyepieces, one at one slide or negative and the other at the second film. Install within the box, under the translucent material, a timer (either mechanical or electronic) that switches the current back and forth between a light under both slides (or negatives). Both of your eyes are not viewing the same object (the film), so it is necessary that you be in a very dark room. The method is crude and not nearly as efficient as the professional blinkers.

Another method that has had a great deal of success was designed by the California amateur astronomer, Ben Mayer. This unique system allows you to compare two color transparencies (or 35-mm black-and-white negatives mounted in slide mounts) alternately as they are projected through two different slide projectors, each of which is aimed so that it centers atop the other as they are projected onto the same screen. The method has the nice distinction of allowing you to sit back and enjoy the view as the slides flash onto and off of the screen. As in the blink comparator, this method makes an object visible on one film but not on the other so that only one film appears to be blinking on and off.

Any method of comparing one film to the other requires exacting care to align the camera on the star field at the time the photograph is taken. The orientation and the center of each photograph should be the same, and of course, you must use the same optical system throughout. The high-speed Ektachrome 400 (Kodak) is preferred over black-and-white film in that the field will be dark when viewed by whatever comparator is used. Black-and-white negatives, when projected, give a stark white background that quickly tires the eyes.

Studying External Galaxies For the Supernovae

Although only one supernova per century on an average is thought to occur in our galaxy, there is a way in which a hundred times that many supernovae explosions can be monitored in the same length of time by astronomers: Look at the events in a hundred galaxies.

There are at least 100 galaxies within reach of modest amateur equipment that are extensive enough to show any supernova that might occur within them. When most supernovae occur in these external galaxies, they appear as very faint stars embedded in the seemingly nebulous outer portions of the galaxy. The brightest are on the order of magnitude 12 or 13, so telescopes of 8 inches (20 cm) or larger are required to see them visually.

To search for supernovae in external galaxies requires that you know the appearance of the galaxies as they should appear in your telescope. Photographs of galaxies are often published that have had the central regions burned in in an effort to expose the outer regions of the galaxies.

Any bright knots or star-like images that might normally be seen visually are thus obliterated. Consequently, it is necessary that you set up a schedule for recording the fields of the galaxies that you plan to observe before you start your actual observations.

Recording the Galaxy Field

You should choose a number of galaxies large enough to allow you to examine 15 galaxies for supernovae each night. With the great number of galaxies visible in our spring skies, there is a predominance of excellent subjects during that season. At the advent of the project, examine each galaxy on your list, preferably in sequence by right ascension. Using a circle about 4 inches (10 cm) in diameter, sketch the field as you see it through the telescope eyepiece, including every star within the field of view.

Because most galaxies appear small to us, draw the galaxy field with moderate magnification, say about 100x to 150x, and only on the very best night. It will take you much longer to make the initial field sketch than it will to search for the galaxies. Draw the fields with the telescope that you will use for the searches.

Once you have drawn the fields, it is a simple matter to examine as many galaxies in sequence as possible or practical for the night. Eventually, you will become so familiar with the fields that referring to the charts you have drawn will be necessary only when you find an unfamiliar object.

Galaxies Recommended For Supernovae Surveillance

The galaxies listed in Table 12-3 are recommended as part of a regular patrol for the possible supernova event. Be aware that many of them might have star-like points near the galaxy. By all means record the points on your initial drawing. Monitor each galaxy on every available night because the detection of a supernova in its early stages is of the utmost importance to the professional astronomer. Notice that only *spiral* galaxies are listed in Table 12-3; the supernova is considerably more common in the spiral than in the elliptical galaxies.

CONCLUSION

The study of variable stars is well suited to the equipment and interests of any amateur astronomer, yet the observations are meaningless unless they are coordinated with the efforts of other observers and recorded with those persons who can use them the best. It is interesting that from 1572 to 1850 only 30 variable stars were known and monitored. Today there are an incredible 25,000 known variable stars. It is easy to see that the history of the variable star—and that what is done today will later become the history of that star—cannot be left to the professional astronomer. Many stars vary within a minute's time, whereas others require years to complete their cycles. Still the results slowly accumulate, and the dedicated amateur astronomer compiles a vast record of the star's life history.

If you wish to organize an efficient program for observing variable stars, I urge you to participate in membership in one of the two clearinghouses that provide information to the professional. It is through these organizations that the valuable history of a star's life cycle is preserved and made available when necessary.

In the United States

American Association of Variable
Star Observers
187 Concord Avenue
Cambridge, MA 02138

In Great Britain

British Astronomical Association
Variable Star Section
12 Taylor Road
Aylesbury, Bucks HP 21
8 DR, England

TABLE 12-3. External galaxies suited for amateur surveillance of supernovae.

Designation	*Right Ascension*	*Declination*	*Constellation*	*Magnitude*	*Size*	*Type*
NGC 224 (M31)	00h 40m	+41° 00′	Andromeda	4.3	160 × 35′	Sb
NGC 7727	23h 37m	-12° 34′	Aquarius	10.7	3 × 3′	S
NGC 772	01h 57m	+18° 46′	Aries	10.9	5 × 3″	Sb
NGC 5248	13h 35m	+ 9° 08′	Boötes	11.3	6 × 4	Sc
NGC 2403	07h 32m	+65° 43′	Camelopardalis	8.9	17 × 10′	Sc
NGC 2655	08h 49m	+78° 25′	Camelopardalis	10.7	5 × 4′	S
NGC 2775	09h 07m	+ 7° 15′	Cancer	10.7	3 × 2′	Sa
NGC 4258 (M106)	12h 17m	+47° 35′	Canes Venatici	8.9	20 × 7′	Sbp
NGC 4490	12h 28m	+41° 55′	Canes Venatici	9.7	6 × 2′	Sc
NGC 4631	12h 40m	+32° 49′	Canes Venatici	9.3	13 × 2′	Sc
NGC 4736 (M94)	12h 49m	+41° 23′	Canes Venatici	7.9	5 × 4	Sbp
NGC 5005	13h 09m	+37° 19′	Canes Venatici	9.8	5 × 2′	Sb
NGC 5055 (M63)	13h 14m	+42° 17′	Canes Venatici	9.5	10 × 5′	Sb
NGC 5194 (M51)	13h 28m	+47° 27′	Canes Venatici	8.1	10 × 6′	Sc
NGC 1068 (M77)	02h 40m	-00° 14′	Cetus	8.9	6 × 5	Sbp
NGC 4192 (M98)	12h 11m	+15° 11′	Coma Berenices	10.7	9 × 2	Sb
NGC 4251	12h 16m	+28° 27′	Coma Berenices	10.2	3 × 1	Sa
NGC 4254 (M99)	12h 16m	+14° 42′	Coma Berenices	10.1	5 × 4	Sc
NGC 4321 (M100)	12h 20m	+16° 06′	Coma Berenices	10.6	6 × 5	Sc
NGC 4414	12h 24m	+31° 30′	Coma Berenices	9.7	3 × 2	Sc
NGC 4450	12h 26m	+17° 21′	Coma Berenices	10.0	3 × 2	Sb
NGC 4501 (M88)	12h 30m	+14° 42′	Coma Berenices	10.2	6 × 2	Sb
NGC 4559	12h 34m	+28° 14′	Coma Berenices	10.6	11 × 5′	Sc
NGC 4565	12h 40m	+26° 16′	Coma Berenices	10.2	14 × 2′	Sb
NGC 4571 (M91)	12h 34m	+14° 28′	Coma Berenices	10.9	3 × 2′	S
NGC 4725	12h 48m	+25° 46′	Coma Berenices	8.9	10 × 6′	Sb
NGC 4826 (M64)	12h 54m	+21° 57′	Coma Berenices	8.8	7 × 3	Sb
NGC 6946	20h 34m	+59° 58′	Cygnus	9.4	9 × 7′	Sc
NGC 6503	17h 50m	+70° 10′	Draco	9.6	5 × 1	Sb
NGC 5236	13h 34m	-29° 37′	Hydra	10.1	10 × 10	Sc
NGC 2903	9h 29m	+21° 44′	Leo	9.1	11 × 5′	Sb
NGC 3351 (M95)	10h 41m	+11° 58′	Leo	10.4	6 × 4′	Sb
NGC 3368 (M96)	10h 44m	+12° 05′	Leo	9.1	5 × 4′	Sbp
NGC 3521	11h 03m	+00° 14′	Leo	9.5	7 × 4′	Sc
NGC 3623 (M65)	11h 16m	+13° 23′	Leo	9.3	8 × 2′	Sb
NGC 3627 (M66)	11h 18m	+13° 17′	Leo	8.4	8 × 3	Sb
NGC 2859	09h 21m	+34° 44′	Lynx	10.7	5 × 4′	SBa
NGC 3344	10h 41m	+25° 11′	Lynx	10.4	8 × 6′	Sc
NGC 7331	22h 35m	+34° 10′	Pegasus	9.7	10 × 2	Sb
NGC 628 (M74)	01h 34m	+15° 32′	Pisces	10.2	11 × 9′	Sc
NGC 598 (M33)	01h 31m	+30° 24′	Triangulum	6.7	65 × 35′	Sc
NGC 2681	08h 50m	+51° 31′	Ursa Major	10.4	2 × 3	Sa
NGC 2841	09h 19m	+51° 12′	Ursa Major	9.3	6 × 2	Sb
NGC 3031 (M81)	09h 52m	+69° 18′	Ursa Major	7.9	21 × 10	Sb
NGC 3034 (M82)	09h 52m	+59° 56′	Ursa Major	8.8	9 × 4′	P
NGC 3556 (M108)	11h 09m	+55° 57′	Ursa Major	10.7	8 × 2′	Sc
NGC 3941	11h 50m	+37° 16′	Ursa Major	9.8	2 × 1	Sa
NGC 3992 (M109)	11h 55m	+53° 39′	Ursa Major	10.8	6 × 4	Sb
NGC 5457 (M101)	14h 01m	+54° 35′	Ursa Major	9.6	22 × 22′	Sc
NGC 4216	12h 13m	+13° 25′	Virgo	10.4	7 × 1	Sb

TABLE 12-3. (cont'd)

Designation	Right Ascension	Declination	Constellation	Magnitude	Size	Type
NGC 4303 (M61)	12h 19m	+04° 45′	Virgo	10.1	6 × 4	Sc
NGC 4569 (M90)	12h 34m	+13° 26′	Virgo	10.0	8 × 2′	Sc
NGC 4579 (M58)	12h 35m	+12° 05′	Virgo	9.2	5 × 4′	SB
NGC 4594 (M104)	12h 37m	-11° 21′	Virgo	8.7	6 × 3′	Sb
NGC 4699	12h 47m	-08° 24′	Virgo	9.3	3 × 2′	Sa
NGC 5566	14h 18m	+04° 11′	Virgo	10.4	6 × 1′	Sb

REFERENCES

American Association of Variable Star Observers, *Report 30: Light Curves of Long Period Variables*. Cambridge, MA, August, 1975. Paperback.

Campbell, Leon, and M.W. Mayall, *Studies of Long-Period Variables*. Boston, 1975.

Chandrasekhar, S., *Principles of Stellar Dynamics*. Chicago: University of Chicago Press, 1942.

Clark, David H., and F.R. Stephenson, *The Historical Supernovae*. London: Pergamon Press, 1977.

Gebbie, K.B., and R.N. Thomas *Wolf-Rayet Stars*. Washington, DC, National Bureau of Standards. 1968.

Kopal, Z., *The Computations of Elements of Eclipsing Binary Systems*. Cambridge, MA: Harvard Observatory Publications, 1950.

Glasby, John, *The Dwarf Novae*. New York: American Elsevier, 1970.

——, *The Variable Star Observer's Handbook*. New York: Norton, 1971.

Merrill, P.W., *The Nature of Variable Stars*. New York: Macmillan, 1938.

NASA. *The Supernovae*, EP-126, Washington, DC: Government Printing Office, 1976. Paperback.

Page, Thornton, and L.W. Page, *The Evolution of Stars*. New York: Macmillan, 1968.

Payne, C.H. *The Stars of High Luminosity*. New York: Harvard College Observatory, 1930.

Payne-Gaposchkin, Cecelia, *The Galactic Novae*. New York: Dover Publications, 1964. Paperback.

Rosseland, Svein, *The Pulsation Theory of Variable Stars*. New York: Dover Publications, 1964. Paperback.

Sandage, Allan, *The Hubble Atlas of Galaxies*. Washington, DC: Carnegie Institute of Washington, 1971. Paperback.

Sherrod, P. Clay, Variable Star Section, 1975. Project. Publ. #002 of the Midsouth Astro. Res. Society. Little Rock, 1976.

Strohmeier, W., *Variable Stars*. Vol. 50, International Monographs on Natural Philosophy. London: Pergamon Press, 1972.

13

AN INTRODUCTION TO PHOTOELECTRIC PHOTOMETRY

Consider what we do as amateur astronomers: *everything* we see, all the data we record, and the very instruments we depend on for our studies are governed by the entity we recognize as *light*. The delicate colorations of the planets Mars and Jupiter, the bursting of a comet's nucleus, or the violet train in the wake of a swift meteor—in all cases, we examine the light of the subject for our evaluations.

Occasionally the light from our subjects changes. If the change is great enough in intensity, our eyes might detect the change. But if it is only slight, we will not notice it. Sometimes a star or an asteroid might be subtly shaded in coloration, a characteristic that is quite important in our understanding of the physical nature of these objects. But human eyes do not see many hidden hues or changes, and our minds are not always objective in their definition of color. Many people perceive celestial objects as being red, whereas others see the same object as yellow. Such are the problems encountered in interpreting data recorded by visual observers.

Until the 1970s, amateur astronomers have not had available to them the sophisticated instrumentation necessary for the delicate measurement of the qualities of light. The professional spectroscope—the valuable tool capable of detecting unseen stars in orbits around other known stars and used to study the velocities of great galaxies as they fly away from our vantage point—is not suitable for amateur research. The tiny spectroscopes that *are* available to amateurs are capable of nothing but diffracting the light of a star into a rainbow of color. The spectroscopes used by the professional astronomer in the studies of the lights from celestial objects are so massive that the spectroscope alone weighs more than the amateur's telescope. Large-aperture professional telescopes are necessary, not only to carry the large spectroscopes, but also to define and resolve the fine absorption and emission lines superimposed on an object's spectrum. Such definition is beyond the reach of amateur equipment, particularly for the faint celestial objects suitable for research projects. It is these tiny lines within the spectrum that tell the story of star evolution, the duality of a faint close double star, and the rotation of a great spiral galaxy.

There is a better way that is adaptable to the amateur's efforts and means. Rather than measuring precise increments of a complete spectrum, as is done when the spectroscope is used, you can take measurements of the intensity of the light in about five or six selected regions and can compare them to present a good representation of the object's total spectrum. Such a method is much like looking at five pieces of a puzzle and imagining in your mind or sketching on paper what the puzzle would look like when completed—but without putting the pieces together physically. Suppose as well that this instrument has the capability, unlike the spectroscope, to detect and indicate subtle changes in the light that otherwise would be invisible to the human eye. An instrument of these characteristics is capable of the following:

☆ Establishing the exact magnitudes of stars, using the total integrated light of all wavelengths, as compared to a standard star.

☆ Detecting and monitoring slight changes in the brightness of variable stars, changes from which more information is needed to interpret the physical processes of that object.

☆ Measuring the regions of the spectrum—the ultraviolet, the blue, yellow, red and infrared regions—to reach better insight into the temperatures and constitutions of celestial objects.

This instrument is the *photoelectric photometer*.

THE NATURE OF THE PHOTOELECTRIC PHOTOMETER

Early attempts were made to compare the light of something known to that of something unknown. Instruments that could compare magnitude of stars' light were made late in the nineteenth century. Early stellar photometers used the light of a burning lamp as a comparison for stars in telescope fields. By using prisms to converge the beams of light of both the wick and the star into the same field of view, astronomers could compare a star by simply reducing the burner's light until it properly matched the intensity of the star. This was usually done by inserting into the photometer a disk into which very small holes of various diameters had been drilled. By rotating the disk in sequence, the light was reduced as the size of the hole in the disk became smaller.

However, immediate problems arose when using such a device. Similar *comparison photometers* are used by amateurs although today, rather than using a kerosene burner as a source of light, the modern device uses a small electric lamp. Regardless of the source of light, the observer is at the mercy of the temperature—and thus the *color*—of the comparison light. It is difficult to compare a bright white star with a dull red or orange source, as the flame of a wick burner would be. Similarly, the intensity of such comparison sources cannot be adequately controlled. If the wick of the burner is allowed to burn down, the intensity will likewise diminish.

Modern photometers are improvements on the basic theme of the need and function of an instrument to measure color and brightness. Only recently has the development of electronic detectors and amplifiers made it possible for weak signals from faint stars to be magnified in such a way that their light can cause the deflection of a needle (i.e., the pen) of a galvanometer. The instrument consists of five to six basic units, each closely matched to record the subtle light and color of celestial objects.

The Telescope For Photoelectric Photometry

As in most studies in astronomy, there is not any one telescope particularly suited solely for the study of photometry. Some telescopes, however, are better suited for the study than others. Avoid using refractor, or lens, telescopes for serious photoelectric astronomy, primarily because of the great degree of *chromatic aberration* present in even the best refractor lenses. Because the index of refraction increases with the frequency of light, violet (of all visible colors) comes to a focus closest to the lens, and red does so the farthest, with the other colors in between. Hence, although you may be *exactly* focused for yellow, the blue and red images will be far enough out of focus to appear as annoying colored fringes. Consequently, the color-sensitive photoelectric photometer will record best the color of light that is best focused into the detector. It will not record all of them equally well. Achromatic (i.e., two-element) refractor lenses reduce somewhat the color problems of the telescope but do not eliminate them. Similarly, apochromatic (i.e., four elements, usually) refractors are somewhat better but not perfect in color focusing.

A Newtonian or Cassegrainian telescope is best suited for photometry because the light is only reflected, not refracted, light. Even some of the commercial catadioptic, or compound (Schmidt–Cassegrain or Maksutov), telescopes with glass correction lenses, suffer from a greater degree of chromatic abberation than do the Newtonian or Cassegrain reflectors.

Even with the most sophisticated photoelectric photometers affordable by the amateur astronomer, the magnitude limit attainable in modern size telescopes can be at first somewhat disappointing; as is shown in the following table.

Telescope Aperture	*Magnitude Limit (V)*[a]
6 in. (15 cm)	9.0
8 in. (20 cm)	10.0
10 in. (25 cm)	10.5
12 in. (30 cm)	11.1
14 in. (35 cm)	11.7
16 in. (40 cm)	12.4

[a]Magnitude limit given is for light at 500–650 nm, or V filter, which is the closest comparison to visual estimations.

If you do colorimetry work, using the U, B, V, R, and I sequence, you will find that somewhat brighter stars are the limit in the blue and ultraviolet regions, whereas the red and infrared transmissions allow for an extended limit in magnitude threshold.

The ideal telescope for photoelectric photometry has a long focal length so that the increased scale of the image means that less field needs to be centered in the region of the star being measured. This results in considerably less scatter from background illumination (e.g., faint stars not seen visually, background sky glow, etc.), making the readings considerably more consistent and reliable. For this reason, a Cassegrain telescope (with a focal ratio usually between f/12 and f/20) is preferred over a Newtonian (f/5 to f/8).

Another consideration in your choice of telescope for such work is the mounting. A heavy detector mounted onto the eyepiece rack-and-pinion focuser will tend to cause severe balancing problems in German equatorially mounted instruments, particularly if they are of the Newtonian design. This is because the focuser is mounted onto the side of the telescope tube, which results in severe offset of the dynamic balance. A fork mounting, equipped with slow-motion controls in both axes, is the preferred mounting for photoelectric photometry, because the balance of equipment on the bottom of the instrument can easily be offset (see Figure 13–1).

The ideal telescope for photoelectric study is a large-aperture Cassegrain reflector of f/16 or f/20, on a fork mounting, complete with slow motions for centering the suspect star in the detector unit. This is not to say that other types and sizes cannot be used. They can *all* be used to varying degrees for stellar and cometary photometry. But you must be aware of the problems inherent in physically attaching the photoelectric photometer to other types, and you must know what aspects of correction must be applied when you use telescopes in which refraction and chromatic aberration are evident.

The Telescope Coupler and Detector Unit

The heart of the photometer system is that portion which receives the light and changes that light into some slight electrical current. This device serves a dual purpose: first to couple the photoelectric instrument to the telescope, and second, to transform light energy into electrical energy. You can think of this unit simply as a replacement for the eyepiece, as well as for the human eye. It consists of a variety of component parts, as discussed in the following paragraphs.

The aperture diaphragm. The aperture diaphragm is located at the focal plane of the telescope objective, which is simply a tiny hole

FIGURE 13-1. A commercially made photoelectric photometer mounted to a Schmidt–Cassegrain telescope with fork mounting. The telescope coupler–detector is affixed to the telescope, whereas the wire feeds the weak current generated in the detector to an amplifier below the telescope. A chart recorder (at lower left) records the signal from the celestial source. (Photograph by the author)

FIGURE 13-2. A much simplified illustration of the telescope coupler–detector unit of a photoelectric photometer. (a) Standard mount to be affixed to telescope; usually mounts in focuser. (b) Flip mirror to divert image to viewing eyepiece for centering. (c) Diaphragm for isolating subject and eliminating stray light. (d) Fabry lens. (e) Filters for standard UBVRI indexing. (f) Photomultiplier tube (which requires cooling chamber), or photodiode. (g) Viewing eyepiece, which is not usable during measurement.

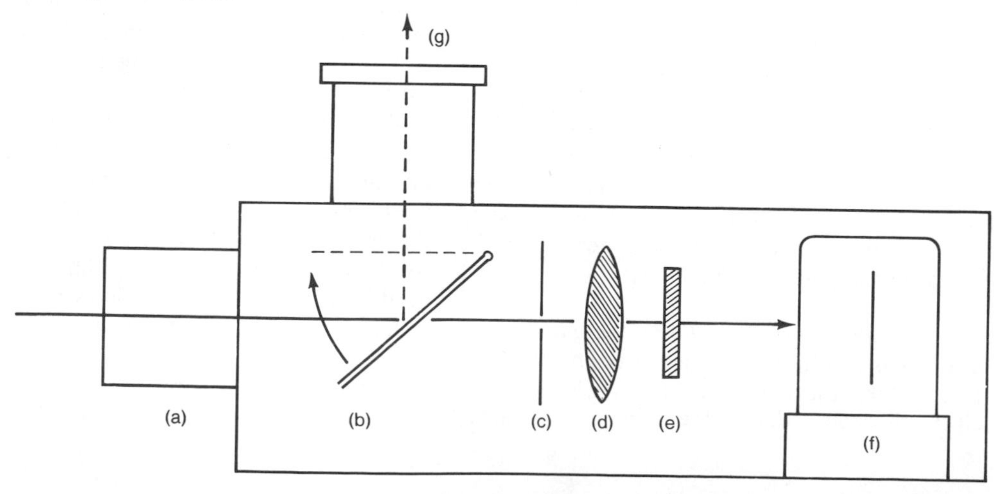

that allows only the immediate area of the star to pass into the detector, thus eliminating background illumination. Its purpose is to make sure that only the light of the object to be measured is transferred to the photocell or photodiode. It is important that the diaphragm opening be large enough that no light from the star's Airy disk is interrupted, yet not so large that an excessive amount of sky glow is also detected and transmitted to the amplifier. Most apertures of diaphragms are not less than 5″ (angle of sky entered). In general, the fainter the subject you wish to measure, the smaller the aperture opening you should choose.

The Fabry lens. The Fabry lens is directly behind the tiny aperture diaphragm. This lens takes the light formed by the objective and forms a focused image on the photocell, or the photodiode. Because not all photocells and photodiodes are perfectly uniform in the amount of current that is generated from different parts of their surfaces, something must smooth out the star's light and average it out over the entire surface of the detector. Otherwise, some fluctuations in the readings of the star's light will occur as the light falls on differing areas of the photocell. This Fabry lens also has the effect of reducing scintillation effects that would cause the star's reading to flicker.

Filters. Most, but not all, photometric equipment is equipped with at least one filter located between the Fabry lens and the photocell. If only one filter is used, it is the "V," or yellow, filter, matching closely the sensitivity of the human eye. Many photometers are equipped with a wheel by which a series of filters can be rapidly interchanged, and five or six color readings of one star can be taken in rapid succession. The filters are not simply colored glass. Each one is a selected narrow pass-band filter, allowing only a narrow portion of the spectrum of the star's light to pass through.

Photocells, or photodiodes. Light-sensitive photocells or photodiodes are used in the detector to turn the energy of the star's light into electrical energy that is measurable on a gauge, once the tiny current is amplified. A common photocell used in many small photometers is the RCA 1P21, which is comprised of an antimony–cesium alloy that is uniformly sensitive to most wavelengths of light. These photocathode tubes' sensitivity to color varies depending on the temperature, and the color of the light changes as the temperature slowly rises. Consequently, serious photometrists cool the tube by encasing it in dry ice to a temperature of −79° Celsius. During such cooling, however, the cell becomes increasingly more sensitive to the blue end of the spectrum, and unsuitable for measuring in red and yellow wavelengths.

More modern (but less efficient to date) detectors to substitute for the photocathodes are the silicon PIN photodiodes. These units, although not as blue-sensitive as a comparable photocathode, have the distinct advantage of not being temperature-variable. Because they are solid-state components, little heat is accumulated for normal photometric work. Many new photodiodes are entering the market, and soon these units will replace the tube-type detectors. During the manufacturing process, many of the photodiodes are given built-in extra blue sensitivity. In most cases, a well-chosen photodiode will equal the sensitivity and response of inexpensive photocathodes through all wavelengths from 300 to 1100 nm, when operated at a normal 25° Celsius.

Viewing eyepiece. Some provision must be incorporated into the telescope coupler unit so that the precise angular circle that will be registered by the detector can be centered in the telescope beforehand. A simple viewing eyepiece is normally employed, using a flip mirror in the light path. When you orient the telescope on the proper subject, the mirror intercepts the light path and diverts it into the viewing eyepiece. The eyepiece, normally a Kellner or Ramsden design, is etched with a small circle that represents the field of view of the detector. To activate the photocell or photodiode, merely flip the mirror out of the light path, thereby allowing the starlight to fall onto the detector element.

The signal amplifier. The current generated by a tiny photodiode or even by the most advanced photocathode is incredibly small, amounting to less than half a microampere. Thus, some form of amplification is necessary to turn this weak signal into a usable current. The amplifier of a photoelectric photometer can operate on DC or AC current, the most popular for amateur work being the DC unit.

The amplifier is normally contained in a unit consisting of a selector for controlling the gain, as well as offset controls for reducing "dark current" from the electronics of the system and reducing the noticeable effects of atmospheric glow.

A selector for the brightness of the star to be monitored should be included in any photoelectric photometer design to adjust the gain of the amplifier unit. Low gain would be used for very bright objects giving maximum deflection; very high gain would be for stars near the threshold of the telescope.

There must be some provision for zeroing the galvanometer dial so that the meter can be set to zero at the beginning of each measurement.

An offset control is also frequently installed in amateur photometers, allowing the dark current in the electrometer (detector) to be nulled. Also the glow of the nighttime sky near the subject to be measured can be cancelled with such an offset.

Some provision for transferring the current to an external read-out, such as a strip chart recorder, should be incorporated into any control unit. This allows a graphic display of the nature of a variable star's light changes, as well as a permanent record of activity.

The strip chart recorder. Most strip chart recorders are of the ink type, in which a tiny stylus records on moving strips of paper the activity as recorded by the detector. Such recorders are expensive yet reliable. Recently, some reasonably priced pressure-sensitive chart recorders have come to the market. Those made by Rustrak are reliable and can be equipped with interchangeable gear trains for speeds of up to 1 in. per minute. Using paper that darkens from the heat of friction, these units simply tap a pointed stylus onto the paper once per second, or some corresponding rate, depending on the speed of the gear train. These small units are perfectly suitable for the amateur astronomer's needs.

FIGURE 13-3. The light path through the photoelectric photometer system.

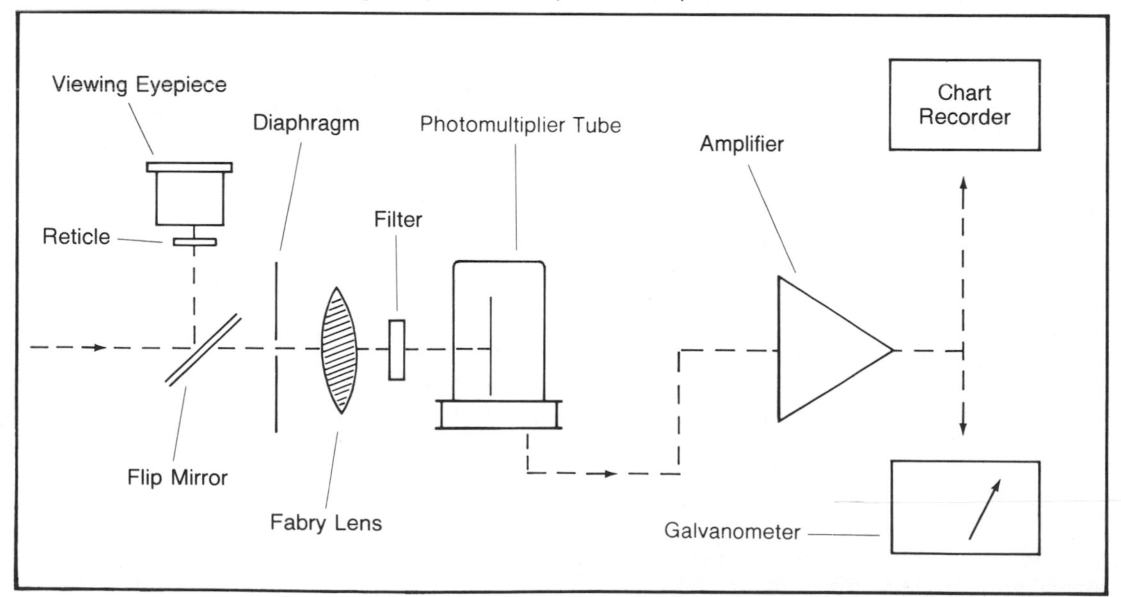

PROJECTS IN PHOTOELECTRIC ASTRONOMY

You can accumulate a great deal of actual research data if you are equipped with a reliable photoelectric photometer, provided that the research protocol is a systematic effort and that you do not do it haphazardly. Sporadic observations are of little value. The areas of endeavor discussed in the following paragraphs are best suited for the amateur's efforts. It is best to choose one area that most interests you and concentrate heavily on that study.

Meteorological Monitoring

A better understanding of local meteorological conditions, particularly the degree of atmospheric absorption and diffusion of starlight, could be an important by-product of serious photoelectric studies. The closer an object is to the horizon, the greater the thickness of air its light must pass through. Therefore, there is a greater absorption of the light. Also, scattering of the light by the air's molecules and other particles, notably dust and moisture, disperses blue more than red (Rayleigh scattering). Consequently, objects will seem to dim considerably as they approach the horizon, and will seem to noticeably redden as well.

Using known standard stars that are constant in brightness, you can conduct tests from your observing site to determine the degree of extinction for various angles from the zenith. For minimum refraction, choose a comparison star at or near the zenith. Choose another star at, say, 30° from the zenith, another at 60°, and one very near the horizon, and determine the proportional amounts of brightness of each. In each case, the comparison star for each angle from the zenith should be somewhat dimmer than predicted, increasing in order by the number of degrees from the zenith. In other words:

Sky Extinction Coefficient = Predicted Magnitude minus Observed Magnitude

Such a study over a year's time can provide valuable insight into the yearly condition of the air in your area. For example, you can probably expect a greater concentration of moisture in spring than in late summer. If so, then some colorimetry work with V magnitudes minus R magnitudes could be of value.

Even if you do not make the effort necessary for meteorological monitoring but if you do keep careful records extinction coefficients (which must be determined anyway for photoelectric observations to be of true value), your records will begin to show seasonal trends valuable to your future endeavors. Determining extinction coefficients for star magnitudes is discussed in a following section.

Photoelectric Photometry Of Minor Planets

Chapter 11, "Visual Photometry of Minor Planets," makes it evident that much valuable data can be obtained by using the photoelectric photometer in place of the eye for measurements of minor planets. In many cases, the brightness of an asteroid can appear constant to the eye, but subtle changes in its magnitude might exist because of irregularities in its shape or albedo.

Of the six known variable asteroids listed in Table 11-1, only 15 Eunomia, 44 Nysa, 216 Kleopatra, and 433 Eros are within range of amateur photometers. Except when it is near a favorable opposition when it attains a brightness nearly visible to the naked eye, 433 Eros is difficult. However, you should not limit your studies to only those asteroids known to be variable. Because of varying tilts of the axis of a minor planet, it is possible that subtle light changes of many of the brighter asteroids have gone unnoticed.

Colorimetry work on the minor planets is of considerable value, also, providing data on the reflectance in various wavelengths, which gives clues as to the makeup of the body. It has been found that the majority of the asteroids are quite red in color, sparking theories that they contain much iron and that they perhaps have surface characteristics not unlike those of Mars.

Colorimetry is a relative value for the minor planet and is expressed as the color index of the asteroid. The standard indexes used could be V-R, R-I, and U-V. Most combinations will provide valuable information. For such work, you need no comparison star.

If you wish to determine the apparent magnitude of a minor planet, use some known comparison object, preferably one with a magnitude known to be constant to within 0.01 magnitude. If the comparison star is close to the asteroid (not more than 2° distant), you do not need to apply an extinction coefficient. However, instances when a comparison object is located so conveniently close to the asteroid are rare, and thus you must take extinction into account.

If the purpose of your study is merely to monitor the asteroid for light changes—perhaps the most meaningful amateur undertaking—you do not need a comparison object. However, it is good practice to compare the asteroid's magnitude with a known star at the beginning of each observing session so that you can establish a standard. It becomes increasingly important when examining for subtle light changes, however, to take into consideration the extinction that results from the earth's atmosphere. In such a program, it is best to monitor the minor planet within two hours before it crosses the meridian (30° east) and two hours after it crosses (30° west) to allow for minimal extinction. You must still apply some extinction correction, or the minor planet will appear to brighten as it approaches the meridian and dim as it passes.

Photometry of Comets

No good method for photoelectric determinations of comets has yet been devised. Most photoelectric photometers are equipped with aperture diaphragms of around 0.5 mm and cover only about 5″ of sky, so you can evaluate only a portion of the comet's coma at any one time. If you incorporate some provision for substituting or varying the size of the diaphragm in the design of the photometer, you can use it with some success.

Observations centered on the apparent nucleus or condensation of a comet can be quite valuable if you perform them with photoelectric instruments. The evaluation of magnitudes less than 10.0 are difficult to make visually unless the comet passes through some variable star field for which you have exact magnitudes. When you estimate the magnitude of a comet's nucleus, you can expect some amount of scatter from the diffuse cloud surrounding the immediate vicinity of the nucleus. Determine the magnitude using the V filter.

Of greater importance than apparent magnitude, however, is the determination of color indexes, such as B - V, V - R, and R - I, which can provide insight into the possible constituents and relative temperatures of the nuclear regions. Likewise, possible rapid fluctuations in brightness as well as color are changes that you should concentrate on when using photoelectric equipment.

Because most bright comets are seen only when they are low on the eastern or western horizon, extinction is a particular problem that you must consider when studying these objects. Particularly when determining color index, you must have some means of reducing the degree of extinction when you make your final report.

Planetary Satellite Photometry

An interesting project for the amateur astronomer is the study of the brightness of Saturn's many satellites. As least one, Iapetus, is known to vary in brightness substantially. Very little colorimetry work has been done in coordination with these brightness changes, and thus this is fertile ground for amateur studies. Chapter 10 discusses this interesting satellite.

Photoelectric Photometry Of Saturn's Rings

Sometimes the rings of Saturn appear to have areas of concentrated brightness (see Chapter 10). This can be noted both when the rings are presented edge-on to earth and sometimes during a phenomenon known as the "bicolored aspect" of the rings.

During the bicolored aspect, one ansa of the rings will be noticeably brighter than the other. Two explanations for this phenomenon are the following:

1. The material which makes up the Saturnian rings can move dynamically through the ring plane and clump in sections owing to the influence of the gravity of Saturn's satellites;
2. The effect is caused by atmospheric refraction.

The second assumption is based in part on the fact that the bicolored aspect is most pronounced when Saturn is near the horizon and when it is viewed in a red filter. Using a blue filter makes both ansae appear equal in brightness. However, the bicolored aspect has been viewed and photographed well even when Saturn is high in the sky, so the effect might be a combination of both factors. Because refraction may be a key factor during some ring brightenings, colorimetry with photoelectric photometers is a valuable tool for determining the degree of refraction.

When the rings are presented edge-on, one ansa will often be evident as a very thin line, while the other is absent. Likewise, knots or enlargements within the thin rings can frequently be seen. Any photoelectric determinations that you can make of the magnitudes of the rings and any swellings are of the greatest importance.

As for studies of Iapetus, the comparisons of the rings can be made using the satellites previously listed as comparison objects.

Monitoring Known Variable Stars

The most fruitful of all amateur photoelectric studies is monitoring variable stars. Hundreds of thousands of estimates of variable stars are submitted each year by amateurs, but very few derive from photoelectric determinations (see Chapter 12). Since most known variable stars have associated charts giving comparison stars, you can rapidly determine the comparison star. If you use the narrow-field charts, you need not apply extinction coefficients.

Many of the standard AAVSO charts provide some photoelectric magnitudes on each chart for stars that are possible candidates for amateur work. If possible, choose a star whose magnitude is known to within 0.01 accuracy for your determinations. Amateurs who have five-color sequential filters on photoelectric photometers can provide valuable information on the colors of these stars as they cycle through their light changes. In most cases the colors will significantly change if there is a corresponding temperature change. This is especially valuable data for the nova and supernova events. In the case of irregular variables or cataclysmic stars, such as the nova, you should monitor the star photoelectrically at every opportunity. The nova and possible supernova should be monitored hourly during maximum and transition phases.

If you are concentrating on the variable stars, choose the type (i.e., irregular, nebular, eclipsing) of star that most interests you and develop a program suited for the study of several stars of that type. You will always study the appearance of a new nova, of course. It is important that you do not choose too many stars in one season. Otherwise, one night's observing could be too rushed, and you would sacrifice quality observations. Anywhere from six to 10 stars a night is a good number for the experienced observer, making for a more relaxed observing regimen, so that you do not become "burned out" on the program.

Long-period Variables

Long period variables, whose light curves are well known, are best ignored by the amateur with a photoelectric photometer. If anything, these predictable variables are overobserved by zealous visual observers who attempt quantity, rather than quality, observations. If such variables are of interest to you, monitor them at most once a week.

Semiregular Variables

Semiregular variables are somewhat less predictable than the long-period stars, and you should monitor them more often. By following such stars through at least 10 complete cycles of variation, you can obtain a fairly good determination of the stars' periods. However, these stars are also closely followed by visual observers, and the value of photoelectric observations is not as great as for stars of other types.

Short-Period Variables

Short-period variables include many of the brightest stars in the sky. Beta Cassiopeiae is one of the best examples. Because the light changes of these stars are very slight, they are prime candidates for the photoelectric photometer. The changes in these stars are usually quite rapid—as a rule, the greater the magnitude of change, the longer the period of the star. Many of the peculiar M- and A-type stars are short-period variables, and many stars suspected of being variables have not been identified as such.

Dr. Dorrit Hoffleit of Yale Observatory proposes that the stars listed in Table 13-1 are possible candidates for slight variation, all within range of photoelectric photometers and amateur equipment. These stars can be quickly located on any good star atlas, as can be the recommended comparison stars.

Flare Stars

Flare stars are like our sun in that they occasionally exhibit an outburst of both energy and visible light. Amateur astronomers who have photoelectric photometers equipped with a chart recorder can contribute valuable data about the frequency and magnitude of these stellar outbursts. Flare stars should be checked at every opportunity for an outburst; the normal flare star has a rapid outburst that may last for only 5 minutes or less, increasing by perhaps 1 magnitude. It is not yet known if the color of the star significantly changes during an outburst, so some colorimetry is greatly needed. However, such five-color determinations must be made rapidly in the event of a detected outburst, because a maximum of only 5 minutes can be expected. It is possible that slight color changes *precede* the outburst. Therefore, you should attempt colorimetry each time you examine the star. It is not uncommon for amateurs to observe a flare star frequently during a night's session at the telescope.

Irregular and Nebular Variables

It is best to monitor irregular and nebular variable stars once each night. In the case of variables of the SS Cygni or dwarf novae type, monitor the star photoelectrically for long periods of time during an outburst. Likewise, monitor the stars of the Z Camelopardalis type and any other known cataclysmic stars.

Eclipsing Variable Stars

Eclipsing variable stars belong to a classification of stars that can be eagerly and rewardingly attacked by the amateur astronomer who possesses a photoelectric photometer. Monitor these stars according to their predicted periods of eclipse. By following these stars over a long period of time, it will become evident to you about when an eclipse will begin. At that time, train the telescope on the star and monitor at 5-minute intervals for as long a time as possible. It is equally important to monitor the midpoints between deepest eclipses because it is during those times that secondary minima usually occur. It is now believed that periods of the eclipsing variables do not necessarily remain constant. The bright star Algol is a good example of a star with changing period. The reasons behind such changes are a result of the physical nature of the orbit of the secondary eclipsing star, yet the exact cause is not known. Thus, as an amateur astronomer, you can provide valuable knowledge of the proportional changes of many stars.

Searching for Undiscovered Variable Stars

All astronomers yearn for discovery, and the photoelectric photometer can certainly provide a tool by which the discovery of unknown variable stars can become a reality. All stars in the sky vary somewhat, but most of them vary unpredictably or by only very small amounts. By selecting only a few stars to monitor over a long period of time, you might well find that one of those stars varies by some small amount. For such a project, first choose from the bright stars listed in Table 13-1 because these stars are prime suspects for intrinsic variability and they are well within reach of amateur instruments.

TABLE 13–1. Possible variable stars in the Yale Bright Star Catalog.*

HR No.	*GC No.*	*RA (1950)*	*Dec*	*Magnitude*	*Spectrum*	*Peculiarity*
Northern Peculiar A-Stars						
128	634	$0^h29^m44^s$	$+43^\circ13.1$	6.43	A0	Si
	606	0 28 20	+43 5.9	8.1	F8	comparison
	608	0 28 30	+43 40.2	6.64	B8	comparison
682	2813	2 17 38	+49 55.4	5.56	A0p	Si
	2795	2 16 48	+50 43.8	6.72	F5	comparison
	2864	2 20 27	+49 06.7	8.5	B8	comparison
1643	6288	5 06 02	+73 53.1	5.38	A0p	Si
	6196	5 01 50	+73 31.5	8.0	A0	comparison
	6405	5 12 02	+73 12.9	5.76	A0	comparison
2362	8430	6 27 21	+ 9 3.8	6.48	A0p	Si
	8379	6 25 33	+10 20.2	6.19	K0	comparison
	8455	6 28 24	+ 9 58.6	7.65	B8	comparison
4041	14132	10 15 21	+27 39.9	6.46	B9	
	14091	10 13 38	+28 56.0	6.51	G0	comparison
	14130	10 15 16	+27 46.5	7.9	F2	comparison
4751	17007	12 26 14	+26 10.5	6.69	A3	m*
	17005	12 26 08	+26 30.2	6.48	A3	comparison
	17012	12 26 24	+26 11.4	5.38	A0p	comparison
5422	19553	14 27 41	+32 0.7	5.96	B9	Si
	19505	14 25 38	+33 10.4	8.4	A3	comparison
	19650	14 32 04	+32 45.2	6.28	F2	comparison
5982	21580	16 01 14	+46 10.5	4.64	B9	Mn
	21547	16 00 01	+47 16.8	7.40	K0	comparison
	21684	16 05 30	+47 38.2	6.58	A0	comparison
6958	25308	18 29 37	+ 3 37.3	6.34	B9	Si, Cr
	25256	18 27 36	+ 4 1.8	6.50	B5	comparison
	25271	18 28 13	+ 4 28.5	6.80	A2	comparison
7147	25997	18 53 51	+17 55.7	6.41	B9	Si
	26052	18 56 01	+17 17.5	5.37	F5	comparison
	25999	18 53 53	+18 2.5	5.72	K2	comparison
7395	26846	19 24 20	+36 13.0	5.15	A0p	Si
	26792	19 22 18	+36 21.1	6.45	K0	comparison
	26927	19 27 42	+36 10.9	6.62	F0	comparison
7870	28642	20 32 16	+46 31.3	5.59	B9	Si
	28595	20 30 51	+48 2.7	6.82	B2	comparison
	28631	20 31 58	+45 0.2	6.62	A0	comparison
8933	32719	23 29 14	+28 7.7	6.23	A0	Si
	32710	23 29 00	+28 23.4	6.68	K0	comparison
	32817	23 33 49	+28 37.3	8.8	F2	comparison
Southern Peculiar A-Stars						
11	114	0 05 10	$-2^\circ49.6$	6.33	B8	
	124	0 05 38	− 2 43.6	6.32	K0	comparison
	132	0 06 09	− 2 30.1	7.31	G5	comparison

TABLE 13–1. (cont'd)

HR No.	GC No.	RA (1950)	Dec	Magnitude	Spectrum	Peculiarity
1100	4305	3 34 00	−17 37.9	5.32	A0	
	4216	3 29 08	−16 56.9	7.56	F5	comparison
	4368	3 37 20	−17 31.6	7.06	K0	comparison
1240	4801	3 57 47	−24 09.4	4.69	A0	
	4755	3 54 22	−25 19.3	6.93	G0	comparison
	4829	3 59 02	−22 24.9	7.19	F0	comparison
1300	5055	4 09 24	−20 29.1	5.80	A0	
	5006	4 06 58	−20 50.1	8.7	K0	comparison
	5015	4 07 39	−19 25.0	7.09	K0	comparison
2195	7875	6 08 35	− 6 44.5	5.97	A0	
	7876	6 08 36	− 7 16.3	7.55	B8	comparison
	7899	6 09 25	− 6 32.2	5.09	B3	comparison
2320	8274	6 22 02	−56 20.5	5.72	A0	
	8148	6 17 38	−56 58.7	8.00	A0	comparison
	8408	6 26 17	−57 58.2	5.73	K0	comparison
2414	8577	6 32 57	−22 55.4	4.54	A0	
	8592	6 33 24	−22 04.0	6.53	K0	comparison
	8626	6 34 35	−22 34.3	6.23	B8	comparison
4263	15014	10 52 44	−41 59.0	6.30	A0	
	15024	10 53 11	−42 45.2	6.66	A0	comparison
	15069	10 55 41	−41 46.3	7.46	A2	comparison
4776	17095	12 30 00	−13 34.9	5.70	F0	
	17036	12 27 30	−13 07.0	6.41	G0	comparison
	17095	12 30 01	−13 35.0	5.70	F0	comparison
4944	17778	13 04 18	−59 35.6	6.06	B9	
	17733	13 02 07	−60 10.4	8.32	A0	comparison
	17761	13 03 42	−59 10.6	8.19	A3	comparison
4965	17839	13 07 58	−52 18.1	6.29	A0	
	17781	13 04 28	−52 29.0	7.76	A5	comparison
	17783	13 04 39	−53 11.6	5.96	B9	comparison
5049	18141	13 22 09	−70 22.0	5.84	A0	
	18058	13 18 29	−70 17.1	7.42	B9	comparison
	18113	13 21 01	−70 10.1	7.22	K0	comparison
5069	18206	13 25 42	−64 25.0	6.24	A0	
	18204	13 25 41	−63 36.9	7.19	B3	comparison
	18266	13 28 29	−65 03.9	6.62	A3	comparison
5158	18555	13 41 06	−50 45.7	6.46	A0	
	18521	13 39 46	−50 32.3	6.29	K0	comparison
	18622	13 44 18	−50 00.0	6.06	A3	comparison

M-Type Stars to be Monitored

HR No.	GC No.	RA (1950)	Dec	Magnitude	Spectrum	Peculiarity
259	1096	$0^h52^m34^s$	$+24^\circ17.2$	6.36	Mb	M4III
	1091	0 52 17	+23 21.5	5.60	K0	comparison
	1093	0 52 19	+23 49.9	7.38	G5	comparison
2146	7725	6 03 11	+29 31.1	6.32	Mz	M3II
	7673	6 00 52	+27 34.4	7.8	A0	comparison
	7683	6 01 22	+30 14.3	8.2	A	comparison

TABLE 13-1. (cont'd)

HR No.	*GC No.*	*RA (1950)*	*Dec*	*Magnitude*	*Spectrum*	*Peculiarity*
2802	9809	7 19 01	-25 47.8	6.10	Mz	M4III
	9790	7 18 30	-26 38.4	6.59	F0	comparison
	9805	7 18 53	-26 52.1	5.84	B3	comparison
3153	10873	7 58 46	-60 26.9	5.06	Mz	M1.5IIA
	10859	7 58 16	-60 26.8	7.61	F0	comparison
	10874	7 58 47	-60 4.2	6.41	B8	comparison
7442	27045	19 32 19	+49 9.2	6.19	Mb	M4.5III
	27007	19 30 43	+50 4.2	8.07	F8	comparison
	27078	19 33 14	+48 3.3	6.70	A5	comparison
7475	27195	19 37 10	+16 27.3	6.58	K5	K4Ib + M0+IIab
	27191	19 37 01	+18 34.1	7.6	B5	comparison
	27215	19 37 52	+17 53.8	4.37	G0	comparison
7997	29139	20 51 07	-28 6.9	6.46	Ma	M4III
	29124	20 50 20	-28 12.8	8.18	F8	comparison
	29154	20 51 36	-28 7.3	7.62	K2	comparison
8062	29388	21 00 37	+44 35.6	6.38	Mb	M4III
	29371	20 59 59	+43 59.4	6.72	A2	comparison
	29398	21 00 53	+45 41.1	7.6	K2	comparison
8164	29860	21 17 53	+58 24.7	5.79	—	M1 Iabep+B
	29804	21 15 56	+58 24.1	6.41	B3	comparison
	29866	21 18 01	+58 6.8	7.05	K2	comparison
8378	30746	21 55 57	-21 25.3	6.23	Mb	M4III
	30715	21 54 19	-19 25.7	7.73	G5	comparison
	30759	21 56 29	-23 6.7	7.39	F5	comparison
8637	31680	22 39 35	-29 37.4	6.44	Ma	M5III
	31639	22 37 35	-30 55.0	5.98	K2	comparison
	31682	22 39 38	-30 54.8	8.72	G0	comparison
9099	24	0 01 16	+66 26.0	6.62	Ma	M4III
	39	0 02 05	+66 53.3	5.84	K0	comparison
	33331	23 59 24	+66 9.6	7.30	B9	comparison

*As suggested by Dr. Dorrit Hoffleit, Yale University Observatory.

MULTICOLOR FILTER OBSERVATIONS

The utility and research worthiness of any photoelectric photometer is increased manyfold when a series of narrow pass band filters are allowed to intercept the light before it reaches the photocell. By using selected filters, bands of the star's (or object's) spectrum are let pass and all others are blocked. In this way, brightness at one part of the spectrum may be compared to that at another.

Worldwide, astronomers are now using a more or less standard set of filters whose wavelengths are coincident. It is important that you as an amateur astronomer choose a set of filters

TABLE 13-2. Recommended filters.

	Range (nm)	*Recommended Source*	*Peak Transmittance (nm)*
Ultraviolet	320–430	Corning 7-54	360
Blue	370–520	Corning 5-56	450
Visual	480–680	Schott GG14	550
Red	540–870	Schott RG5 + Corning 3965	640
Infrared	690–1070	Corning 2540	820

that closely matches those now in use by the professionals. Those recommended wide wavelengths are given in Table 13-2.

The Color Index System

The most valuable contribution of the photoelectric photometer and the desirable characteristic that sets it apart from the spectroscope or spectrograph is its capability of determining the *color index* of a star, and from that the *color excess* and the *color temperature*. Each of the properties of stars is a vital link in our knowledge of the physical processes of stellar evolution.

The color index is defined as *the apparent magnitude of the source (star) in one wavelength compared to the source's apparent magnitude in another wavelength.* The difference between the two is the *color index*. Because the index depends on the spectral dispersion (reddening) of the light of a star, or object, it indicates whether that source is predominantly blue, red, or whatever color.

In the now-accepted indexing system, the standard color index is determined from UBV photometry, being simply B – V, where B is star's apparent magnitude measured at 420 nanometers (blue) and V is the star's apparent magnitude measured at 600 nanometers (visual). Other indexes can be used, such as U – B and R – I, depending on what system one is working with.

All stars falling within basic spectral types are subclassified as to their luminosities within those classes, to give the *intrinsic color index*. The color index of a star is quite easily determined photoelectrically, and thus it is used more frequently when drawing a graph of the relation to the other stars than is the spectral class or the temperature.

The *color excess* is simply the amount of starlight that is absorbed or reddened by passing through the interstellar medium. If the index is negative, it indicates an excess of radiation from the blue or ultraviolet range, and thus is indicative of hotter stars. A positive index represents cooler stars, with higher color excesses.

FIGURE 13-4. Spectral response for UBVRI filters (peak transmission at indicated wavelengths).

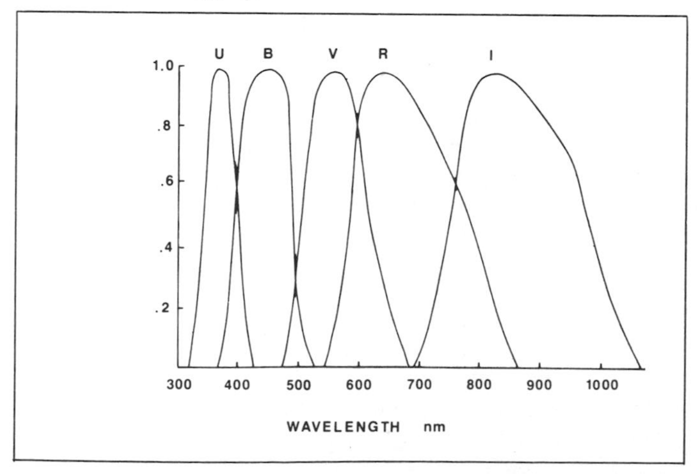

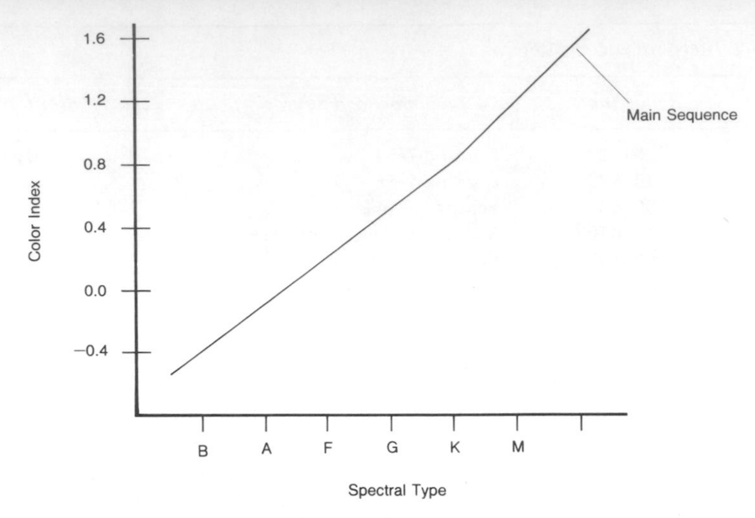

FIGURE 13-5. Graph showing color index versus spectral type.

Because the proportion of blue and ultraviolet radiation is known to increase with increasing star temperature, there is a relationship between the color index (B - V) and the corresponding temperature of the star. Using a slight deviation on Planck's law of radiation of a black body, the temperature—or *color temperature*—of a star can be determined once the B - V index is established. The color temperature is defined as the surface (i.e., photospheric) temperature of a star expressed in terms of a perfect radiator (i.e., black body) whose energy distribution and output throughout all wavelengths of light correspond to that of the star. The temperature is determined through the formula:

$$T_c = 7300/\,[\,(B-V) + 0.6]\ \text{kelvin degrees}$$

where T_c corresponds to the color temperature, and the $(B-V)$ represents the color index of the star. This temperature is always higher than the given effective temperature, which is dependent on the radius of the star. For example, the sun's color temperature is 6500° K, whereas its effective temperature is 6000° K.

ATMOSPHERIC EXTINCTION

The one factor that even the most sophisticated photoelectric photometers cannot cancel is that of atmospheric extinction. The amount of reddening and dimming of an object's light is dependent on the displacement in degrees from the zenith. At the zenith, we might well assume the atmospheric extinction to be zero as a standard for measure, even though some extinction certainly exists through that column as well.

First, determine the approximate displacement in degrees from the zenith at the time of your observation. This can be approximated by thinking in terms of 30° increments and fractions thereof from the zenith to the nearest horizon from the object, or it can be defined accurately by elements of spherical trigonometry, using the formula:

$$\cos Z = (\sin a)(\sin b) + (\cos a)(\cos b)(\cos H)$$

where,

Z = distance from zenith to geometric horizon nearest object

a = observer's latitude on earth

b = declination of object being measured

H = hour angle of the object being measured

Once the correct zenith distance has been estimated or determined, the amount of atmospheric extinction can be determined by the formula:

$$M = 0.35\ (\sec z_1 - \sec z_2)$$

where,

M = extinction coefficient

z_1 = zenith angle of highest star

z_2 = zenith distance of star closest to horizon

REDUCTION OF PHOTOMETER DEFLECTIONS TO ACTUAL MAGNITUDE DIFFERENCE

The conversion to actual magnitudes of deflections observed on the readout of the photometer control is a quite simple calculation. It is given by the formula:

$$M_1 - M_2 = 2.5 \log \frac{L_2}{L_1}$$

where,

M_1 = Magnitude of comparison star

M_2 = Magnitude of star being measured

L_1 = Deflection of comparison star

L_2 = Deflection of star being measured

For example, suppose that we are measuring two stars, L_1, *the comparison star,* and L_2, *the suspected variable*. The comparison star is known to be nonvarying in light, with a photometric V magnitude of 8.80, and it is known to be within 0.5° of the suspected variable star (to minimize the effect of extinction). We read the photoelectric photometer and find that the deflection of the suspected variable is 0.10, whereas the deflection for the comparison star is 0.14. Then,

$$2.5\ (\log \frac{0.14}{0.10})$$

determines the difference of 0.37 magnitude between the two. Because we see that the deflection of the comparison star is higher, then that star is the brighter of the two, and the value 0.37 must be added to 8.80 to obtain a magnitude of 9.17 for the suspected variable.

REFERENCES

Fredrick, L.W., and R.H. Baker, *An Introduction to Astronomy*, 8th ed., pp. 85-91. New York: Van Nostrand, 1974. Paperback.

National Bureau of Standards, Vol. 7, No. 300, *Precision Measurement and Calibration Radiometry and Photometry*. November, 1971.

Parkhurst, J.A., *Researches in Stellar Photometry During the Years 1894 to 1906 Made Chiefly at the Yerkes Observatory*. Washington, DC: Carnegie Institute of Washington, 1906.

Roth, Gunter D., *Astronomy: A Handbook*, pp. 122-124, 439-471. Cambridge, MA: Sky Publishing Corp., 1975.

Sidgwick, J.B., *Amateur Astronomer's Handbook*. London: Faber and Faber, 1958.

Thackery, A.D., *Astronomical Spectroscopy*. London: Eyre and & Spottiswoode, 1961.

Wood, Frank B., *Photoelectric Astronomy for Amateurs*. New York: Macmillan, 1963.

14

ASTROPHOTOGRAPHY FOR THE AMATEUR ASTRONOMER

Perhaps the greatest challenge to the amateur scientist is that of recording on film the images of celestial objects—astrophotography. But waiting down the road to good celestial photography are many pitfalls and obstacles that tend to discourage many would-be photographers.

Even the newest beginner can take astrophotographs. It is just that the more complex the photograph (i.e., the more you want to show), the more difficult is the undertaking. Any camera can take an astrophotograph if the shutter can be locked open, but certain types have distinct advantages.

THE ASTROPHOTOGRAPHER'S CAMERA

Your choice of camera for use in astrophotography depends chiefly on the type of photograph you want to make—moon, planets, deep-sky, or whatever. Another factor, of course, in choosing a camera is your budget. For the first two methods (fixed and piggyback) of astrophotography described in the following pages, any simple camera will do. For prime focus and extended (projection) photography, however, only certain cameras are suitable.

The best all-purpose camera type for the astrophotographer is one of the many 35-mm single lens reflex (SLR) cameras on the market today. Almost all of them are suitable for this type of photography, but they should have certain features if you plan to invest in such a camera. Those features are as follows:

1. The lens should be removable.
2. The camera must have a bulb, or time-exposure setting.
3. There must be provisions on the camera for attaching a cable release.
4. The focusing screen should be as clear as possible, such as those screens supplied for extreme telephoto work.
5. The mechanics of the camera should operate smoothly, with little shutter slap and mirror vibration.

The 35-mm SLR camera will suffice for almost all types of astrophotography. The movement today is toward sophisticated cold camera set-ups and other complex apparatus designed to speed up the accumulation of light onto the film. Such devices are ideal for the *experienced* amateur, but the beginner should not use or even consider them. Learn the nature of astrophotography before you move to such equipment. There is more information in the following pages about the proper use and choice of cameras and lenses.

TYPES OF ASTROPHOTOGRAPHY

There are basically four types of astrophotography. Each is specifically suited for recording a particular type of astrophotograph. The four methods are the following:

1. A camera mounted on a tripod or other stationary support with no driving provision (see Figure 14–1).
2. Piggyback astrophotography, or simply an equatorially driven camera (Figure 14–2).
3. Prime focus photography, which utilizes the telescope optics as a huge telephoto lens on the camera.
4. Projection photography. Table 14–1 summarizes the applicability of each.

Fixed (Tripod-Mounted) Camera Photography

A tripod-mounted camera is the simplest way to get into astrophotography. All that is required is having a dark sky, setting your camera on a tripod or any other fixed mount, aiming the camera at a particular sky area, and opening the shutter for any length of time you desire—usually a minimum of 5 seconds and a maximum of about 30 minutes. The best camera set-up is usually a normal or wide-angle lens that has been set to its maximum speed, usually at f/1.2 to f/2.8.

This method can be used to photograph the brighter stars in a constellation with a 5- to 10-sec exposure on Tri-X (Kodak) film. Such short exposures greatly reduce the amount of star trailing that is caused by the earth's west-to-east rotation.

The fixed camera method is a favorite for the photography of meteor showers, also. The camera can be aimed at a point in the sky from which it is known that the meteors will originate. Dark skies will require 20- to 30-min exposures on fast black-and-white or color films. This interval is usually ample time to record one or more meteors from active showers. A word of caution here: monitor the sky area that you are photographing, and after one or two meteors appear in the area, stop the photograph and start another. The dimmer meteors can become "washed out" on the film by background skylight if you expose the film for a considerable time after the passage of the meteor. As in any type of astrophotography, make careful records of each exposure so that any discovery can be properly timed and placed in the sky. In addition, your records provide systematic methods of cataloging a growing collection of astrophotos.

A tripod-mounted camera can also be used to produce beautiful photos of circumpolar star

FIGURE 14–1. A camera mounted on a tripod for simple astrophotography. Notice that the lock on the cable release is necessary to lock open the shutter because most SLR cameras do not have any method of locking open the shutter with the bulb setting other than holding the button down with a finger.

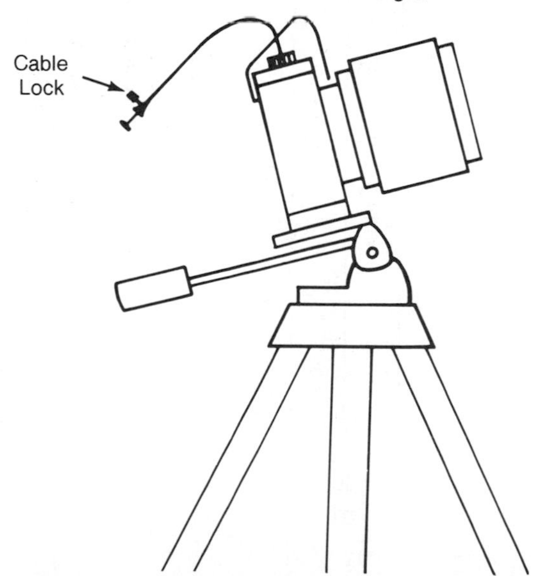

TABLE 14–1. Astrophotography systems.

Method	*Advantages*	*Disadvantages*	*Use*	*Maximum Exposure*
Fixed camera	Speed	Star trails	Bright comets	5 min
	Wide-field	Limited magnification	Circumpolar stars	1 hr (depending on sky)
	Small investment	Limited light accumulation	Meteors	10 min
	Ease of operation	Short exposures	Constellations	1 to 3 min
Piggyback	Longer exposures	Increased investment	Comets to mag. 10	1 hr
	Great light accumulation	Clock drive required	Deep-sky objects	1 hr
	Magnitude gain	Must use darker sky	Milky Way	Up to 45 min
	Increased resolution (with telephoto)		Planet conjunctions	5 min
	Guiding not necessary		Meteors	10 min
Prime focus	Great resolution	Accurate guiding required	Comet close-ups	30 min
	Larger image scale	Substantial investment	Deep-sky objects	2 hr (if necessary)
		Wind & breeze impairment	Some double stars	10 min
			Planetary satellites	About 2–5 min
			Lunar disk	1/250 to 1 sec
Projected	Optimum resolution	Perfect tracking required	Lunar close-ups	Time[a]
	Largest image scale	Excellent optics required	Planets	
	Relatively short exposures	High-quality eyepieces	Double stars	
		Wind or breeze impairment		
		External vibrations		
		Requires excessive film		

[a]Projected photography requires a variety of precise exposure times. These times depend on several factors discussed in the text. Of these, focal length of the telescope and projection system, film type, and sky conditions are the primary factors. Exposure times for the different objects are calculated in the Henry Exposure Graph included at the end of this chapter.

trails in the northern sky. A 15- to 30-min exposure aimed at the polar area (centered near Polaris, the North Star), with some trees, buildings, or distant city in the foreground will become an artistic and spectacular photograph, as the stars describe circular paths caused by the earth's rotation.

In all types of fixed camera photography, always be sure to focus carefully on the stars and frame the desired area prior to starting each picture. This will ensure pinpoint star images or sharp star trails. Most 35-mm SLR cameras have an *infinity* (∞) setting, which is usually the proper setting for star photos. However, do not assume that the infinity setting is precisely in focus. Improper adjustment at the factory can often cause the infinity setting to be slightly off—enough to notice in astronomical photography. If the focusing barrel of the lens cannot be moved in far enough to reach sharp focus of the stars in the viewfinder, find another lens.

Piggyback Astrophotography

A second and more rewarding method of astrophotography is the piggyback system (Shown in Figure 14-2), which is an extremely versatile method of photographing the stars. A camera with lens is mounted to ride along the telescope tube. The film is exposed for long periods of time while a motor drive or hand controls enable the telescope to track the stars. As in the fixed camera method, you keep the shutter of the camera open by putting the camera on the bulb (b) setting, which is on the timing ring, and locking the shutter with a cable release clamp. The system is incredibly versatile because you can use any size (i.e., focal length) lens on the camera, from a wide-angle to a supertelephoto. If you choose the wide-angle lens, you can use the piggyback set-up to photograph wide sky areas, such as the great star clouds of the summer Milky Way. The telephoto lenses allow you to zoom in on the larger, deep-sky objects such as the Andromeda Galaxy (M-31), the Lagoon Nebula (M-2), or the North American Nebula. The list of objects is almost limitless. The longer the focal length of the lens, the more detail you can record, and the greater the number of deep-sky objects within your grasp.

When you use this method, you must consider several factors in order to get pleasing results. These considerations include polar alignment, guiding, drive-motor inaccuracies, sky darkness, and sturdiness of the mounting. Polar alignment and guiding techniques are discussed in Chapter 2. For now, it is necessary only that you realize that the longer the focal length of the lens you use, the more accurate the polar alignment and the guiding must be. Simple lining up by sighting on Polaris for alignment is not accurate enough except for wide-angle or normal lenses, and even then only for exposure under 5 minutes. The guiding of the telescope/camera combination also becomes more difficult with increasing focal lengths. A few examples serve best to indicate what is needed for good results in piggyback photography.

First, let us assume that the telescope is well aligned to celestial north and is equipped with a clock drive. For lenses with focal lengths of 28 mm to 55 mm, 30-min exposures are possible without much guiding other than having the clock drive running. Lenses with focal lengths in the range of 100 mm to 200 mm require some guiding. A high-power eyepiece

FIGURE 14-2. Piggyback mounted camera.

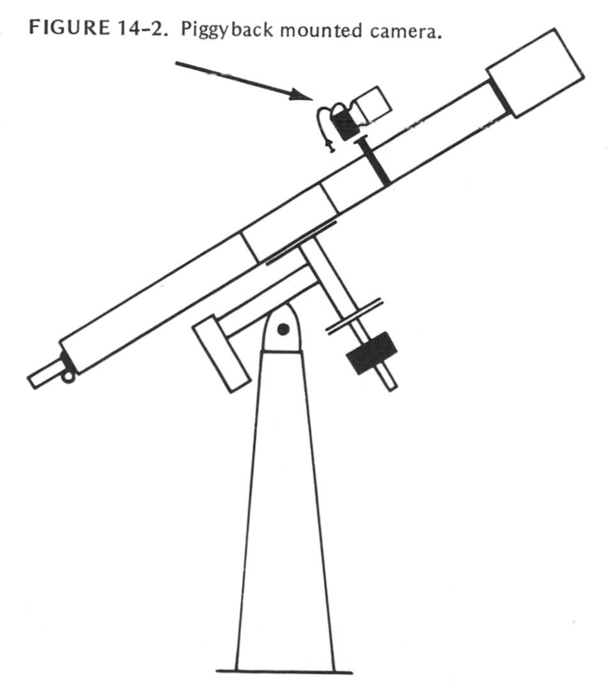

providing about 150x in the telescope is sufficient if you can visually maintain a star image close to the center of the field of view by using the telescope's slow-motion controls. For lenses of focal lengths of 300 mm and more, you can guide best by using a crosshair eyepiece with 100x or more, keeping the star at the intersection of the crosshair web at all times. Figure 14-3 gives a good idea of the approximate sky coverage available using lenses of five different focal lengths.

Two mechanical considerations must be kept in mind to ensure satisfactory piggyback astrophotographs. First, make sure that the telescope mounting is as sturdy as possible so that external vibrations and breezes will not ruin a long exposure. Check to make sure that the mounting is exactly balanced to provide for better tracking by the motor drive (or hand control) and to eliminate a lot of "backlash" and looseness in unbalanced systems. The second mechanical consideration of piggyback photography is the *location* of the camera to be mounted on the tube.

Trial-and-error viewing through the camera viewfinder will show you just how far back from the front of the telescope tube you should mount the camera so as not to photograph the end of the scope in each picture. Notice that this positioning must be done with the widest-angle lens that you may be using to ensure that every lens later used will keep the telescope tube out of the picture (see Table 14-2). Ideally, the camera should be as far back toward the *middle* of the tube as possible to help stabilize the unwieldy tube, and it should be a couple of inches above the surface of the tube. But, of course, the farther back the camera, the greater the possibility of photographing the end of the telescope tube. The equation for finding the angular fields of telephoto lenses is:

TABLE 14-2. Telephoto lenses commonly used with 35-mm format film (fields of view).

Focal Length	*24-mm Dimension*[a]	*36-mm Dimension*[a]
50	28.0	40.0
85	16.0	24.0
105	13.0	19.0
135	10.0	15.0
150	9.2	13.8
200	6.9	10.3
250	5.6	8.2
300	4.6	6.9
400	3.4	5.2
500	2.8	4.1

[a]In degrees.

$$\text{Field coverage in degrees} = \frac{57.3}{\text{Lens EFL (mm)}} \times \text{(frame size)}$$

For example, using a 35 mm camera (24 mm X 36 mm) with 200 mm lens and applying the equation, the following are derived:

$$\text{(vertical)} \quad \frac{57.3}{200 \text{ mm}} \times 24 \text{ mm} = 6.9^\circ$$

$$\text{(horizontal)} \quad \frac{57.3}{200 \text{ mm}} \times 36 \text{ mm} = 10.3^\circ$$

Prime Focus Astrophotography

Prime focus photography can be a real challenge, but it also can be quite discouraging if you are not prepared for its many pitfalls. However, detailed studies of faint galaxies and star clusters are possible with a little care and a lot of patience and practice. I recommend that you practice the piggyback method first, if possible, and when you have mastered it, move to the more exacting prime focus method. I suggest this because all the things that can go wrong and ruin a picture when you are using the piggyback method are only amplified in prime focus photography.

The camera, with lens removed, is attached directly to the eyepiece holder of your telescope, thereby making the telescope itself a huge telephoto lens. Adapters for attaching the camera to the telescope are available through commercial suppliers to fit almost any camera with removable lens. Besides the usual requirements for accurate guiding and precise polar alignment, which are discussed in following pages, a sturdy mounting and a high-quality clock drive are mandatory. As with the piggyback method, there are additionally a few mechanical factors to consider.

Balancing the telescope is an important

factor. In particular, the right ascension axis must be precisely balanced to facilitate proper motion of the driving motor. However, to properly balance a telescope in right ascension, first the declination must be balanced. This is best done by attaching the camera apparatus as it will be used, turning the telescope tube to a horizontal position, and unclamping the declination clamp. Either the tube must be moved laterally to compensate for the added weight of the camera, or weights must be added to equalize the imbalance. Once the declination axis has been suitably balanced, the right ascension axis can be balanced. In many of the modern Newtonian reflectors on the market, there are no clamps to release. Instead, there are clutches that provide tension against the shafts of the declination and right ascension axes. With such arrangements, it is difficult to judge the amount of balancing that is necessary. You must do one of the two following steps:

1. Leave the clutches adjusted and rebalance by *feel*. That is, move the telescope to one side, let go, and note any desire of the scope to move back toward the original direction. Both axes must be balanced in this manner, and the axes must be pushed in both directions until no "backlash" is seen or felt.
2. Unclamp (usually by loosening three screws on each clutch) both clutches and balance as though the axes were unclamped.

Another critical consideration before beginning prime focus astrophotography is *focus*. Once the system is focused, be sure that the mechanical workings of your focuser are tight enough to ensure that this focus will be accurately maintained throughout the exposure. The weight of the camera alone is sufficient to change the focus. If the object to be photographed is fairly dim, or if it is a diffuse object such as a galaxy or nebula, first focus on a bright star. Usually, three or four stars in the same field of view aid in the accuracy of focusing. It is essential to check your focus after each exposure and before the next, and correct it only if necessary.

In prime focus astrophotography, there are

FIGURE 14-3. Approximate fields of view of various lenses.

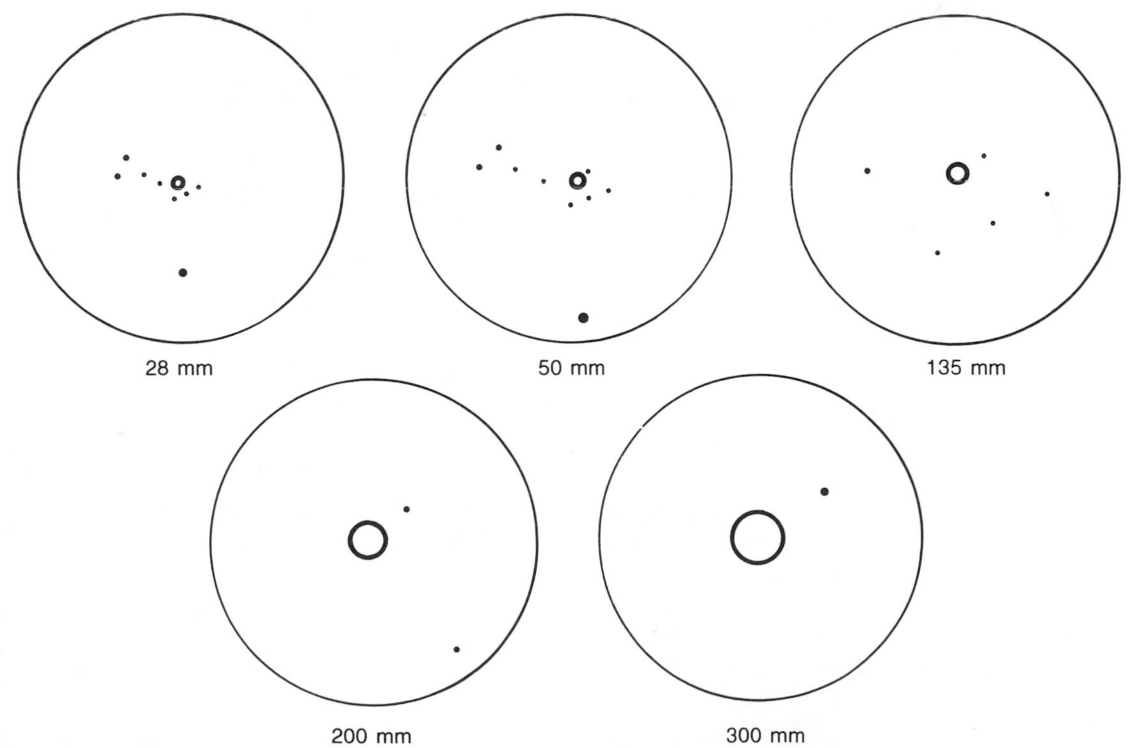

four factors that affect any given system of recording celestial objects:

1. The size of the primary mirror or lens (objective) of the system.
2. The focal length of the objective.
3. The focal ratio of the system.
4. The length of time the film is exposed.

Generally, the larger the mirror or lens of the telescope, the more that can be recorded, but the requirements for such a statement vary somewhat. For instance, to record more star images, a primary objective of larger diameter is needed, but to record more light from an extended field object such as a galaxy or nebula, a *fast* system is needed (or a system with a small focal ratio). As an example, a 6-inch f/5 system can record more light than an 8-inch f/8 system in the same exposure time. This is because the light-recording power of any system is directly proportional to the square of the radius of the objective, and inversely proportional to the square of the focal ratio. Thus, our 6-inch f/5 has a ratio of

$$\frac{[r]^2}{[1]^2} = \frac{[3]^2}{[5]^2} = \frac{9}{25}$$

or 0.36, whereas the 8-inch f/8 system has a ratio of

$$\frac{[4]^2}{[8]^2} = \frac{16}{64}$$

or 0.25. This gives the 6-inch f/5 system a speed of recording 44% greater than that of the 8-inch f/8 system.

However, the system is further complicated by the fact that, in order to resolve fine detail in objects, a longer focal length is more advantageous than a shorter one. For the amateur, this usually means a compromise between system speed and *resolution*. If you want to record faint objects in as short a time as possible, a fast system is needed, but if you want to increase the clarity of detail in those objects, a longer focal length is best. In either situation, it is best to have as large a lens or mirror objective as possible because this increases both the light gathering *and* the resolution of a telescope.

Electronic Cameras In Prime Focus Astrophotography

Many new 35-mm cameras with electronic shutters are being marketed, but they are not very good astrophotographic cameras because long exposures require the shutter to be open for the length of the exposure. This is a considerable drain on the battery. On a cold night, for example, a battery may last only 5 to 10 minutes; then the shutter will close.

Chilled Emulsion Cameras For Deep Sky Astrophotography

Cold cameras are becoming more popular now that some of them use roll film. Formerly, only the one-frame-at-a-time cold cameras were available. A cold camera has one big advantage: reciprocity failure of films is almost nonexistent at temperatures below -50°F. A cold camera usually uses dry ice (-110°F) to cool the film being used. Tri-X 400 then becomes a good astrofilm, being very sensitive, having finer grain than the 103a films, and being less expensive. Color films can be used for longer exposures and better color saturation. Because the films continue to record at their peak values, exposure times are much reduced, usually by a factor of 4, to get the same image density on film. Operation of a cold camera takes practice to get good results. You should work on perfecting the other methods of deep sky photography before attempting to use a cold camera. Cold cameras can produce outstanding photos, both in color and black and white.

Guiding Prime Focus Astrophotographs

Guiding while making prime focus photographs is a critical requirement. Guiding is necessary for several reasons. The air of our atmosphere is constantly boiling, mixing warm and cold air currents. This mixing causes the stars to seem to the naked eye to twinkle, and through the telescope, the stars can either move rapidly back and forth by small amounts, or they actually appear to drift slowly because of the refractive quality of the atmosphere. To keep up with such drifting, guiding is essential during long-exposure astrophotographs. Another problem that necessitates guiding is the inaccuracies of clock drives. Even the best drive systems can and do have small deviations in them. Thus, constant guiding must be maintained. In addition, power fluctuations are common in most areas and can create a significant problem if you are near a large city or using power from the same line that provides power for a refrigerator or air conditioner. When such appliances kick on during exposures, the clock drive can slow considerably.

Regardless of what causes the image to drift in the field of view, the drifting can be corrected with slow-motion controls, and it will be negligible if it is corrected as soon as it starts. Guiding can be greatly simplified if the polar alignment is as accurate as possible. If you must make adjustments in the declination axis more than once every 10 minutes, you should improve the alignment of the polar axis if you want to save yourself some work. Some techniques for aligning the mounting to celestial north are discussed in Chapter 2.

It should be obvious at this point that, in order to guide a telescope, slow-motion controls must be available in both the declination and right ascension axes. Such controls can be either manual or electric, but they must be smooth and easy to use in the dark. Declination slow motion is something that either must be bought or built into a telescope mount. Right ascension slow motion is best obtained by using a *drive corrector*, an electronic system into which the telescope motor is plugged. Drive correctors are supplied with pushbutton controls to either speed up or slow down the motor drive as necessary. Such a system can guide much more accurately than can any manual system. In addition, the remote-control box of a drive corrector frees your hands from the telescope, resulting in noticeably less vibration.

There are several methods of guiding long exposures, each suited to a distinct type of telescope system and each suitable for differing pocketbooks. Some of the advantages and disadvantages and the applicability of each system are described in the following paragraphs.

Guiding telescopes. The most economical method of guiding a prime focus astrophotograph is by using a guide scope mounted parallel to the tube of the main telescope. However, this method is in many cases the least efficient of the others. Guiding should be done with a crosshair eyepiece, either illuminated or not. The illuminated reticle provides you with greater ease of guiding on extremely faint stars, thereby giving greater accessibility to more stars on which to guide. Dark crosshairs can also be used effectively by a simple trick: using a fairly bright star. Simply throw the star out of focus until it appears to be a small disk of light, thus making the crosshairs (unilluminated) appear silhouetted against the disk of the out-of-focus star.

The higher the magnification used for guiding with this system, the better. This increased magnification allows for a slight drift of the star image to be detected in the eyepiece and corrected before it is recorded on film. A minimum of 75x magnification should be used for effective guiding. The larger the aperture of the guide scope, the easier the guiding on fainter stars and the more stars visible in the eyepiece. The minimum aperture for any guide system is 2.4 inches.

Flexure is a problem in large guide telescopes. After some time in one position, optical sag occurs between the main scope and the guide scope. Substantial supports along the guiding telescope and/or shorter exposures are remedies for optical sag.

Off-axis guiding. Some telescopes now come equipped with provisions for guiding through the main telescope at the same time you are taking photographs through the telescope. This has the unique advantage that you can guide on the same object you are photographing, using the same optical system. There are problems with such systems, however. The main drawback of off-axis guiders is inherent in the nature of their design. A small prism deflects some of the light that normally would pass straight through to the camera. The total area of the light cone that is interrupted by this prism is quite small, affording only a tiny field of view in the off-axis guider. Consequently, finding a guide star is sometimes impossible.

Off-axis guiding systems are expensive and difficult to add to a telescope/photography system not manufactured to accept them. However, guiders are available for those systems now manufactured for this application (most catadioptic telescope systems) that are much cheaper than if you were to add a guide telescope.

Electronic guiding. Most of the fun in astrophotography comes from the actual taking of the picture yourself. Thus, it seems somewhat unfortunate that a trend is rapidly progressing toward using electronic guiding systems that make all the necessary corrections for you. These systems are now available commercially at a modest cost and are extremely precise. The telescope using such a system must have on both axes electric slow-motion controls that are connected to the electronic unit. A guide star is picked up in the guiding eyepiece and followed electronically, subject to the tiniest deviations. You merely need to set the unit initially, start the exposure, and sit back while the telescope, camera, and guider do the rest.

Eyepiece Projection Astrophotography

Projected image photography is used for close-up views of the moon and planets or for high-resolution photography of, for example, double stars. The principle behind such photography is similiar to that of a slide projector casting an enlarged image onto a screen—the farther the screen is from the projector the larger the image will appear on the screen. In our case, we use an eyepiece to project an enlarged image of a celestial object onto film (see Figure 14–4).

The use of an eyepiece between the telescope and camera increases the effective focal length (EFL) of the system, as shown in the computations on page 254 and in the Henry graphs shown in Figures 14–5 to 14–9. Consequently, the system is many times slower compared to that of prime focus photography. Because of the slow nature of projected photography, only the moon, planets, and brighter objects are usually photographed. Such objects require exposures ranging from ¼ to 15 sec.

Computing Effective Focal Ratios

The effective focal length for any given combination of telescope, eyepiece, and camera

FIGURE 14–4. Demonstrating the positive (or eyepiece) projection system of astrophotography.

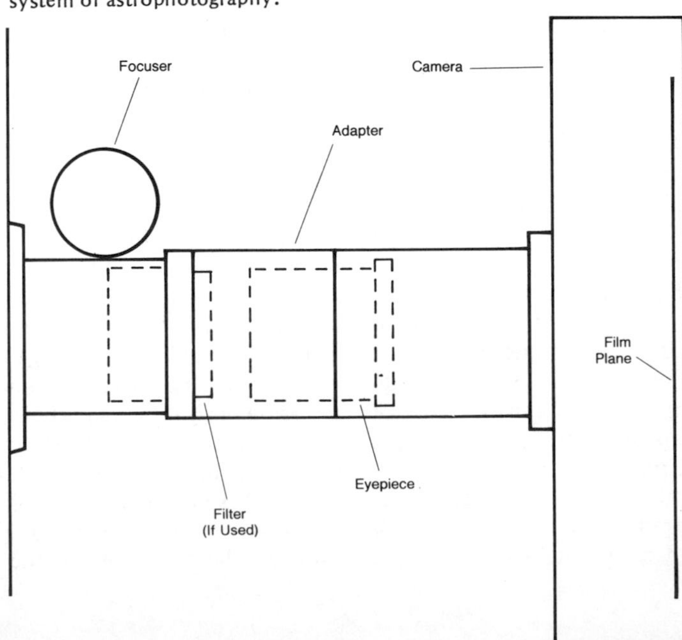

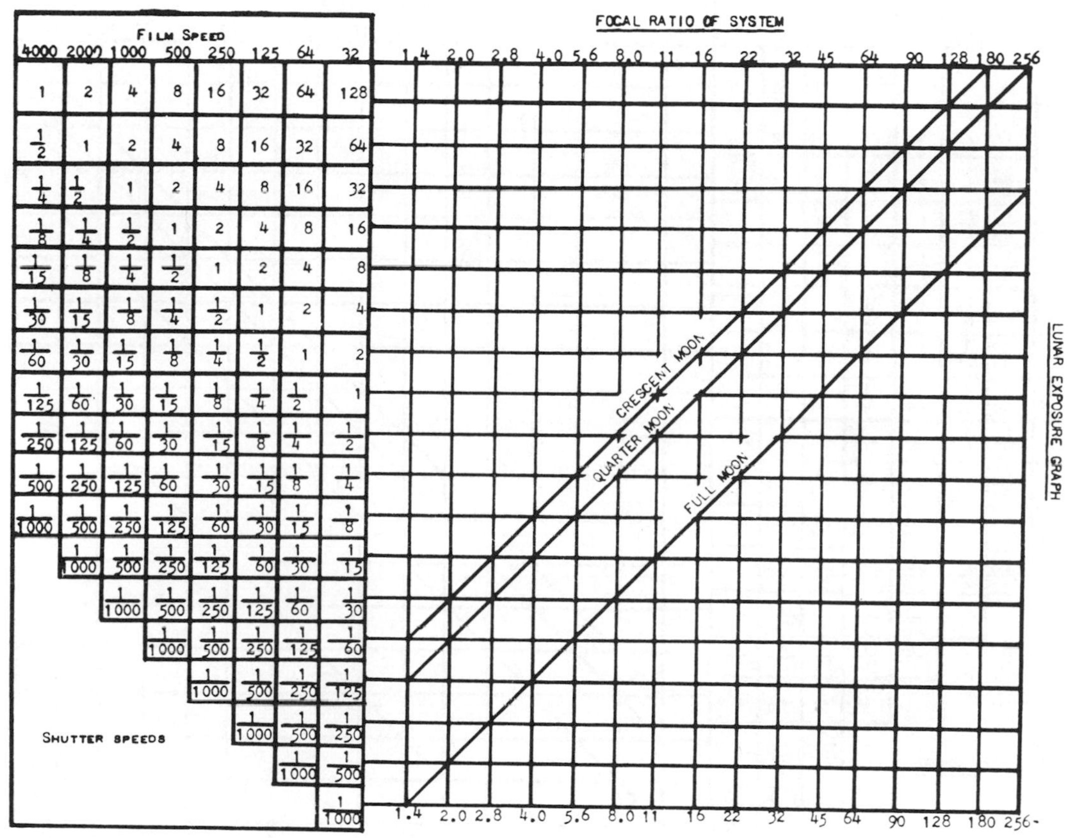

FIGURE 14-5. Lunar exposure graph. (Courtesy, Midsouth Astronomical Research Society)

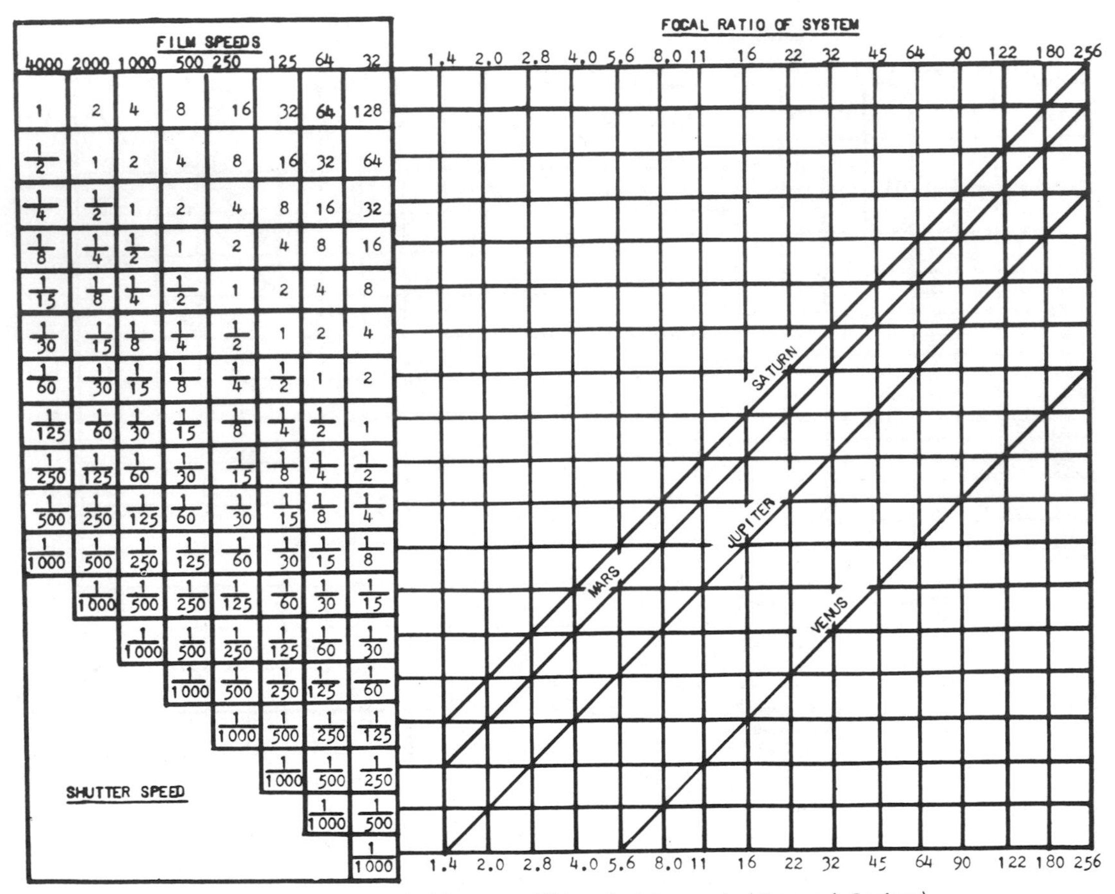

FILM SPEEDS 4000	2000	1000	500	250	125	64	32
1	2	4	8	16	32	64	128
1/2	1	2	4	8	16	32	64
1/4	1/2	1	2	4	8	16	32
1/8	1/4	1/2	1	2	4	8	16
1/15	1/8	1/4	1/2	1	2	4	8
1/30	1/15	1/8	1/4	1/2	1	2	4
1/60	1/30	1/15	1/8	1/4	1/2	1	2
1/125	1/60	1/30	1/15	1/8	1/4	1/2	1
1/250	1/125	1/60	1/30	1/15	1/8	1/4	1/2
1/500	1/250	1/125	1/60	1/30	1/15	1/8	1/4
1/1000	1/500	1/250	1/125	1/60	1/30	1/15	1/8
	1/1000	1/500	1/250	1/125	1/60	1/30	1/15
		1/1000	1/500	1/250	1/125	1/60	1/30
			1/1000	1/500	1/250	1/125	1/60
				1/1000	1/500	1/250	1/125
					1/1000	1/500	1/250
						1/1000	1/500
							1/1000

FIGURE 14-6. Planetary exposure graph. (Courtesy, Midsouth Astronomical Research Society)

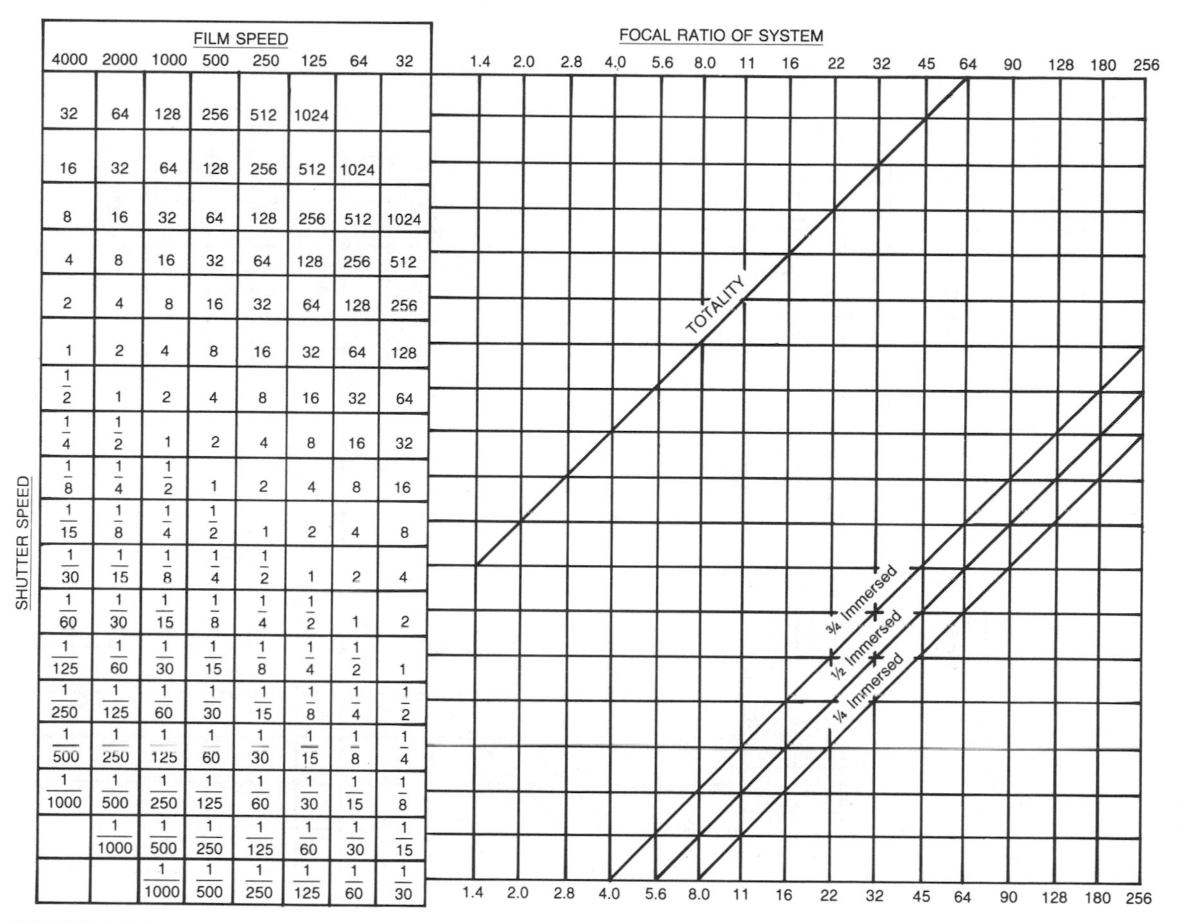

FILM SPEED							
4000	2000	1000	500	250	125	64	32
32	64	128	256	512	1024		
16	32	64	128	256	512	1024	
8	16	32	64	128	256	512	1024
4	8	16	32	64	128	256	512
2	4	8	16	32	64	128	256
1	2	4	8	16	32	64	128
1/2	1	2	4	8	16	32	64
1/4	1/2	1	2	4	8	16	32
1/8	1/4	1/2	1	2	4	8	16
1/15	1/8	1/4	1/2	1	2	4	8
1/30	1/15	1/8	1/4	1/2	1	2	4
1/60	1/30	1/15	1/8	1/4	1/2	1	2
1/125	1/60	1/30	1/15	1/8	1/4	1/2	1
1/250	1/125	1/60	1/30	1/15	1/8	1/4	1/2
1/500	1/250	1/125	1/60	1/30	1/15	1/8	1/4
1/1000	1/500	1/250	1/125	1/60	1/30	1/15	1/8
	1/1000	1/500	1/250	1/125	1/60	1/30	1/15
		1/1000	1/500	1/250	1/125	1/60	1/30

FIGURE 14-7. Lunar eclipse exposure graph.

FILM.SPEED							
4000	2000	1000	500	250	125	64	32
1	2	4	8	16	32	64	128
1/2	1	2	4	8	16	32	64
1/4	1/2	1	2	4	8	16	32
1/8	1/4	1/2	1	2	4	8	16
1/15	1/8	1/4	1/2	1	2	4	8
1/30	1/15	1/8	1/4	1/2	1	2	4
1/60	1/30	1/15	1/8	1/4	1/2	1	2
1/125	1/60	1/30	1/15	1/8	1/4	1/2	1
1/250	1/125	1/60	1/30	1/15	1/8	1/4	1/2
1/500	1/250	1/125	1/60	1/30	1/15	1/8	1/4
1/1000	1/500	1/250	1/125	1/60	1/30	1/15	1/8
	1/1000	1/500	1/250	1/125	1/60	1/30	1/15
		1/1000	1/500	1/250	1/125	1/60	1/30
			1/1000	1/500	1/250	1/125	1/60
				1/1000	1/500	1/250	1/125
					1/1000	1/500	1/250
						1/1000	1/500
							1/1000

SHUTTER SPEED

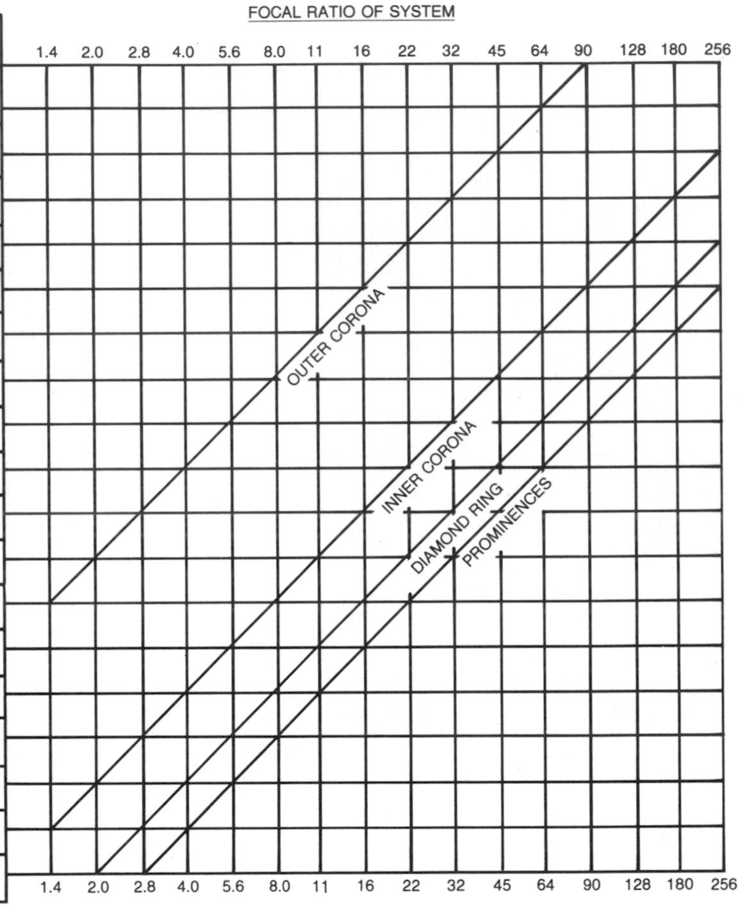

FOCAL RATIO OF SYSTEM

1.4 2.0 2.8 4.0 5.6 8.0 11 16 22 32 45 64 90 128 180 256

OUTER CORONA

INNER CORONA

DIAMOND RING

PROMINENCES

1.4 2.0 2.8 4.0 5.6 8.0 11 16 22 32 45 64 90 128 180 256

FIGURE 14-8. Solar eclipse graph.

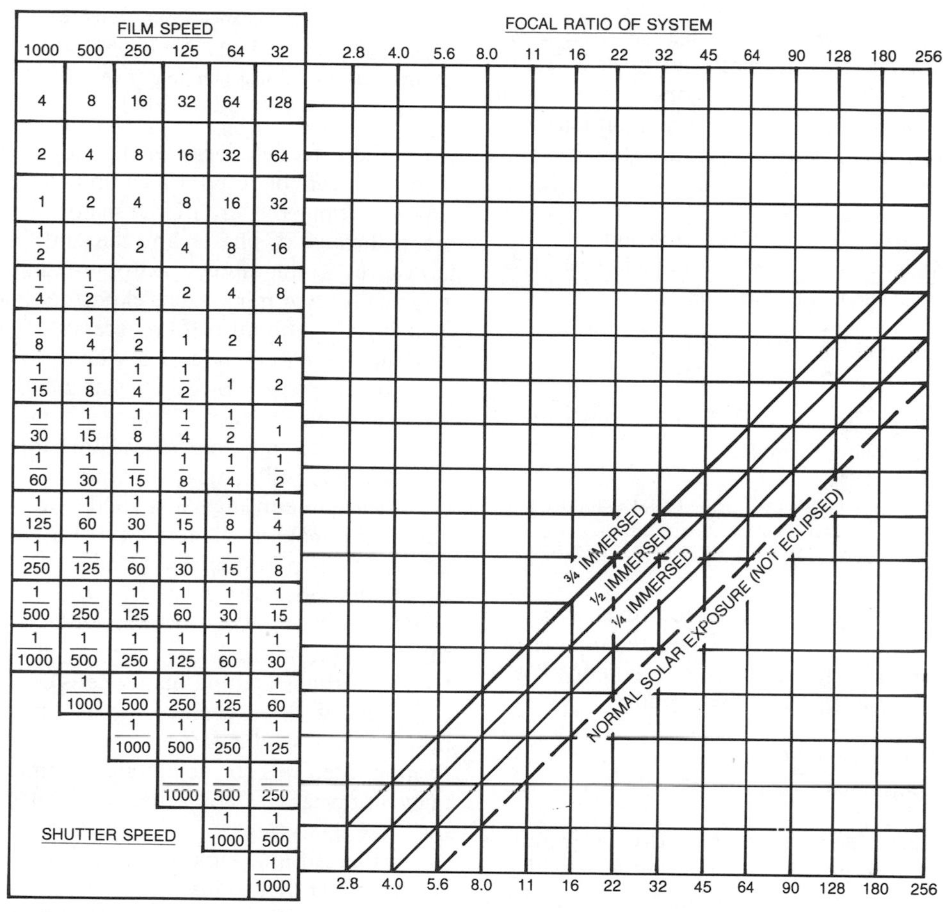

FIGURE 14-9. Solar exposure and partial solar eclipse graph. Exposures given on this graph are for No. 4 neutral density filters. For No. 2 density filters, decrease 6 stops. For No. 3 neutral density filters, decrease 3 stops. For No. 5 neutral density filters, increase 3 stops. For No. 6 neutral density filters, increase 6 stops. To use high-contrast copy film (1045), decrease 3 stops.

can be calculated as shown in the following discussion and by Figures 14-5 to 14-9.

Camera with lens. When using only a camera with lens, the focal ratio (EFL) is whatever f/stop one chooses on the lens barrel.

Afocal method. The afocal method uses a camera with lens attached (set at infinity) to photograph through a telescope projecting an image with eyepiece. Mount camera on telescope, or better, on a sturdy tripod. The following formulas apply:

$$\text{EFL} = \frac{\text{Scope Magnification} \times \text{Focal Length of Camera Lens (in inches)}}{\text{Objective Diameter (in inches)}}$$

$$\text{Scope Magnification} = \frac{\text{Objective Focal Length}}{\text{Eyepiece Focal Length}}$$

When using a Barlow lens in the afocal method, use the following formula:

$$\frac{\text{Barlow Magnification} \times \text{Scope Magnification} \times \text{Focal Length of Camera Lens}}{\text{Objective Diameter}}$$

Prime focus. This method uses a camera without lens that is mounted on the focuser of a telescope. The telescope is not equipped with eyepieces. Use the following formula:

$$\text{EFL} = \frac{\text{Objective Focal Length}}{\text{Objective Diameter}}$$

Negative projection. Negative projection is similar to the prime focus method except that a Barlow is used in the system, according to the following formula:

$$\text{EFL} = \frac{\text{Objective Focal Length}}{\text{Objective Diameter}} \times \text{Barlow Magnification}$$

Positive (eyepiece) projection. This method uses a camera without lens that is mounted on the telescope fitted with an eyepiece. A photographic adapter should be used to mount the camera to the telescope, which also should hold the eyepiece. It is helpful if the distance from the eyepiece to the film plane can be changed and also if filters can be added to the adapter if desired.

$$\text{EFL} = \frac{\text{Magnification (by projection)} \times \text{Objective Focal Length (in inches)}}{\text{Objective Diameter (in inches)}}$$

Magnification in the preceding equation is derived as follows:

$$\text{Magnification (by projection)} = \frac{\text{Distance from eyepiece to film plane (in inches)}}{\text{Focal Length of eyepiece (in inches)}}$$

Once the effective focal length has been determined, it can be used to compute the proper exposure time for any of the major planets, the moon through its phases, and the sun. The Henry Exposure Graph allows astrophotographers to graphically determine exact exposure times, thus reducing the amount of film necessary to bracket exposures to ensure at least one proper exposure.

The eyepiece projection system has some of its own unique problems that are not encountered with the other methods of astrophotography so far described. Still critical requirements, however, are a sturdy mounting and a good clock drive. To obtain the maximum capability of a system and to assure crisp resolution of planets and lunar detail, the clock drive must be capable of converting to a *planetary tracking rate*, instead of the common sidereal rate of most clock drives. Usually a rheostat type of device is coupled into a frequency drive corrector that can vary the power input to the drive, thereby changing its turning rate. This corrected rate becomes increasingly important as the effective focal length of the system increases.

Projected image photography has two major enemies: (1) vibrations in the telescope system and (2) atmospheric turbulence. Vibrations can be caused by such things as the camera shutter, a slight breeze blowing on the telescope tube, the shutter-release cable, rough clock drives, or simply someone walking on your concrete pad or back porch while you are trying to take the picture. Some of these things may seem minor, but consider the magnifications at which this system is capable of working. Remember that prime focus photography generally works at 10x to 40x magnification over the normal 50-mm camera lens. Momentary vibrations or disturbances in the atmosphere are usually not too noticeable at this magnification. Projected photography, however, is capable of magnifications 500x and greater. At this power even the slightest vibration can subtract from the resolution of an otherwise excellent photograph.

The atmosphere also presents a few problems. Those beautiful, clear, dark nights that you might be accustomed to thinking of as good "astronomy nights" are often a disappointment when it comes to high resolution photography. Quite often, the best nights for projection photography are those that are slightly hazy as a result of the atmospheric inversion layers that put a cap on the boiling mixture of warm and cold air. Best results are usually obtained in the predawn hours, around 2:00 AM, after most of the earth's escaping heat has equalized with the surrounding air. Remember, if you cannot see the detail visually, you certainly cannot photograph it. Nothing can be accomplished on a night when the stars twinkle wildly and the image of a planet looks as though you are seeing it through a glass of water.

Let's now look at the problem of *vibration* specifically. The best way to check your system for vibration is to set it up as you would use it for projection photography. Choose an eyepiece that gives you a focal ratio of f/150 or higher if possible—the higher the f/number the better. Next put a bright planet (for example, Jupiter, Saturn, or Mars) in the field of view. Use your camera viewfinder to monitor the planet. This critical evaluation will give you an idea of the usefulness of your total system. First, check out your focusing mechanism. Is it smooth? Will it maintain your focusing once it is set? Is the drive operating correctly? Check to see how long it takes the image to drift one way or the other. If the drift is obvious, make fine adjustments in the driving rate until you notice only unpredictable deviations in tracking. Be sure to check for vibrations introduced by the clock drive assembly, a problem not easily remedied. Next, check to see how long it takes the image to settle down after you have closed the shutter of the camera. Up to 5 sec can be expected in any system. Anything longer can mean that the mounting of your scope must be beefed up a little. A simpler remedy is to try a smoother technique in shutter release. Sometimes an *air-release* shutter cable can help, or you can try using a 2- to 3-inch cable, releasing it entirely after opening the shutter. Because the most useful f/ratio for projection photography is around f/80 to f/100, any vibrations detected and eliminated at this higher f/ratio will be considerably less in actual use.

The Henry Exposure Graph gives the exposures needed for a given focal ratio/film combination. For the shorter exposures (less than 1/15 sec), you must rely on a smooth shutter and good technique to give a sharp photograph. For longer exposures, try the old card or hat trick. Hold a large dark card in front of the telescope tube while you click open and lock the shutter. Wait about 2 to 5 sec for all the vibrations to die down and then remove the card for the desired length of exposure time. Simply move the card back in front of the telescope to end the exposure, and then be sure to release the cable release, thus closing the camera shutter. This is an excellent technique and will eliminate most common vibrations that can ruin your photograph. Exposures from 1/8 sec to 2 sec are a little hard to estimate, but those of 2 sec and more can be estimated with fair accuracy after some practice. The rule of projection photography is to take lots of pictures and hope that a few are good; the good ones make up for all the wasted film.

CHOOSING FILMS FOR ASTROPHOTOGRAPHY

Many types of film, both black and white and color, are available to the amateur astrophotographer today. Of these, you can choose several color and many black and white to make sure of taking successful astrophotographs. In fact, some black-and-white film emulsions are made specifically for astrophotography.

Let's consider the ideal astrofilm. Such a film will have met several requirements, which are listed on the following page.

☆ *Speed*. The most obvious requirement is film speed, which is measured by its ASA rating. The higher the ASA number, the faster is the film, therefore the shorter is the exposure.

☆ *Grain*. Grain size is another element that must be considered for astrophotographs. Grain is determined by the physical nature and makeup of the film emulsion (e.g., silver grains in black-and-white films). To make enlargements of the negative and give satisfactory results with little loss in resolution, a fine grain film is desirable.

☆ *Contrast*. Another requirement is contrast, which can also play an important role in astrophotography. Bringing out the faint spiral arms of a galaxy or defining the ring divisions of Saturn can be enhanced by high-contrast film.

☆ *Color sensitivity*. Color sensitivity is an important consideration, even in black-and-white films. Distant galaxies and reflection nebulae emit light in the blue range of the spectrum. Emission nebulae and planetary nebulae are often red. Planets, have almost the entire range of color. Thus, *panchromatic* films are desirable for solar system photography because they possess the capability of recording all colors equally well, except at the extreme ends of the spectrum.

To summarize our "perfect" astrofilm, we need a fast, fine grain, high-contrast, high-resolution, panchromatic film. Unfortunately, this film, whether black and white or color, does not exist. Fortunately, for any given astronomical subject, only a few of these many characteristics are important. Therefore, different films suitable for photographing different objects.

Breaking down the astrosubjects into three categories helps to determine which films can be used for each purpose. The categories are (1) constellation, (2) lunar and planetary, and (3) deep sky.

Constellation Photography

Constellation photography (or meteor photography) is usually performed by a fixed camera on a tripod and is suitable for recording the brighter stars in a constellation, or meteors as they streak by. Because about 30 sec is the upper limit for exposures without showing star trails, a *fast* film is the basic requirement. Films with ASA ratings of 400 or higher are recommended, such as Kodak's Tri-X (ASA 400) or Kodak's 2475 Recording Film (ASA 1200). Both of these films can be push-processed in the darkroom to even higher ASA values than those given, if you use Ethol Blue extended developer.

Several new color films have emerged on the market that also meet this speed requirement, such as Fujichrome 400 (Fuji) and Ektachrome 400 (Kodak). There is also a fairly good color slide film, High Speed Ektachrome (Kodak), which can be push-processed to an ASA of 800.

Lunar and Planetary Photography

Lunar and planetary photography has requirements much different from those previously described. The moon and the planets are fairly bright, even if you use increased magnifications (effective focal lengths). Generally, exposures of less than 1 sec for the moon and less than 10 sec for the planets are sufficient even with some of the slow films (low ASA numbers). Therefore, high speed is definitely *not* a necessity for such photographs. However, to record fine details in the turbulent cloud belts of Jupiter, or close-up views of the lunar craters, *high resolution, contrast*, and *fine grain* are a steadfast must.

Many black-and-white films are suitable for recording the moon and planets, meeting the requirements stated, but the selection of color films is rather limited and quite disappointing. The most readily available black-and-white film for lunar and planetary photography is Kodak's Plus-X. This is a fine grain, moderately high resolution, panchromatic film—an excellent film to experiment with. You will find yourself coming back to this film many times after experimenting with some of the more specialized films.

Two other widely used black-and-white films for lunar and planetary shots are Kodak's SO-410 (now 2415), originally manufactured as a photomicrography film, and High Contrast Copy Film. SO-410 (or 2415) is a faster, very high resolution, very fine grain film with panchromatic and extended red sensitivity. It is highly recommended for photographing Saturn or Mars and is used with much success on Jupiter and lunar close-ups. High Contrast Copy Film is slower, but it has virtually undetectable grain, and possesses extremely high resolution capability and sharp contrast. This film is often used on Jupiter, because of its brightness and fine detail. Both 2415 and High Contrast Copy can be push processed to 3.5 times their normal speeds using Ethol UFG, which maintains contrast and increased effective speed.

Color films are not very well suited for high resolution photography. However, there are some advantages to color planetary photography. Quite often what contrast does exist on the planet is due to slight color differences and not to contrast variations. For such subtle detail, Kodak's High Speed Ektachrome works fairly well, but it becomes somewhat fuzzy if enlarged very much. Kodachrome 64 and Kodachrome 25 have much finer grain, but they are insensitive to many of the spectral regions necessary for successful planetary photography. Kodak's Ektachrome 400 (ASA 400, slide film) is an excellent choice for planetary photography.

Deep-Sky Photography

Recording deep-sky objects (e.g., nebulae, galaxies, clusters) is an entirely different realm of photography. Long exposures are almost always required, usually of at least 10 min but more often 20 to 30 min. At first glance, a high-speed film would appear to be the answer to such sky shooting, but such is not always the case, particularly when it comes to color film.

Many of the conventional films do not continue to record an image after the first few minutes of exposure to the light of that image. For example, 2475 Recording Film, with an ASA of at least 1200, is insensitive to light after only several seconds of exposure. Kodak's Tri-X film, the usual answer to low light photography, is efficient for only about 10 min. This lack of further light accumulation after an interval of time is known as *reciprocity failure* and is present to some degree in all films.

Because of the long exposures needed in astrophotography, Eastman Kodak developed a film that is low in reciprocity failure, commonly known as the 103a films. Originally developed for professional observatory programs, this excellent film is now available to the amateur astronomer in the popular 35-mm size to fit all common SLR cameras. The cost is about six times that of a comparable roll of black-and-white 35-mm film. Although rated at about the same ASA as Tri-X film (ASA 400), the 103a films can record three or four times more light in a 30- to 40-min exposure than Tri-X, and will continue to record for at least three hours. However, the recording rate does gradually decrease to a slower ratio after the first hour of use. The 103a films have the disadvantage of having coarse grain. However even when they are compared to Tri-X for enlargements up to 8 X 10 inches, the grain is acceptable. These films are available in several different color sensitivities, hence are usually referred to as having *spectroscopic emulsions:*

103aE—emulsion with extended sensitivity into the red region of the spectrum; this is an excellent choice for emission nebulae such as the Orion Nebula and the Lagoon Nebula.

103aO—blue-sensitive only and good for reflection nebulae such as that surrounding the Pleiades star cluster; also good for the outer arms of distant galaxies and general galactic photography. Comets with strong dust tails are recorded well by this emulsion.

103aF—an all-purpose spectroscopic film, because it is panchromatic and sensitive to all ranges of the spectrum. Good choice if several different types of objects are to be photographed in the same evening, or if your budget precludes purchasing all three types.

Color films are limited in deep-sky photography. High Speed Ektachrome (Kodak) is a fairly good choice for the very bright objects and if exposures of 15 or less are to be made. The Kodachrome (Kodak) films are virtually a waste in deep-sky photography; they have very low recording power in low light situations. Two popular films, GAF 500 and GAF 200, are quite good for deep-sky work, being usable up to one hour. However, these two films are in short supply and will no longer be produced.

The best commercially available color film for deep-sky photography is Ektachrome 400. This film has much greater recording power (sensitivity) on exposures up to 15 min. This limited exposure range should not be a serious deterrent to this film's use, however. By using special mailers available at any photo store, in which you send your film to Kodak, Ektachrome 400 can be push-processed by Kodak up to ASA 800, thereby increasing its effective recording power with a low-reciprocity range. Ektachrome 400 is also red sensitive, which makes it great for red objects such as nebulae, but it does not respond well to blue light. Likewise, no color film on the market at this time is blue-sensitive since the discontinuing of GAF's films. Ektachrome 200 (ASA 200) works fairly well in the blue end of the spectrum if it is used on conjunction with a cold camera, which lowers the film's reciprocity failure.

Film Recommendations

As must be apparent by now, astrophotography is quite often a trade-off when it comes to film choice. Trial and error with a particular system of camera/telescope/film is often the answer to finding the best film.

For the beginning astrophotographer, Kodak's Tri-X and Plus-X are the best films to start with. Although not perfect for anything, they are satisfactory for just about everything. Tri-X, with its fast speed, can be used for deep-sky photography and push-processed to give some good photographs from a 10-min exposure. Plus-X, with its fine grain, is good for the moon and planets. Another advantage of using these two films is their availability at almost any store where film is sold. In addition, these two films can easily be processed (developed and printed) at home. The best black-and-white photographs will be the ones you have done yourself—from beginning to end. Commercially processed black-and-white films and prints rarely show the detail and contrast that can be obtained by processing these films yourself. So, for the best and the most rewarding astrophotos, try to process them yourself.

Processing color film is a different story. Color slide film seems to work better, with the added advantage that you are able to enlarge the image of the slide on a large screen to show off to fellow astronomers. Color slide film is best developed commercially, unless you are really into film developing. Slides are much more difficult, expensive, and time consuming than the black-and-white processing. In addition, the chances of doing as good quality a job on the processing as a commercial firm can do are really very slim. Prepaid processing mailers will have your processed slides back to you in a week or less—far less time than is usually required to process the slides yourself.

When you have slide film commercially processed, I strongly recommend that you shoot the first two frames on ordinary daylight "earth" subjects. This allows you to check the film for quality or flaws after it is returned from the processor. It also gives the photo lab a good starting reference to use for cutting the film into individual frames. If the first few shots are dark photos, the cutting machines used by commercial processors cannot have a distinct cutting reference, and the lab may have to guess where to start cutting. Such guesswork could ruin some really great photographs that took much time and work.

Table 14-3 provides a summary of black and white color films and their characteristics for astrophotography. Color films are summarized in Table 14-4.

New Films

Two new color films have recently come on the market—the Ektachrome 200 and 400. My tests show that the 400 film has as good grain and better resolving power than the 200. Also, the 400 can be push-processed to 800 by Kodak with

TABLE 14–3. Recommended black-and-white films.

Film Name	*ASA No.*	*Color Sensitive*	*Resolution*	*Contrast*	*Advantages*	*Disadvantages*	*Use*	*Maximum Exposure*
Tri-X	400 to 1600	All (Ethol Blue)	Fair	Fair	Speed; moderate resolution.	Excessive grain.	Constellations; deep sky; beginning photography of planets; meteors.	10 min
Plus-X[a]	125	All	Good	Fair	Speed; resolution.	Moderate grain.	Planets; moon.	30
High contrast	25	Red	Fair	Excellent	Contrast; no grain.	Extremely slow.	Planets; moon.	10
SO–115[a]	100 to 380	All (Ethol UFG)	Good	Excellent	Speed; contrast; no grain.	—	Planets; moon.	10
2475 Record.	1200+	Red	Poor	Fair	Speed only.	Limited exposure; excessive grain.	Stars; meteors.	3
103aE	400	Red	Good	Good	Long exposures; red objects; city use.	Limited color response; grainy.	Emission nebulae.	120
103aO	400	Blue	Good	Good	Long exposures; blue objects.	Limited color response; cannot use near city lights; grainy.	Galaxies; clusters; comets.	120
103aF[b]	400	All	Good	Good	Recording power; all colors	Grainy.	All-purpose deep sky.	120

[a]Recommended for lunar and planetary projection photography.
[b]Recommended for deep-sky prime focus or piggyback photography.
Note: All films listed are by Kodak. Each may be easily processed in the home by readily available darkroom chemicals. Processing your film yourself provides better contrast, resolution, and rendition of celestial objects, than can be obtained through commercial black-and-white processing, and with considerable savings.

TABLE 14-4. Recommended color films for astrophotography.

Film Name	*ASA No.*	*Color Sensitive*	*Resolution*	*Contrast*	*Advantages*	*Disadvantages*	*Use*	*Maximum Exposure (min.)*
Ektachrome	200+	Red	Good	Fair	Slide film; rapid recording under 10 m.	Reciprocity.	Meteors; constellations; others under 10 min.	15
Ektachrome	400	Red	Good	Fair				10
Kodacolor 64	64	All	Good	Fair	Very low grain.	Limited recording power.	Moon; bright objects.	5
Kodacolor 400[a]	400	All	Good	Good	Speed; low grain.	Print film.	Planets; moon; deep sky.	30
GAF 200[a]	200	Blue	Good	Fair	Resolution; sensitivity.	Not available.[c]	Planets; deep sky.	60
Fuji R100[b]	100	Red	Good	Good	Use in city; sensitivity; extended recording power.	Slow.	Deep sky.	60
Fuji 400	400	Red	Fair	Good	Speed.	Poor resolution.	Constellations; deep sky.	30

[a]Recommended for lunar and planetary projection photography.
[b]Recommended for deep-sky prime focus or piggyback photography.
[c]Try K-Mart Focal 400.
Note: Slide film is listed as an "advantage" simply because it can be processed readily through commercial firms with good results, whereas print film (color) does not render quality images when processed commercially.

only a little loss in the final outcome. This is better film than the old High Speed Ektachrome 160. Exposure can be up to 15 min. Longer exposure records little extra. Also, this film is very red sensitive, so it is a good choice for emission nebulae. Ektachrome never has been very blue sensitive, so the 200 and 400 are not great performers on reflection nebula and galaxies. Stick to the GAF films for these objects if they can be found. (K-Mart Focal 400 is the same film as GAF 500.)

BLACK-AND-WHITE FILM PROCESSING

The best way to have black-and-white film developed is to do it yourself. Commercial labs do not know how to handle astronegatives. You need to use special developers and techniques to bring the most out of your negatives. However, all the materials and chemicals can be obtained at your local photo store. The actual developing can be done in a small bathroom.

The common black-and-white films, such as Plus-X and Tri-X, have developing directions included in the packaging. The development of the special astrofilms is covered here. The 2415 is most often developed in D-19 for 4 ½ minutes at 68°F. Stop bath is then used for 30 to 60 sec, followed by 3 to 5 min of fixing in rapid fixer. Wash for 10 to 15 min. Soak in Photo-Flo for 1 min before hanging to dry. This produces high-contrast negatives with very fine grain and a speed rating of about ASA 160. This film can also be developed in HC 110 Dilution D, for 8 min at 68°F. This produces lower contrast but reduces grain and results in an ASA of about 80. Pushing 2415 to ASA 400 can be done in Ethol UFG at full strength for 8 min. This still maintains the film's high contrast, high resolution, and fine grain characteristics.

Kodak's 103a series films are usually used on low-contrast objects such as nebula and galaxies. Therefore, this film should be developed in high-contrast developers to bring out the faintest details in the photo. D-19 is usually used with the 103a film. Standard development is for 4 min in D-19, with a 60-sec *water* stop bath. *Do not use an acid stop bath with this film*. Fixing is normal—3 to 4 min in rapid fixer. Wash for 10 to 15 min, followed by a 1-min soak in Photo-Flo. The 103a film can be given a 50% push be developing for 6 min in D-19 instead of 4 minutes. This produces slightly more grain, but it can bring out some very faint detail.

One more step needs to be added to this process. The 103a film has an antihalation backing built into the film to help prevent light from a bright star bouncing off the back of the film and scattering back through the emulsion. This backing partly dissolves in the developing process, and the remainder can be removed after the soaking in Photo-Flo. The negative will appear slightly dirty. The backing can be gently rubbed off with a cotton swab soaked in Photo-Flo. The negative is then rinsed one more time in Photo-Flo before hanging it to dry.

REFERENCES

de Vaucouleurs, G., *Astronomical Photography*. London: Faber and Faber, 1961.

Miczaika, G.R., and W.M. Sinton, *Tools of the Astronomer*. Cambridge, MA: Harvard University Press, 1961.

Paul, Henry, *Outer Space Photography*. New York: Amphoto, 1960.

Rackman, Thomas, *Astronomical Photography at the Telescope*. London: Faber and Faber, 1972.

Selwyn, E.W.H., *Photography in Astronomy*. Rochester, N.Y.: Eastman Kodak Co., 1950.

Sidgwick, J.B., *Amateur Astronomer's Handbook*. London: Faber and Faber, 1971.

GLOSSARY

A

absolute magnitude A standard by which the actual luminosity of a star can be compared to other celestial objects. It is the magnitude that the star would appear if located at a distance of 10 parsecs (one parsec = 3.26 light years).

acceleration The eastward drift against the normal current of Jupiter's clouds, resulting in a decrease in the observed longitude of some feature.

accretion The process by which the mass of a celestial object is gradually increased by bombardment of that object by other celestial objects. The process is increased rapidly as the larger object accumulates additional mass.

accuracy The estimated error, in seconds or fractions thereof, of an observer's timing of an occultation. If the observer feels that the occultation timing might have been affected by poor seeing and could be 0.5 sec off, the accuracy of that event is given ±0.5 sec. Each occultation event has its own accuracy rating; the observer has a *consistent* accuracy predetermined as his or her personal equation, which remains unchanged for all observations.

achromatic A two-element lens common in most refractor telescopes. The addition of a second (flint) element reduces serious chromatic (color) aberration where various colors focus at varying distances and not at one point. The achromatic lens combines one positive and one negative lens, each of which nearly cancels the chromatic aberration of the other.

Airy disk A bright central image, appearing as a tiny disk, from a point source of light such as a star. The disk is formed by the diffraction patterns of the objective and is surrounded by many faint concentric rings of light. About 87% of the star's light is concentrated in the disk itself, whereas less than 13% remains in the concentric rings. A residual amount of light is lost through scatter. Defined by Sir George Airy, an Astronomer Royal of England, who served from 1835 to 1881.

albedo The ratio of the total amount of light a celestial body receives in comparison to the amount reflected from its surface in all directions. A perfect reflector has an albedo of 1.0, which means that it reflects 100% of the light it receives. A theoretical black body, which cannot reflect any light, would have an albedo of 0.0.

altazimuth A simple mounting allowing a telescope to move horizontally (in azimuth) and vertically (in altitude). Not particularly useful for serious astronomy studies, other than for sweeping for comets.

altitude A rectangular coordinate measure of the vertical; the angular distance of a celestial object above the horizon at sea level.

amplitude The total range, expressed in magnitude, of the light variation of a minor planet. The amplitude generally increases when the axis of spin of the asteroid is tilted near 0° relative to the earth.

amplitude The total variation of a star, ranging from maximum to minimum, expressed in units of magnitude. The measure is from crest to trough of the light curve, and is the range of magnitude change in that interval.

Ångstrom unit (Å) A unit of length, 10^{-10} meters. Numerical designations for spectral emission or absorption lines give each one's wavelength in Angstroms.

ansae The extensions of the ring system on either side of Saturn as seen from earth.

antitail A lesser, secondary tail of a comet, usually following the comet in its path and thus suspected to be composed of heavier dust particles.

apparent magnitude The magnitude of a star (or other celestial object) as it appears to the eye or telescope. This value is that recorded by the variable star observer.

ASA A numerical representation of the light sensitivity of a particular film; the higher the ASA number, the more sensitive to light. The abbreviation for American Standards Association, this is the equivalent to the European DIN rating.

azimuth A rectangular coordinate measure of the horizontal. It is measured from north, toward east, with east representing 90°, south 180°, and west 270°.

B

backlash Looseness or improper meshing of drive assembly gears so that the system requires several moments to catch up and begin tracking once initiated.

belt A dark horizontal linear feature that is continuous through several degrees of longitude of Jupiter and Saturn. The belts are thought to be valleys in the clouds, the result of convective cooling that plummets the Jovian gases toward the interior.

bifilar micrometer An instrument again becoming readily available that is used for measuring the size of the coma, the position angle of the comet tail(s), and the spacing of nucleus components should splitting occur.

binary star Two or more stars that share and orbit a common center of gravity. The brighter star normally is designated as the *primary* star, and the fainter is known as the *secondary* star.

bolide A fireball that breaks apart, sometimes with recognizable sound. Many times the breakup of a bolide results in a change of the color of the object.

C

catadioptic A compound telescope, using the properties of refraction and reflection, each correcting for inherent problems of the other. Common catadioptic instruments are the Schmidt-Cassegrain (using a corrective plate) and the Maksutov (using a meniscus lens).

celestial equator A great imaginary circle in which the extension of the earth's equatorial plane is represented against the celestial sphere. The coordinates of right ascension and declination are measured in reference to this extension.

center transit The instant at which a feature of a planet is centered on the central meridian.

central meridian An imaginary north-south line dividing a planet into two equal halves. Any feature crossing this line can be timed to within 1 minute, thereby facilitating the determination of longitude. On Mars, unlike on earth, longitude begins at a designated point (Sinus Meridiani) and progresses continually through 360° back to that point. On Jupiter timing of any feature as it crosses the central meridian (CM) can place the feature to within 0.6° in correct longitude.

chromatic aberration A problem common in any optical system through which light must be passed through glass. The problem prevents varying wavelengths of light from focusing at the same distance from the objective, and thus all wavelengths cannot be measured simultaneously.

class The group in which a particular type of variable star is placed, according to its characteristics that most closely match the other stars in that group (e.g., regular, semiregular, irregular).

collimation As used regarding telescopes, the centering of all optical elements on one another, thus achieving maximum performance of the instrument.

color excess An excess over the expected ratio of the color index U–B, where U is the ultraviolet reading from the star and B is the blue measurement. If the amount is higher than the intrinsic value expected, it indicates that the star was reddened as its light passed through the interstellar medium.

color index The slight difference between the apparent magnitude as measured in one wavelength of light and the apparent magnitude in another. The index indicates a predominance of a particular wavelength in the star's spectrum.

color (star spectral class) A star's spectral type is based on its characteristic color. A star of solar type (G-type spectrum) is nearly the same color as the lunar disk since the light from the moon is reflected sunlight of the same class. Thus, during occultation observing it is more difficult to observe such a star near the bright edge of the moon than it is to observe one of B-type (bluish) or M-type (very red).

color temperature A method by which the actual temperature of a star can be determined through the color index B–V, in relation to the temperature of a standard black body (perfect radiator) whose energy distribution through all wavelengths of light corresponds closely with the star.

coma A great cloud of gas that can measure several hundred thousand kilometers in diameter. It emanates from the nucleus of a comet.

comet Seeker A wide-field (2°+), altazimuth-mounted telescope used in the search for comets. It is convenient to use for its portability, generous field of view, and straight-line motion for scanning in a horizontal path.

condensation The graduated brightening of a comet near its center, or the intensity of the nucleus. Some comets show no condensation within the coma, whereas others exhibit intense stellar-like nuclei.

convection The rapid rising of heated gases within Jupiter's and Saturn's clouds, resulting in the detail visible on their globes. As the rising gases cool, their velocities slow. The decreased kinetic energy results in gas compression and eventual drops in altitude.

crater wall The rim, or edge, of a lunar crater formed by the expulsion of lunar crustal material at the time of meteorite impact. The wall's edge can be quite sharp when new, but erosion caused by solar wind will deteriorate it in time.

culmination (of Polaris) The passage of the North Star across the celestial meridian, at which time that star will be exactly south or exactly north of the true pole. Lower culmination occurs when Polaris is at a minimum altitude above the observer's horizon.

D

deceleration The westward drift of a feature on Jupiter, against the normal current of Jupiter's clouds, resulting in an increase in the observed longitude of that feature.

defocusing A procedure by which the lights of the comparison star and the variable star are examined out of focus in the same field of view. A diffused disk of light is easier to examine for subtle differences in magnitude than is a point.

deserts The name given to the seemingly barren red plains of Mars. The reddish coloration apparently is a result of the oxidation of iron in the deposits in these plains.

Doppler effect A phenomenon in which the spectrum of an object rapidly receding from the point of measurement is shifted toward the red wavelength in proportion to the velocity of recession, and shifted to the blue similarly, if the object is approaching the point of measurement.

drive corrector An electronic device into which the electric clock drive of the telescope is plugged. The corrector allows changes in the voltage output, thereby facilitating speeding up or slowing down of the telescope for astrophotography guiding.

dust tails Strongly curved comet tails containing solid molecular particles. These tails have no intrinsic luminosity and are seen only by reflected sunlight.

E

early spectra Spectral-type stars that exhibit spectra preceding that of our sun on the spectral scale O, B, A, F, G, K, M. Because the sun is a G-type star, early spectral types are O, B, A, and some of the hotter F stars.

ecliptic circle A great circle where the plane of the earth's orbit intersects the celestial sphere. The sun's apparent path through the sky appears to be nearly coincident with this circle, and all the planets except Pluto can also be found near the ecliptic circle.

effective focal length The resulting focal length of an optical system after the addition of converging or diverging lenses, or convex or concave secondary mirrors.

elevation For an observatory or observing site, elevation is a measure of the height above sea level. This value is necessary for the reduction of occultation and other data.

ephemeris A table that lists a comet's or planet's orbital elements for a given period of time, usually the geocentric (earth) distance, heliocentric (sun) distance, exact position on the celestial sphere, and magnitude.

equatorial mounting A mounting designed to follow the natural motions of the celestial objects as a result of the rotation of the earth's axis. This mount allows the telescope to move in arcs rather than in straight paths, as in the altazimuth, thereby facilitating use of clock drives and setting circles. An equatorial mounting can be of many varieties; the two most common are the German equatorial and the fork.

erosion The process by which bombardment of the lunar surface by wind—particle radiation—and micrometeorite bombardment results in erosion much like that the earth experiences through wind and water erosion. Solar wind erosion requires millions of years to become evident, although a great deal of the lunar dust is a result of eons of such breakdowns of the lunar highlands.

extended object Any celestial object that is not a point source (i.e., a star). It can be a planet, diffuse nebula, comet, and so forth.

extinction The reduction and reddening of starlight as a result of refraction and scattering of that light as it passes through our atmosphere. Extinction coefficients increase as the angular distance from the horizon decreases.

extrinsic variability Variations in a star's light as seen from earth that are not caused by a physical change in the star. Cases of extrinsic variability are eclipsing variables and the nebular variables.

F

festoons Very faint and infrequent linear features on Jupiter and Saturn that appear to bridge two belts. Most festoons are visible in the equatorial zone, between the north and south equatorial belts.

filter, eyepiece Small section of dense glass, similar to welder's glass, which attaches to the eyepiece of the telescope. These filters are very dangerous and should not be used.

filter, full-aperture An ideal solar filter that fits on the outside of the objective lens or mirror, and thus blocks solar radiation before it enters the telescope. It allows the entire surface of the objective to be utilized, thereby increasing resolution and brightness.

filter, Mylar Very inexpensive aluminized film that reflects most of the sunlight. It is used in two layers for optimum brightness level. These filters transmit a great deal of blue light and are recommended only if some better material is not available.

filter, neutral density A filter that can be used at the eyepiece to minimize brightness, provided that some filtration has already taken place as described above.

filter, off axis A filter of smaller size than the telescope objective, usually located off center of the optical path so that secondary mirror holders (in the case of reflectors) do not disperse the light. Much less expensive than full-aperture filters, but brightness decreases with size.

filter (V) In photoelectric astronomy, a *V* filter closely matches the visual magnitude, as determined with the human eye, for determining the magnitude of the comet.

fireball An extremely bright meteor, with a brightness equal to the brightest planets. Many fireballs leave a wake of fire and are so bright that they cast shadows.

five color index A measure of the color of a celestial object in selective wavelengths representative of a wide range of the spectrum, including ultraviolet (U), blue (B), visible (V), red (R), and infrared (I).

focal length The distance between the objective of a telescope, as measured from its center, to the point of focus where all rays of light converge.

focal length For prime focus photography, the focal length of a system is merely the distance required for the light rays, after passing through or reflecting from the objective, to converge to focus.

focal ratio The ratio f/d of a system, where f is the focal length and d is the diameter of the objective lens or mirror. Focal ratios of f/5 and less are considered "fast" systems in regard to their ability to expose an emulsion to a given total amount of light; ratios of f/10 and above are considered "slow."

following end transit The time at which the west edge of any feature passes across the central meridian of a planet.

G

galvanometer A meter capable of reading weak current. In the photoelectric photometer, the current is normally expressed in milliamperes (mA).

gas tail Known also as the "ion tail," this straight and narrow tail is composed only of ionized molecules from the comet. Solar wind and light pressure are responsible for the elongation of this tail.

geodetic (topographical) survey maps Issued by the United States Geological Survey Commission, these maps are available for any land area within the United States. The scale and visibility of obvious landmarks on these maps allows one accurately to determine the location of the observing site.

grain The emulsion of the film onto which an astrophotograph is recorded. The tiny silver deposits are not distributed uniformly or smoothly on most emulsions, thereby giving a "grainy" appearance when the photograph is excessively enlarged.

grazing occultation An event that occurs when the moon in its orbit overtakes a star or planet and partially occults (eclipses) it on either the north or south pole. Such occultations are particularly important in that the topography of the lunar poles can be ascertained by the passage of the star through the lunar valleys and behind unseen peaks (see also Chapter 8).

group Any seemingly associated number of sunspots, either connected or so close that they appear to have a physical relationship with one another. All group members are positioned in the same general latitude.

G-type Stars with spectra resembling that of our sun; they are named for their placement on the spectral scale.

I

infinity A setting on a camera at which all objects, usually greater than 100 feet distant, are in focus. For practical purposes of astronomical photography, all celestial objects are placed at infinity.

integrated light The light of a celestial object concentrated to a point, as in a star. The integrated magnitude of a celestial diffuse object is the magnitude at which that object would appear if all light were concentrated into a point at infinity.

interaction The drifts of two or more features in the clouds of Jupiter that eventually bring the features within the same longitude if they are located in concurrent latitude. The interaction usually results in some visible disruption or change in the normal current of Jupiter's rotation.

intrinsic variability A characteristic of the majority of all variable stars. In these stars the light changes are the result of a physical change (e.g., temperature or size).

J

Jovian Of, or pertaining to, Jupiter.

Julian day A sequential time reference suitable for computer programming and graphic representation over long intervals of time. All variable star reports are expressed in this time, which began at noon, January 1, 4713 BC and is expressed in elapsed 24-hour days since that date.

K

Kirkwood gaps The gaps between major components of the rings of Saturn. The two most prominent are the Cassini and Encke divisions. They are the results of orbital resonances between the periods of the ring particles and the periods of Saturnian satellites. They appear dark and empty in earth-based telescopes, but Voyager I in 1980 revealed that there are many rings within the apparent gaps, and that what appear from earth to be only three rings are actually more like a thousand.

L

latitude A measure on earth, in degrees, starting at the equator (0°) and progressing north or south to the poles (90°). Exact placement in latitude is expressed in degrees (°) minutes (′) and seconds (″) and decimal parts thereof.

light curve A graph exhibiting the change in magnitude over an interval of time so that determinations of period, shape, and amplitude can be derived.

light grasp The ability of a telescope to collect light. The greater this ability, the greater the potential to discern faint objects. Light gathering is in proportion to the square of the diameter.

limiting magnitude The threshold of a telescope, described by the faintest object discernible. The greater the light grasp of an instrument, the fainter the threshold that is achieved.

longitude A measure, in degrees, of east and west on earth. The beginning of longitude is arbitrarily set at Greenwich, England (0°) and progresses as either east or west longitude to 180°. Longitude is accurately placed in degrees (°), minutes (′) and seconds (″) of arc.

longitudinal drift The change in longitude of any feature as it rotates faster or slower than the normal current of rotation of Jupiter's clouds. Such drift can allow for the determinations of rotational periods of the Jovian features.

longitudinal expanse The total breadth of a feature, as measured from east or west, usually expressed in degrees.

long-period comet A comet that requires more than 200 years to complete one orbit about the sun.

luminosity The intrinsic or absolute brightness of a star, equal to the total energy radiated in 1 sec from the star.

lunar limb Any edge of the lunar disk. The eastern edge is frequently called the preceding limb because the moon moves eastward; the western edge is termed the following limb.

M

magnification The enlargement of celestial objects as viewed through a telescope.

magnitude, absolute The perceived brightness of a star if placed 10 parsecs away (32.6 light years). For meteors, the standard distance is 100 km altitude and, to allow for atmospheric extinction, this distance is measured toward the zenith.

magnitude, apparent The magnitude that is actually perceived of any object.

main sequence The main sequence consists of

90% of all known stars ranging from high temperature/high luminosity stars to low temperature/low luminosity stars, as expressed.

maria Dark areas, or "seas," on the moon. Maria are less-cratered areas of ancient lava flow covered with great depths of lunar dust.

maria The dark markings on Mars whose locations are constant. The somewhat darker coloration is perhaps a result of the albedo (reflectance) of these features, or a concentration of moisture within certain minerals.

melt line A distinct dark band surrounding the periphery of the Martian polar caps during thawing. The line is thought to be the result of mineral discoloration caused by increased moisture from the melting caps.

meridian A great imaginary line on the celestial sphere that passes from exact south, through the zenith, to exact north. The astronomical coordinate system of declination is based along this line. The intersection of the meridian and the celestial equator constitutes 0° declination. Measurements of declination north of that point are positive, whereas those south are negative.

meteor The entry of a usually quite small particle from space into the earth's atmosphere, causing a bright streak or train across the sky.

meteoric Of meteoroid origin; of or like a meteor.

meteorite A meteoroid that has impacted on another celestial object. Many of the meteorites that have struck the moon remain beneath the lunar crust as *mascons* (i.e., concentrated gravitational areas distributed throughout the lunar globe).

meteoroid A particle, much smaller than an asteroid in size, moving through space. Perhaps all meteoroids are the results of a breakup of a comet.

milliampere A small unit of measure of current, representing 1/1000th of an ampere (A).

N

nanometer A small unit of measure, used similarly to the Ångstrom (Å) to measure the position on the spectrum of a particular wavelength of light. One nanometer is equivalent to 10^{-9} meter.

north celestial pole (NCP) or south celestial pole (SCP) The two points on the celestial sphere where the earth's axis, if extended, would intersect. The declinations of these points are +90° and -90°, respectively.

nova A star that suddenly and unpredictably increases its brightness by 10 or more magnitudes. Such stars are usually of late spectral type. The frequency of occurrence of these stars is 10 to 15 per year per galaxy. The increase in brightness is rapid, followed by a less rapid decline and long transition phase, at which the decline in magnitude becomes progressively slower.

nuclear magnitude The magnitude of the nucleus, or central condensation of the comet, as determined by the eye or a photometer.

nucleus The core, or physical bulk, of a comet. Although every comet must originate from some central body, not all comets exhibit a distinct nucleus.

O

objective The primary lens or mirror of an optical assembly, responsible for collecting and converging the light.

occultation The event at which one celestial object occults, or eclipses, another, such as the moon covering a star. Occultations of stars by comets are extremely rare, yet they are quite important for astronomers to monitor, revealing varying densities of the comet gases through the dimming of starlight.

opposition The point in the orbits of the earth and a planet at which the earth is between the sun and the planet.

opposition (Mars) *Aphelic* opposition occurs when the earth is between the sun and Mars, although—because of the elliptical orbit of Mars—the planet is considerably farther away than its nearest possible approach. Aphelic opposition occurs roughly every 26 months. *Perihelic* opposition occurs once every 17 years average. This opposition brings the planet considerably closer to both the earth and sun than during the aphelic oppositions. Meteorological activity is greater during perihelic oppositions because of increased radiation received from the sun.

opposition In the study of comets this term refers to a comet that is approaching the sun from the opposite side of the earth, thereby being in opposition. Such a comet is near the meridian at midnight, opposite the sun.

orbit The true path of the meteor or comet through space when it is plotted with reference to the sun rather than to the earth or stars.

P

panchromatic A film emulsion nearly equal in sensitivity to virtually all wavelengths of the visible spectrum.

path The apparent track of a meteor through the stars, as seen by the observer and recorded on star charts.

penumbra The surrounding halo of a sunspot, often thought to be the convection wall of the sunspot phenomenon. Always lighter than the umbra, and many times streaked with fine linear markings.

perihelion The point in an object's orbit at which it is closest to the sun.

period The interval of time required for a variable to complete a cycle; the time from maximum to maximum, or minimum to minimum.

period The total length of one complete rotation of a minor planet, as determined from one maximum to the following third maximum, or from a minimum to the following third minimum.

personal equation A realistic evaluation by the observer of his or her delay time from the time of actual occultation until the event is realized and responded to. A personal equation of 0.3 sec is normal, and this value is subtracted from each timing to allow for this delay.

phase The event of variation of a star; a description of the cycle of the variable star as expressed from the height of maximum to the depth of minimum.

photometer An electronic (or mechanical) device that is capable of measuring light intensity, much as does the eye, but much more precisely.

photometry The study of the nature and intensity of an object's light, either through visual means, or with instrumentation such as the photoelectric photometer.

photosphere The visible portion of the sun; the layer at which the sunspot phenomenon occurs. This layer, several hundred kilometers thick, is the transition layer at which solar energy is accumulated and radiated into space.

planetary tracking rate The apparent rate at which the planets appear to traverse the sky, as compared to the sidereal rate of stars, which is somewhat faster.

polar alignment Aligning the mechanical axis of an equatorial mounting so that the right ascension axis and motion are parallel to the axis and turning of the earth.

polar caps Frozen material at the Martian north and south poles. The Mars caps are known to be a thin layer of water ice over which is deposited a thicker layer of frozen carbon dioxide.

position angle The angle of a comet's tail relative to due north. The angle is always measured from north (0°) to east (90°). South is 180°, west is 270°.

position angle The PA is measured from the north pole of the moon (0°) eastward (90°). It serves as a reference primarily for star reappearances on the dark limb of the moon.

preceding end transit The passing of the easternmost (front) edge of a feature across the central meridian.

preconception The tendency of the occultation observer, who knows from the predictions the approximate time of occultation or reappearance, to allow his actual timing to be influenced by this foreknowledge. Preconception can result in widely varying accuracies through one night's observations.

presentation The angle to earth at which Saturn's rings are viewed as a result of the changing angle of the orientation of the earth with respect to the plane of Saturn's rings. This presentation can be at a maximum of 26° or a minimum of 0°, at which time the rings are viewed exactly edge-on from earth.

prime values The mean rotational rates for the currents of Jupiter. For visual observers there are two: System I at 9h 50m 30s and System II at 9h 55m 40s. All rotational period determinations for features are based on these values.

Purkinje effect The ability of the eye to increase its sensitivity to light in red wavelengths after exposure to the source for a period of time; the phenomenon with which some observers record objects that have reddish coloration as being brighter than others as a result of the increased sensitivity of their eyes to those wavelengths.

push process The processing of film to higher-than-normal ASA ratings by the use of special processing techniques or chemicals.

R

radiant The point in the sky from which the meteors of a shower appear to originate.

reaction time The time required for the observer to see an occultation and record it by whatever means employed. Reaction time varies with the method used.

reciprocity The ability of film to record faint light over an extended period of time. Reciprocity failure is the falling off of a film's sensitivity to light after a short time, thereby causing it to fail to record the light falling on the emulsion.

red dwarf A star of very late spectral type, lying on the main sequence and of very small surface area.

red giant A star that has cooled to 2000° to 3000° Centigrade and has a diameter 10 to 100 times greater than the sun. The red giant stars are perhaps in their final evolutionary stages, exhausted of their hydrogen fuel. Because of the great instability of the great stars, they comprise a vast number of the known variable stars.

reflector (Cassegrain) A "folded" optical system similar to the Newtonian design, except a convex hyperboloid secondary mirror intercepts the light focused from the paraboloid primary, thereby amplifying the light cone and increasing the focal ratio, which is usually from f/16 to f/24.

reflector (Newtonian) An optical assembly utilizing a parabolic front-surface mirror as an objective. The light path is diverted by means of a smaller secondary flat mirror at a right angle to the observing position. Normal focal ratios are from f/4 through f/8.

refractor A telescope utilizing a convex lens as an objective. It is often preferred by the visual planetary observer. Focal lengths are common from f/12 to f/20, with f/16 preferred.

resolution The ability of an optical system to discern very fine detail, as measured through the resolving power. Resolving power is defined as the smallest angle between two point objects (i.e., stars) that produces two distinct images not in contact.

reticle Usually refers to a transparent disk inserted into the focal plane of an eyepiece. The reticle is graduated into a set scale for the purposes of measurement.

retrograde motion A temporary apparent westward motion of a planet as the earth swings between it and the sun. This apparent motion is realized only when viewing planets with orbits greater than earth's. This aspect appears to make a planet (as seen from earth) slow down in its normal eastward path through the sky and eventually reverse that path in a loop. As the earth rounds its orbit relative to the planet, the planet will appear to resume its normal drift direction and rate.

rice grain A common name for granulation on the surface of the sun. Granulation is often visible when small telescopes are used under ideal conditions. Looking somewhat like cells under a microscope, the "rice grains" are actually convective cells latticed in a network throughout the photosphere. Each granule is about 1000 km in diameter and is separated from another by a thin, dark lattice suspected to be vertical walls of rising gases.

rippling Similar to the Kirkwood gaps, and of the same origin, but minor resonances with the satellite's gravity and orbital velocity results in intensity thinnings of the ring particle material, rather than complete separations.

Roche's limit The minimum distance from the center of a major mass (planet) at which an orbiting body can remain in equilibrium. It is thought that Saturn's rings are the result of the breakup of a major satellite that orbited at a lesser distance than Roche's limit.

S

screw value Also known as the *pitch* of a filar micrometer. This is simply the scale on the vernier (dial) that is read after the micrometer is adjusted for the measurement.

seeing A measure of the steadiness of the earth's atmosphere at the time of astronomical observations. The conditions of seeing depend on the steadiness of the air; poor seeing produces rapid motion of the image of a celestial object so that the image appears blurred and enlarged. Seeing can be evaluated on a scale from 1 to 10, with 1 worst and 10 ideal steady seeing.

shower Any group of meteors, all entering the atmosphere in succession, that can be traced back to a radiant. All shower members have the same velocity and orbit.

sky glow The effect of sky lighting that results from the reflection of extraneous artificial lighting at night. Sky glow is reflected normally from particles of water vapor, dust, and industrial pollution.

slumping A gravitational process that results in landslides within lunar craters. Normally sharp crater walls frequently collapse from their weight into the crater floor.

solar type Any star with spectral characteristics similar to our sun.

sporadic meteors Those meteors, perhaps 12 per hour, that cannot be traced to any known shower. They appear seemingly at random and are not associated with any known annual stream of meteors.

spectroscope An instrument used to display the component colors of the electromagnetic spectrum in their proper order and expanse. A *spectrograph* uses film to record the image, thereby facilitating easy measuring of critical wavelengths.

star atlas A set of charts on which the positions and relative magnitudes of stars are plotted.

star catalog A list, usually in book form, that accompanies some star atlases. Such a list is valuable for the comet observer for reference to magnitudes of comparison stars.

sunspot Ranging in size from a tiny pore to over 1 billion sq km, sunspots are somewhat cooler than the surrounding photosphere of the sun, and thereby appear darker. Each sunspot can be comprised of both an umbra and penumbra. Most spots occur not randomly but in groups. Each sunspot is the center of an intense outburst of solar magnetism and can last from one week to several months.

supernova A very rare astronomical occurrence in which a star brightens quite rapidly to an absolute magnitude of greater than -15, over 100 times the brightness increase of a nova. Only three supernovae events have been recorded in our galaxy, but over 400 such explosions have been seen in external galaxies since 1936.

sweephand stopwatch A stopwatch capable of reading in either 0.1-sec increments or 0.01-sec increments. For the first, a mechanical stopwatch is preferred; for the latter an electronic digital stopwatch is necessary.

System I The great equatorial current of Jupiter, rotating faster than the temperature and polar regions, and confined within the two equatorial belts. The period is 9h 50m 30.0s.

System I As on Jupiter, this represents the major average current of features rotating within the equatorial region of Saturn, at a period of 10hr 14m 13.08s.

System II The mean rotational period of features located outside (north and south) of the equatorial zone of Jupiter; it is approximately 5 min slower than features within System I.

System II The mean period of the current for features of Saturn that rotate either north or south of the equatorial region—10h 38m 25.42s.

T

Terby white spots Scintillating bright patches sometimes seen on the ring system of Saturn. Their cause is unknown, but they might be related to reflection from the Saturnian globe as seen from earth.

threshold The limit of light grasp of an instrument as determined by the faintest stellar image discernible in it. It is a factor of the objective diameter, the resolution of the instrument, and the angle from the zenith at which the object is viewed and amount of sky glow.

total magnitude The magnitude of a comet as reported in terms of its brightness on any given day. This magnitude is the total brightness of only the coma, including the nucleus, if a nucleus is present.

train The light or "tail" of a meteor that is left behind in the meteor's wake; not all meteors have trains.

trajectory A three-dimensional path of a meteor considered relative to the observer on earth.

transit The passage of a feature across the central meridian.

transparency Like seeing, this is another state of the earth's atmosphere at the time of observation, representing the clarity of the air. A convenient scale might be from 1 to 6, each number representing the faintest star discernible to the naked eye.

U

umbra The name given to the dark inner portion of a sunspot. Every sunspot must have an umbra but not necessarily a penumbra. The umbral region is about 4000° C cooler than the photosphere of the sun, thus accounting for its dark appearance.

V

vignetting The loss of the periphery of a telescope's field of view, usually as a result of an obstruction in the light path, or—in the case of reflector optics—improper matching of secondary size or placement in the light path.

W

wake Following the meteor; also used to designate a meteor train lasting less than 1 sec, as most do.

waning The phase of the moon either after full or before new when the eastern visible quadrant is illuminated.

wavelength The distance that any periodic wave must travel to complete one cycle (e.g., from crest to crest). The value is normally expressed in nanometers. The ultraviolet has the shortest wavelength, and the infrared has the longest.

waxing The phases of the moon either before full or after new, when the western visible quadrant is illuminated.

Z

zenith The point overhead intersected by the meridian; it is, simply, straight up.

zeroing Adjusting the photoelectric photometer amplifier so that the instrument components do not produce spurious deflection.

zones The alternating bright bands encompassing Jupiter and Saturn. They are representations of high cloud altitudes and increased reflectivity.

Zurich/Wolf number A daily index demonstrating the relative intensity of the sun by means of its only visible phenomena—the sunspots. Expressed by the formula $RSN = [(g \times 10) + n]k$ where g is the total number of groups, n is the number of sunspots seen, and k is a factor determined by the observer to bring his or her daily estimates more in keeping with the numbers derived with the standard Zurich telescope.

APPENDIXES

APPENDIX I

Converting from Universal Time To Standard Or Daylight Time

Universal Time, otherwise known as *Greenwich Meridian Time* is a standardized method of global timekeeping. By converting to such a method, there is no need for time zones, and discrepancies in recent times by two or more observers are kept at a minimum. No matter where an observer is located on earth, when it is 03:00 UT, it is 03:00 everywhere.

This time is based on 24-hour time as noted at the Greenwich Meridian, in England. At midnight at this location, Universal Time is 00h 00m, and a new day begins. Universal Time is measured in 24 hours, from midnight to midnight; for conversion purposes any times after 12:00 UT are PM. Conversion to standard time requires only subtracting the following values: Eastern Standard, subtract 5 hours; Central Standard, subtract 6; Mountain Standard, subtract 7; Pacific Standard, subtract 8. If necessary, add 24 hours before subtracting.

Table I-1 gives quick conversion for major time zones of the United States.

TABLE I-1. Time conversion table.

Standard Time					*Daylight Savings Time*			
Eastern Standard	*Central Standard*	*Mountain Standard*	*Pacific Standard*	*Universal Time*	*Eastern Daylight*	*Central Daylight*	*Mountain Daylight*	*Pacific Daylight*
2:00 AM	1:00 AM	12:00	11:00 PM	07:00	3:00 AM	2:00 AM	1:00 AM	12:00
3:00 AM	2:00 AM	1:00 AM	12:00	08:00	4:00 AM	3:00 AM	2:00 AM	1:00 AM
4:00 AM	3:00 AM	2:00 AM	1:00 AM	09:00	5:00 AM	4:00 AM	3:00 AM	2:00 AM
5:00 AM	4:00 AM	3:00 AM	2:00 AM	10:00	6:00 AM	5:00 AM	4:00 AM	3:00 AM
6:00 AM	5:00 AM	4:00 AM	3:00 AM	11:00	7:00 AM	6:00 AM	5:00 AM	4:00 AM
7:00 AM	6:00 AM	5:00 AM	4:00 AM	12:00	8:00 AM	7:00 AM	6:00 AM	5:00 AM
8:00 AM	7:00 AM	6:00 AM	5:00 AM	13:00	9:00 AM	8:00 AM	7:00 AM	6:00 AM
9:00 AM	8:00 AM	7:00 AM	6:00 AM	14:00	10:00 AM	9:00 AM	8:00 AM	7:00 AM
10:00 AM	9:00 AM	8:00 AM	7:00 AM	15:00	11:00 AM	10:00 AM	9:00 AM	8:00 AM
11:00 AM	10:00 AM	9:00 AM	8:00 AM	16:00	12:00 NOON	11:00 AM	10:00 AM	9:00 AM
12:00 NOON	11:00 AM	10:00 AM	9:00 AM	17:00	1:00 PM	12:00 NOON	11:00 AM	10:00 AM
1:00 PM	12:00 NOON	11:00 AM	10:00 AM	18:00	2:00 PM	1:00 PM	12:00 NOON	11:00 AM
2:00 PM	1:00 PM	12:00 NOON	11:00 AM	19:00	3:00 PM	2:00 PM	1:00 PM	12:00 NOON
3:00 PM	2:00 PM	1:00 PM	12:00 NOON	20:00	4:00 PM	3:00 PM	2:00 PM	1:00 PM
4:00 PM	3:00 PM	2:00 PM	1:00 PM	21:00	5:00 PM	4:00 PM	3:00 PM	2:00 PM
5:00 PM	4:00 PM	3:00 PM	2:00 PM	22:00	6:00 PM	5:00 PM	4:00 PM	3:00 PM
6:00 PM	5:00 PM	4:00 PM	3:00 PM	23:00	7:00 PM	6:00 PM	5:00 PM	4:00 PM
7:00 PM	6:00 PM	5:00 PM	4:00 PM	24:00	8:00 PM	7:00 PM	6:00 PM	5:00 PM
8:00 PM	7:00 PM	6:00 PM	5:00 PM	01:00	9:00 PM	8:00 PM	7:00 PM	6:00 PM
9:00 PM	8:00 PM	7:00 PM	6:00 PM	02:00	10:00 PM	9:00 PM	8:00 PM	7:00 PM
10:00 PM	9:00 PM	8:00 PM	7:00 PM	03:00	11:00 PM	10:00 PM	9:00 PM	8:00 PM
11:00 PM	10:00 PM	9:00 PM	8:00 PM	04:00	12:00	11:00 PM	10:00 PM	9:00 PM
12:00	11:00 PM	10:00 PM	9:00 PM	05:00	1:00 AM	12:00	11:00 PM	10:00 PM
1:00 AM	12:00	11:00 PM	10:00 PM	06:00	2:00 AM	1:00 AM	12:00	11:00 PM

Converting Hours Of Universal Time To Decimals

Example: Suppose we wish to find the decimal day equivalent of 2:32 UT. Going down the scale to "2" and across the horizontal scale to "30," our decimal is 0.103; adding the additional 2 minutes, we derive a value of 0.10426 UT (the decimal for 2 minutes was taken from the table above [.00136]).

	Minutes[a]												
UNIVERSAL TIME (In Hours)	00	05	10	15	20	25	30	35	40	45	50	55	60
1	.041	.044	.048	.051	.055	.058	.061	.065	.068	.072	.075	.078	.083
2	.083	.086	.090	.093	.097	.100	.103	.107	.110	.114	.117	.120	.123
3	.123	.126	.130	.133	.137	.140	.143	.147	.150	.154	.157	.161	.166
4	.166	.170	.173	.177	.180	.184	.187	.190	.194	.197	.201	.204	.208
5	.208	.212	.215	.219	.222	.225	.229	.232	.236	.239	.242	.246	.250
6	.250	.253	.257	.260	.264	.267	.270	.274	.277	.281	.284	.287	.292
7	.292	.295	.298	.302	.305	.309	.312	.315	.319	.322	.326	.329	.333
8	.333	.337	.340	.344	.347	.350	.354	.357	.361	.364	.367	.371	.375
9	.375	.378	.382	.385	.389	.392	.395	.399	.402	.406	.409	.412	.416
10	.416	.420	.423	.427	.430	.434	.437	.440	.444	.447	.451	.454	.458
11	.458	.462	.465	.469	.472	.475	.479	.482	.486	.489	.492	.496	.500
12	.500	.503	.507	.510	.514	.517	.520	.524	.527	.531	.534	.537	.541
13	.541	.545	.548	.552	.555	.559	.562	.565	.569	.572	.576	.579	.583
14	.583	.587	.590	.594	.597	.600	.604	.607	.611	.614	.617	.621	.625
15	.625	.628	.632	.635	.639	.642	.645	.649	.652	.656	.659	.662	.666
16	.666	.670	.673	.677	.680	.684	.687	.690	.694	.697	.701	.704	.708
17	.708	.712	.715	.719	.722	.725	.729	.732	.736	.739	.742	.746	.750
18	.750	.753	.757	.760	.764	.767	.770	.774	.777	.781	.784	.787	.792
19	.792	.795	.798	.802	.805	.809	.812	.815	.819	.822	.826	.829	.833
20	.833	.837	.840	.844	.847	.850	.854	.857	.861	.864	.867	.871	.875
21	.875	.878	.882	.885	.889	.892	.895	.899	.902	.906	.909	.912	.917
22	.917	.920	.923	.927	.930	.934	.937	.940	.944	.947	.951	.954	.958
23	.958	.962	.965	.969	.972	.975	.979	.982	.986	.989	.992	.996	1.00
24	1.00	.003	.006	.009	.013	.016	.020	.023	.027	.030	.034	.037	.041

Table and data compiled by Clay Sherrod.

[a]For accuracy to within a minute's time, add the following:

1 minute - .00068
2 minutes- .00136
3 minutes- .00204
4 minutes- .00272
5 minutes- .00340

APPENDIX III

Coordinates for Bright Stars For Use with Setting Circles, Epoch—1970

Ordinary Name	*Greek Name*	*Mg.*	*RA*	*Dec.*	*Dist. L. Y.*
Aldebaran	α Tau	1.1	04H 34M	+16° 27′	68
Alfard	α Hya	2.2	09H 26M	-08° 31′	94
Altair	α Aql	0.9	19H 49M	+08° 47′	17
Antares	α Sco	1.2	16H 28M	-26° 22′	520
Betelgeuse	α Ori	0.1	05H 54M	+07° 24′	650
Capella	α Aur	0.2	05H 14M	+45° 58′	45
Deneb	α Cyg	1.3	20H 40M	+45° 10′	540
Fomalhaut	α PsA	1.3	22H 56M	-29° 47′	23
Hamal	α Ari	2.2	02H 05M	+23° 19′	76
Mizar	ζ UMa	2.4	13H 23M	+55° 05′	88
Pollux	β Gem	1.2	07H 43M	+28° 05′	35
Procyon	α CMi	0.5	07H 38M	+05° 18′	11.3
Regulus	α Leo	1.3	10H 07M	+12° 08′	84
Rigel	β Ori	0.3	05H 13M	-08° 14′	900
Sirius	α CMa	-1.6	06H 44M	-16° 40′	8.7
Spica	α Vir	1.2	13H 23M	-11° 00′	220
Thuban	α Dra	3.7	14H 04M	+64° 32′	?
Vega	α Lyr	0.1	18H 36M	+38° 30M	26.5
Stars for aligning polar axis of telescope:					Deviation from meridian
Alkaid	η UMa	1.9	13H 46M06S	+49° 26′	4.6′
Almach	γ And	2.3	02H 02M06S	+42° 13′	1.1′
Arcturus	α Boo	0.2	14H 14M05S	+19° 19′	1.6′
Segin	ε Cas	3.4	01H 52M06S	+63° 31′	3.2′
50 Cas	—	4.0	02H 01M03S	+72° 18′	1.4′

APPENDIX IV

Eyepiece Angular Size from Interval of Star-drift Time

Time Interval (min.)	Declination of Object—North or South															
	0°	10°	15°	20°	25°	30°	35°	40°	45°	50°	55°	60°	65°	70°	75°	80°
1	.25°	.24°	.24°	.23°	.22°	.21°	.20°	.19°	.17°	.16°	.14°	.12°	.10°	.08°	.06°	.04°
2	.50	.49	.48	.47	.45	.43	.41°	.38	.35	.32	.28	.25	.21	.17	.13	.08
3	.75	.73	.72	.70	.67	.64	.61	.57	.52	.48	.42	.37	.31	.25	.19	.12
4	1.00	.99	.97	.94	.91	.87	.82	.77	.71	.64	.57	.50	.42	.34	.26	.17
5	1.25	1.23	1.21	1.17	1.13	1.08	1.03	.95	.88	.80	.72	.62	.53	.42	.32	.21
10	2.50	2.46	2.41	2.35	2.26	2.16	2.05	1.91	1.76	1.60	1.44	1.25	1.05	.85	.64	.43
15	3.75	3.69	3.62	3.52	3.39	3.24	3.07	2.86	2.65	2.40	2.15	1.87	1.59	1.27	.97	.64
20	5.00	4.94	4.83	4.70	4.53	4.33	4.09	3.83	3.53	3.21	2.87	2.50	2.11	1.71	1.29	.87
25	6.25	6.17	6.04	5.87	5.66	5.41	5.12	4.78	4.41	4.01	3.58	3.12	2.64	2.13	1.61	1.08
30	7.50	7.39	7.24	7.05	6.79	6.49	6.14	5.74	5.30	4.82	4.30	3.75	3.17	2.56	1.94	1.30
35	8.75	8.62	8.45	8.22	7.92	7.57	7.17	6.69	6.18	5.62	5.02	4.37	3.70	2.98	2.26	1.51
40	10.00	9.85	9.66	9.40	9.06	8.66	8.19	7.66	7.07	6.43	5.74	5.00	4.23	3.42	2.59	1.74
45	11.25	11.08	10.87	10.57	10.19	9.74	9.22	8.61	7.95	7.23	6.46	5.62	4.76	3.84	2.91	1.95
50	12.50	12.31	12.07	11.75	11.32	10.80	10.23	9.50	8.83	8.00	7.17	6.20	5.28	4.20	3.23	2.10
55	13.75	13.54	13.28	12.92	12.45	11.88	11.26	10.45	9.71	8.80	7.89	6.82	5.81	4.62	3.55	2.31
60	15.00	14.78	14.49	14.10	13.59	12.99	12.28	11.49	10.60	9.64	8.61	7.50	6.34	5.13	3.88	2.61

APPENDIX V

Magnitudes and Colors of Standard Stars

Name	*α 1950*	*δ 1950*	*V*	*B – V*	*U – B*	*Sp*
γ Peg	00h 11m	+14° 54′	2.83	–0.23	–0.87	B2 IV
χ Peg	00 12	+19 56	4.80	+1.58	+1.92	M2 III
δ Cas	01 22	+59 59	2.68	+0.13	+0.12	A5 V
τ Cet	01 42	–16 12	3.50	+0.72	+0.20	G8 Vp
α Ari	02 04	+23 14	2.00	+1.151	+1.12	K2 III
HR 875	02 54	–03 55	5.17	+0.084	+0.05	A1 V
α Tau	04 33	+16 25	0.86:	+1.53	+1.90	K5 III
π^3 Ori	04 47	+06 53	3.19	+0.45	–0.01	F6 V
π^4 Ori	04 48	+05 32	3.69	–0.17	–0.80	B2 III
η Aur	05 03	+41 10	3.17	–0.18	–0.67	B3 V
β Eri	05 05	–05 09	2.80	+0.13	+0.10	A3 III
HD 35299	05 21	–00 12	5.70	–0.22	–0.87	B2 V
ι Ori	05 33	–05 56	2.77	–0.25	–1.08	O9 III
β Cnc	08 14	+09 20	3.52	+1.480	+1.78	K4 III
η Hya	08 41	+03 35	4.30	–0.195	–0.74	B3 V
ι UMa	08 56	+48 14	3.14	+0.18	+0.07	A7 V
α Leo	10 06	+12 13	1.36	–0.11	–0.36	B7 V
λ UMa	10 14	+43 10	3.45	+0.03	+0.06	A2 IV
β Leo	11 46	+14 51	2.14	+0.09	+0.07	A3 V
γ UMa	11 51	+53 58	2.44	0.00	+0.01	A0 V
δ UMa	12 13	+57 19	3.31	+0.08	+0.07	A3 V
γ Crv	12 13	–17 16	2.60	–0.11	–0.35	B8 III
80 UMa	13 23	+55 15	4.01	+0.16	+0.08	A5 V
η Boo	13 52	+18 39	2.69	+0.58	+0.19	G0 IV
α Boo	14 13	+19 27	–0.06	+1.23	+1.26	K2 IIIp
109 Vir	14 44	+02 06	3.74	0.00	–0.03	A0 V
β Lib	15 14	–09 12	2.61	–0.108	–0.37	B8 V
α CrB	15 33	+26 53	2.23	–0.02	–0.02	A0 V
α Ser	15 42	+06 35	2.65	+1.168	+1.24	K2 III
β Ser A	15 44	+15 35	3.67	+0.06	+0.07	A2 IV
λ Ser	15 44	+07 30	4.43	+0.60	+0.10	G0 V
ϵ CrB	15 56	+27 01	4.15	+1.230	+1.28	K3 III
τ Her	16 18	+46 26	3.89	–0.152	–0.56	B5 IV
ζ Oph	16 34	–10 28	2.56	+0.02	–0.86	O9.5 V
α Oph	17 33	+12 36	2.08	+0.15	+0.10	A5 III
β Oph	17 41	+04 35	2.77	+1.16	+1.24	K2 III
γ Oph	17 45	+02 43	3.75	+0.04	+0.04	A0 V
α Lyr	18 35	+38 44	0.04	0.00	–0.01	A0 V
α Aql	19 48	+08 44	0.77	+0.22	+0.08	A7 IV,V
β Aql	19 53	+06 17	3.71	+0.86	+0.48	G8 IV
α Del	20 37	+15 44	3.77	–0.06	–0.22	B9 V
ϵ Aqr	20 45	–09 41	3.77	+0.01	+0.04	A1 V
10 Lac	22 37	+38 47	4.88	–0.203	–1.04	O9 V
HR 8832	23 11	+56 54	5.57	+1.010	+0.89	K3 V
ι Psc	23 37	+05 21	4.13	+0.51	0.00	F7 V

APPENDIX VI

Observing Form for Meteors

THE AMERICAN METEOR SOCIETY

Observer...

Place...

Date...

Began...

Ended...

Condition of Sky...

TIME			*No.*	*Class*	*Color*	*Magn.*	*Distance*	*Duration*	*Duration of Train*	*BEGINNING*		*ENDING*		*Accuracy*	*Serial No.*
										α	*δ*	*α*	*δ*		
H	M	S					°	S	S	°	°	°	°		

APPENDIX VII

Observing Form for Comets

COMET ______________________

Observer ______________________ Date (UT) ______________________

Address ______________________ Time (UT) ______________________

______________________ Distance from horizon ______________________

Place of observation ______________________

Approximate R.A. ______________________ Approximate declination ______________________

Sky conditions (Transparency, haze, moonlight, twilight, city lights, faintest naked eye star near comet)

MAGNITUDE ESTIMATES (State methods and telescopes, with type f/no., and power.)

Total magnitude ________ Tel. ____________ Magnitude of nucleus ________ Tel. ____________

Comparison stars, magnitudes, and catalog used. Were they in the same field of view? Give exact time if extinction correction must be calculated.

DESCRIPTION (State methods and instruments used for each observation.)

Tail length ________ Position angle of tail from head. (Plot on an atlas and measure. North is 0°, East is 90°.) ______________________

Remarks on tail (details, color, shape, etc.):

Coma description (diameter by drift method, shape, jets with P.A.):

Degree of coma condensation. Scale: 0 = diffuse, 9 = stellar. ______________________
(A sketch will be of value.)

FIELD SKETCH: Draw on an atlas and recopy.

Field orientation ______________________

Diameter of field ______________________

Instrument, power ______________________

Drawing paper may be used instead.
Please do not fold.

APPENDIX VIII

Observing Form for Occultations

OCCULTATION OBSERVATIONS

PLACE NAME

TELESCOPES AND POSITIONS

	DESCRIPTION	LONGITUDE	LATITUDE	HEIGHT	STATION CODE
A					
B					
C					

OBSERVERS

a	d	g
b	e	h
c	f	i

METHOD OF TIMING

PERSONAL EQUATION APPLIED /NOT APPLIED

YEAR _____

1 DATE AND U.T. MTH. D H M S	2 LUN.	3 Z.C. NO.	4 T X MAGN.	5 OBS.	6 PH.	7 REMARKS *(Accuracy)*	8 METHOD OF TIMING	9 TIME SIGNAL	10 P. EQN.

N.A.O. 1966 MARCH

MORE FORMS REQUIRED *YES / NO*

APPENDIX IX

Observing Form for Mars

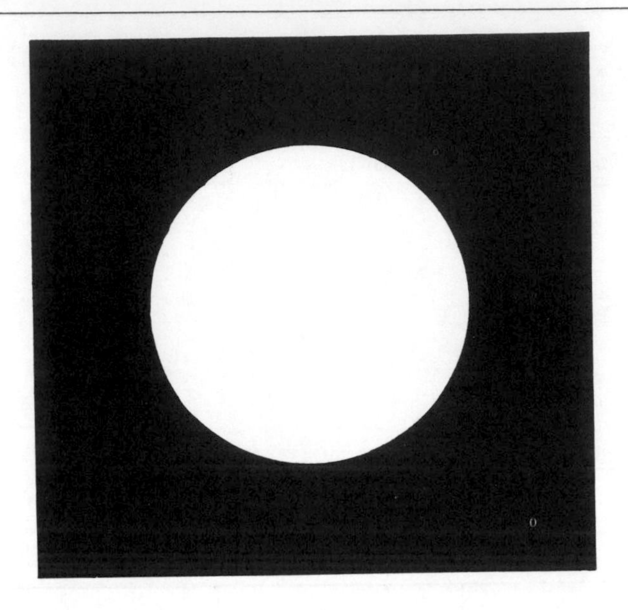

MARS

Drawing Time Begin______ End______ (U.T.)

Diameter______ (seconds of arc)

Seeing (1—5)____

Magnitude______

Aperture____ ″

Magnification________X

Observer____________________

Filter____________

Location____________________________

Comments______________________________________

Photograph: Time (U.T.)______, Exp.______, E.F.L.______, Film______________

Comments:______________________________________

Mid-South Astronomical Research Society
Little Rock, Arkansas

APPENDIX X

Observing Form for Jupiter

WEST

EAST

Date (by U.T.)______
U.T.______
Seeing (0 to 10, 10 best)______
Transparency (1-5, 5 best)______
CM(I)______
CM(II)______
Telescope______
Magnification______
Observer______

Date (by U.T.)______
U.T.______
Seeing (0 to 10, 10 best)______
Transparency (1-5, 5 best)______
CM(I)______
CM(II)______
Telescope______
Magnification______
Observer______

NOTES:		COLOR:
NTeZ		
NTB		
NTrZ		
NEB		
EZ		
SEB		
STrZ		
STB		

NOTES:		COLOR:
NTeZ		
NTB		
NTrZ		
NEB		
EZ		
SEB		
STrZ		
STB		

TIME (UT)	SYSTEM	LONGITUDE	DESCRIPTION

AFFILIATED WITH:

– Association of Lunar and Planetary Observers – Astronomical League – Mid South Astronomical Research Society –
– Smithsonian Astrophysical Observatory – United States Naval Observatory –

APPENDIX XI

Observing Form for Saturn

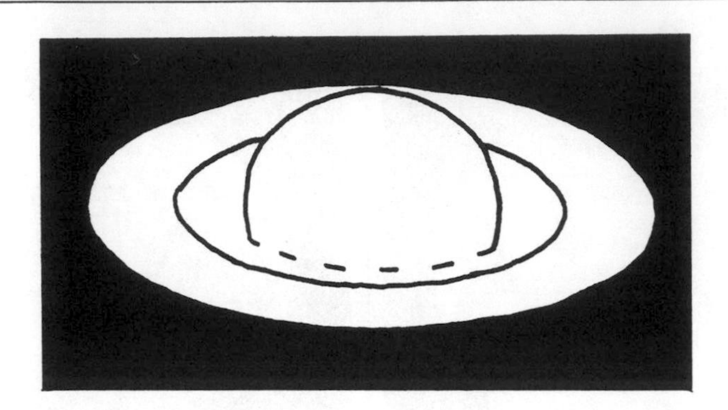

MIDSOUTH ASTRONOMICAL RESEARCH SOCIETY

PLANETARY OBSERVING LOG

OBSERVER ______________________

TELESCOPE ______________________

LOCATION ______________________

TIME (U.T.) ______________________

MAGNIFICATION ______________________

SEEING (1-5) ______________________

WITNESSED ______________________

(B=±26)

OBSERVING NOTES (below and over)

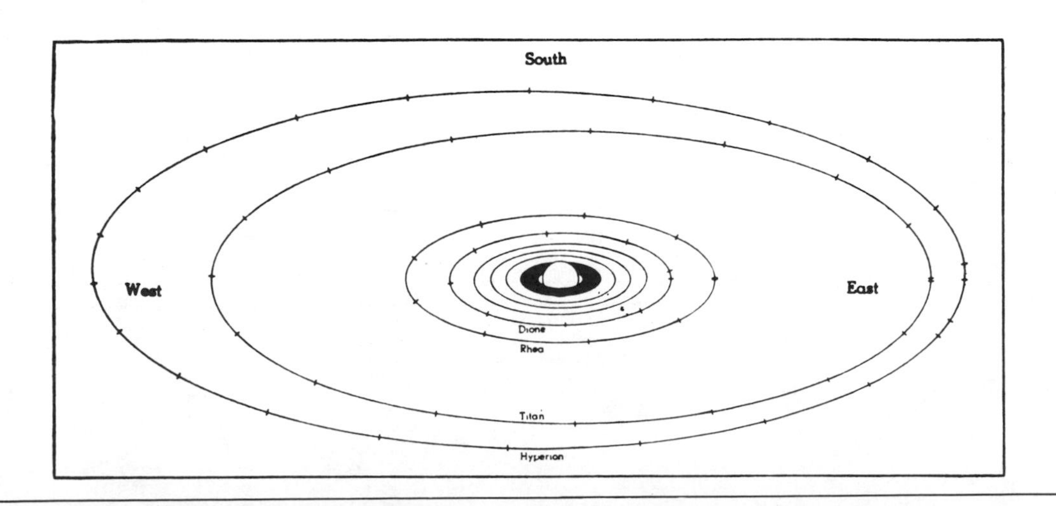

APPENDIX XII

Observing Form for Minor Planets

ALPO Minor Planets Section
Photometry Report Form

PLANET: ______________

Observer ______________ Address ______________

Location ______________

Date (UT) ______________ Telescope ______________ magn. ______________

Time (UT) ______________ Seeing ______________ Transparency ______________

Type of estimate (visual, photoelectric, etc) ______________

Other equipment (e.g., photometer) ______________

ENTER ONLY ONE NIGHT ON THIS FORM
ENTER ONLY ONE PLANET ON THIS FORM

ESTIMATES

time	mag	time	mag	time	mag

LIGHT CURVE
0.1 mag steps

15 minute intervals

Star field used ______________

Sequence used:

star	mag	source of mag

Were the comparison stars and the asteroid in the same field of view?

NOTES:

Complete and return to Alain Porter, Photometry Coordinator,
10 Sea Lea Drive, Narragansett RI 02882

APPENDIX XIII

Observing Form for Variable Stars

VARIABLE STAR OBSERVATIONS

For

THE AMERICAN ASSOCIATION OF VARIABLE STAR OBSERVERS

Report No. . . . Sheet . . of . . .

For Month of . . . . 19 . .

Observer . . . .

Street . . . .

City . . . . State . . . . Zip . . .

Time Used, G.M.A.T., or. . . . .

Instrument . . . .

DO NOT WRITE HERE

Recd. . . . .

Pltd. . . . .

Ackd. . . . .

Posted. . . . .

Ledgd . . . .

DESIGNATION	VARIABLE	JUL.DAY&DEC.	MAGN.	DESIGNATION	VARIABLE	JUL.DAY&DEC.	MAGN.

TOTAL NUMBER OF STARS OBSERVED	TOTAL NUMBER OBSERVATIONS

Observations should be sent to Headquarters, 187 Concord Avenue, Cambridge, Mass. 02138, as soon as possible, after the first of each month.

APPENDIX XIV

How To Organize An Astronomy Club

There is no better way to enjoy astronomy and the night sky than to have others along to share in your experiences. The thrill of discovery—a bright new comet, an active meteor shower, a lunar eclipse—is magnified many times when amateur astronomers have others to participate and discuss with. But many times in both large and small towns, and in rural areas, it is difficult to locate others of similar interests. Many times, several people meet for a few months and then lose interest because their meetings have no organization or intent. This discussion will aid those wishing to form an astronomy club and help those now participating in a failing organization.

The Club Organization

Before we discuss how to organize a club, let's say something about what can keep a club together. Of course, it would be nice if clubs could exist where people get together very informally and have a good time and still manage to get things done. However, such an ideal situation is not realistic. People involved with *any* club, whether it be civic, job-related, or whatever, are participating in that organization *on their own time*. Therefore, they come to meetings expecting a little entertainment, and also to better themselves for the experience.

A club that is too tightly structured with a lot of rules and frowning faces does not invite persons to participate. On the other hand, an organization that is too loosely structured, with no rules and poor conduct throughout the meetings and activities, will eventually fail due to lack of respect for the organization by the members. People involved in scientific organizations such as astronomy clubs come to the meetings primarily to gain some knowledge and share in the common interest. Therefore, the meetings should be structured with those objectives in mind:

1. Use the same agenda (or meeting schedule) each month.
2. Always cover business first and the program last.
3. Try to involve as many of the members as possible with the meeting.
4. Recognize visitors at your meetings—they are potential members of the future.
5. Above all, do not let arguments build during meeting hours; if disagreements emerge, discuss them on your own time, or organize committees to smooth them over.
6. *Stick to the planned meeting every month*.

It is essential to have some type of program at every meeting no matter how short it is. Many younger members and restless adults will eventually tire of nothing but business. In the following pages, I list some activities that might be included in your club's meetings to better inform the members and to entertain visitors and younger folks. *Remember*—a club is not just a monthly meeting; it requires outside activity, planning, and dedicated members who will work between those meetings so that there can be something to meet about!

Sections. A good way to involve as many people as possible is by organizing *sections* within the club, each section devoted to a particular aspect of astronomy which the members seem to like. One person who has proved to be an active observer or student in each area should be appointed by the club president and executive council to fill the position of chairperson of that section.

At each meeting, each section chairperson will present a report of what activity has taken place in that section during the preceding month and recognize observers who have actively worked in that area. Sections that might be included in your club organization are:

1. Solar Section for those who will be observers of the sun.
2. Planetary Section for persons interested in the solar system.

3. Meteor Section to help organize observing parties for meteor showers.
4. Variable Star Section, working closely with the AAVSO.
5. Messier Club for members to strive to observe as many deep-sky objects as possible.
6. Sky Transparency Section whose members keep track of sky conditions throughout the month and compile yearly logs of weather conditions in your area.

Of course, many more sections than listed above might be added. However, each section's report at the meeting might last as long as 10 or 15 minutes, thereby requiring a great deal of your meeting time if there are many sections.

The responsibilities of a section chairperson should be:

1. Organize all people in your club interested in similar observing projects (meteors, planets, etc.).
2. Keep those actively participating within the section informed of new developments that might require observation.
3. Present a monthly report to the meeting about activity within sections, and acknowledge members for their observations.
4. Describe to the full membership at each monthly meeting what activities will be occurring for the coming month (such as describing upcoming meteor showers).
5. Organize all observations made by members into a club file, and submit members' observations to professional astronomers or national organizations (ALPO, AAVSO, etc.).

The executive committee. Every club must have some leaders. The best arrangement for astronomy clubs is to have the following:

President—in charge of conducting meetings and organizing activities.

Vice President—runs meetings when president is absent and works as club publicity agent.

Secretary—keeps records of all meetings and monthly activities. In addition, answers correspondence from potential members.

Treasurer—keeps accounts of the club's finances.

Of course, the duties of each can be expanded if necessary. The executive committee also organizes committee meetings and all club outings.

The Meeting Agenda

If your club is organized into sections, a very good format to follow during meetings is described as follows.

I. Introduction of visitors.

II. Secretary's minutes of last meeting.

III. Treasurer's financial report.

IV. List of coming club activities (described by president).

V. Section reports:
  a. solar section.
  b. meteor section.
  c. variable star section.
  d. planetary section, including comets.
  e. sky transparency report.

VI. Old business.
Discuss any items that were brought up at the last meeting but not fully resolved.

VII. New business.
Discuss any planned activities, new procedures, and items that have been requested by membership.

VIII. Snack break. Everybody brings a snack potluck style.

IX. Program.

X. Discussion. Announcement of place and date of next meeting.

XI. Meeting adjourns.

XII. Observing party if sky is clear.

When using this agenda, the meeting should take about two hours with a 30-minute program included. It is good to begin meetings at about 7:00 or 7:30 PM local time, and the best day seems to be Saturday. Try to have your meeting on a regular schedule. A good practice is to have every meeting on the second or third Saturday of each month, which does not require members to memorize the dates of each meeting.

The Meeting Place

For a small club, meetings can be held in members' homes, moving the meeting from home to home for different months. However, larger clubs might require a regular meeting place, such as a classroom, in which to hold meetings. Contact local schools, banks, and civic recreational centers about the possibility of meeting in their facilities. Explain to them what type of club it is that will be meeting, how many persons to expect, and so on. One person will have to be responsible for ensuring that the facility is left clean with no damage after every meeting. Some meeting rooms require a deposit or a small fee, but the amount is usually less then $10, and the atmosphere of a nice meeting room enhances all meetings.

Activities

Try to plan as many star parties, club outings to dark sites, club picnics, softball games, and so forth as possible so that the membership remains as well-rounded as possible. *Do not expect* every member to participate in such activities; nor should they be frowned upon if they do not. It takes all kinds of people to make astronomy an interesting hobby. Be sure to announce all activities such as these at least one month prior to their dates and send a reminder through the mail to all members. These activities must be fully organized by responsible persons, preferably through committees appointed by the executive committee during a regular monthly meeting. These activities will help to hold your club together. Try to plan several observing parties and Messier Object Observing Nights as time permits. Be sure to get groups together for meteor showers, new comets, eclipses, and similar events.

The Club Newsletter

If the club has more than 10 members it should have a newsletter. The newsletter does not need to be fancy be any means; it can be mimeographed, Xeroxed, or even carbon copies. The main point is communication—let the members know the dates, times, places of all club activities and meetings in the newsletter. Dues collected (see following section) should be applied to cover the cost of mailing, and copying. One capable person who has a reasonable amount of writing experience and who has time, should be designated as editor of the club newsletter. Usually there is enough space in a one-page typewritten newsletter to cover all activities and even have some space left over for recent discoveries and coming sky events.

Dues

No one likes to pay dues, but they are necessary in volunteer clubs. They enable the club to pay for a newsletter, meeting room (if necessary), pay for printing of observing forms (for meteors, planets, etc.), and eventually they may even help to build a club observatory and telescope. It is best to break the dues into two categories: for regular members and for student members. Fees for regular members should either be set at $2 or $1; those for student members at either $1 or 50¢, half that of the regular membership. Dues should be paid each month, with the responsibility of collecting them given to the treasurer. All club money should be kept in a bank account. Many banks provide free checking accounts with 50 free checks for nonprofit organizations, so be sure to ask for such an account. Honorary memberships can be given to members who have exhibited exceptional service to the astronomical community of your area, or outstanding persons within the club. These members, of course, should not pay dues.

Club Publicity

It is the responsibility of the club vice president or the editor to make sure the club has as much publicity as possible. Astronomy clubs need newspaper, radio, and TV coverage as much as possible to obtain. News releases should be sent out every time some occasion might be of public interest. These releases should be typewritten, double spaced, with a name and phone number where a club representative can be reached for interviews or for further information. Be as accurate with your information as possible. The media are depending on *you* to provide them with the truth. The following subjects are some that would warrant a news release to the local media:

1. A new comet that will become visible to the naked eye.
2. A solar or lunar eclipse that will be visible in your area.
3. A nice meteor shower.
4. Club meetings.
5. Coming star parties in which the public can participate.
6. Special honors achieved by some member of your club.

If you want publicity, *do not pester* the newspaper and media people; they will give you coverage if space allows, and if your subject is of public interest. *They will not* give you coverage if it is found that some of your information was inaccurate, so be sure to check things out thoroughly.

Organizing a New Astronomy Club

A new club should be set up as I have described. However, the problem still remains of contacting interested persons when organizing a new group. Again, the media are your best bet. Set up a time, place, and agenda for a charter meeting in your area. Make the meeting something that will be attractive not only to the potential members, but also to the newspapers and other media. This will help get coverage before and during such a meeting.

Let the media know on typewritten, double spaced copy, the place, time, date, and subject of the meeting about two weeks prior to the date. Send out copies of this news release to local junior high, high schools, and colleges in your area. Be sure to include a phone number where potential members can contact you to get additional information. Try to have a public interest program, dealing with popular subjects. It is desirable if a guest speaker from a local university or observatory can present the program. Not everyone who turns out the first night will become members.

On the meeting night, describe what the meeting agenda will be each month, the purpose of the club, and the club's objectives. Let people know what the dues might be (don't set a fee until everyone votes on it that has joined), have your program, and then encourage persons to join up. It also helps if there can be some observing through several scopes after the charter meeting.

Encourage all age groups to join, from age 10 years and up. Many clubs divide their clubs into a main group and a junior group. The junior group has young people from age 10 to 16 who can choose to belong to the main group as well as to the junior group, of who may wish just to participate in the younger set. Usually dues for junior members are only about 50¢ per month, less than regular members. Junior members should be encouraged to participate as well in the main group, but the junior group should be off-limits to people over 16.

Do not try to elect officers on the first night, and *do not* put yourself in the position of "president" or whatever; this will really turn people away. The organizer can act as chairman of the meetings until elections are held, preferably about the third meeting.

Above all, remember that an astronomy club is for fun—it is a way to relax and enjoy your hobby. If you go about organizing a club believing this, then others will follow. It may be a year before 10 members are on the books, but eventually there will be a group all dedicated to the same interest—astronomy.

Constellation Projects

The designs of the sky—the constellations—can provide years of enjoyment to those who study them. Learning the curious legend of some peculiarly named star or the physics of some enormous supergiant star can keep you occupied day and night alike.

These pages are a beginning—an incentive—for persons now interested in the night sky only for the love of its legend and beauty to begin a serious study of the constellations. Not every constellation is included, primarily for lack of space. However, enough of each season is provided perhaps to spark an otherwise latent interest in the sky.

The constellation projects are easy and relaxing; you can accomplish all the tasks with only the naked eye (or binoculars), a good book, and a lawn chair. One or two constellations each week is a good starting pace. After you study those provided here, you can study the remaining constellations at leisure. Good luck, and enjoy the night sky!

CLASSIC GREEK

The alphabet. The Greek language used twenty-four letters.

	FORMS *Italic*	*Roman*	NAMES	SOUNDS
Α	*α*	α	alpha	a
Β	*ϐ β*	β	beta	b
Γ	*γ*	γ	gamma	g
Δ	*δ*	δ	delta	d
Ε	*ϵ*	ε	epsilon	e short
Ζ	*ζ*	ζ	zeta	z
Η	*η*	η	eta	e long
Θ	*θ*	θ	theta	th
Ι	*ι*	ι	iota	i
Κ	*κ*	κ	kappa	k, c
Λ	*λ*	λ	lambda	l
Μ	*μ*	μ	mu	m
Ν	*ν*	ν	nu	n
Ξ	*ξ*	ξ	xi	x
Ο	*ο*	ο	omicron	o short
Π	*π*	π	pi	p
Ρ	*ρ*	ρ	rho	r
Σ	*σ*	σ ς	sigma	s
Τ	*τ*	τ	tau	t
Υ	*υ*	υ	upsilon	u
Φ	*φ*	φ	phi	ph
Χ	*χ*	χ	chi	ch
Ψ	*ψ*	ψ	psi	ps
Ω	*ω*	ω	omega	o long

The letters ϐ and ϑ are rare forms of the delta, and should not be used as symbols.

The letter σ is the sigma used at the beginning or in the middle of a word; ς is used at the end of a word.

All Year/Circumpolar

URSA MINOR

STAR NAMES: α -"Polaris"
β -"Kochab"

CONSTELLATION FACTS: Unfortunately, to most, this constellation is known as "The Little Dipper". Actually this constellation represents the Little "Bear", overshadowed by its neighbor, Ursa Major. This inconspicuous constellation contains POLARIS, our north star. A 3" telescope will show a faint 9th magnitude companion to Polaris.

DATE:____________________TIME:______________U.T.

OBSERVER:________________ OPTICAL AID:________________

α δ ε ζ η β γ

Visual Impressions:

Polaris____________________ Color: __________

companion visible?______power used__________

COMMENTS:

FACTS: Distance of Polaris________
Spectral Type ________
Orbital period of secondary star ________
Magnitude of Polaris_______of secondary star_______

DEFINE:
Circumpolar____________________
Precession of the Pole____________________
____

TRANSLATE: Polaris -
Kochab -

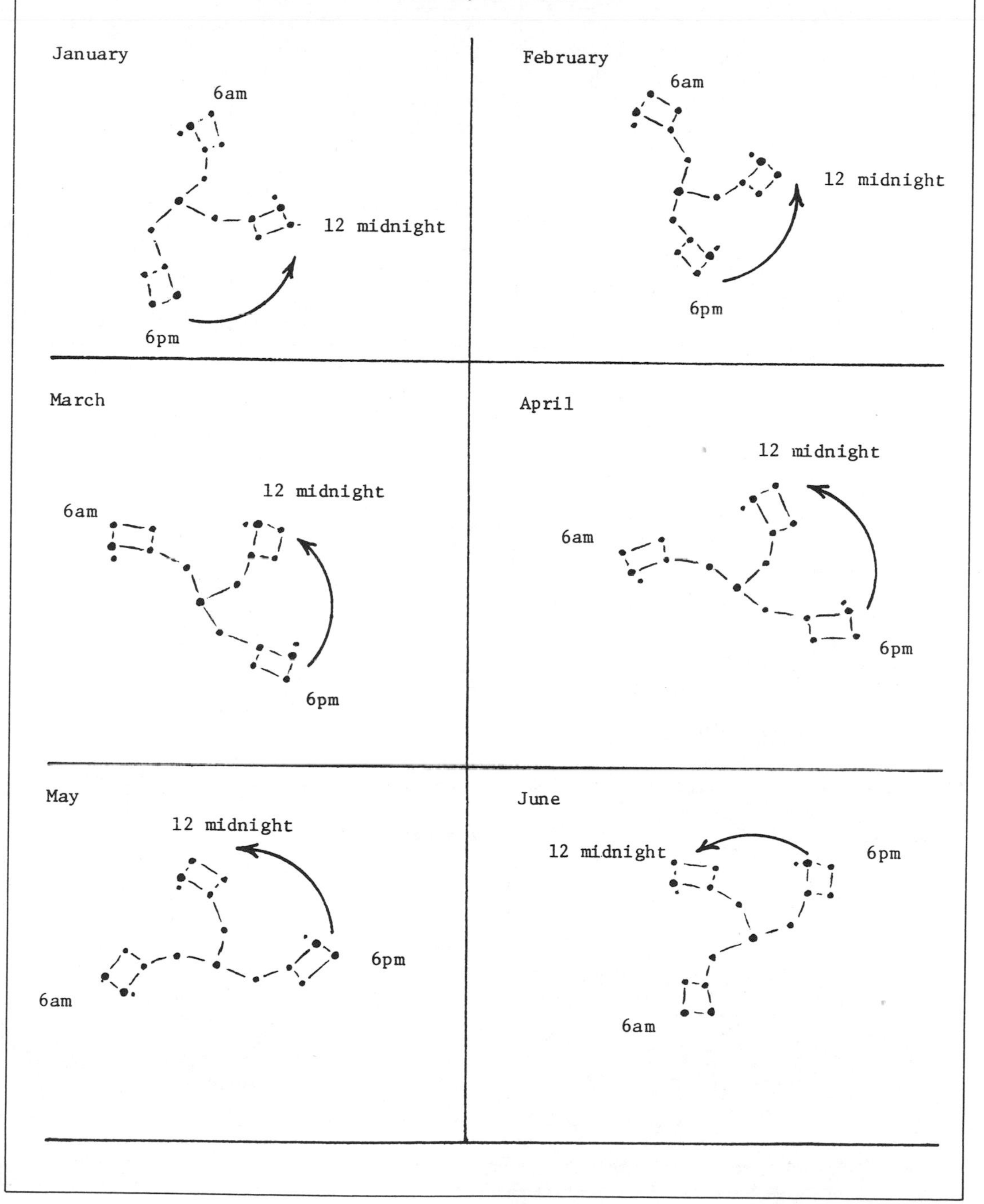
A Celestial time clock using the Orientation of the Little Dipper
Based on 6:00 p.m., 12:00 midnight, and 6:00 a.m.
15th day of Month
January
6am
12 midnight
6pm
February
6am
12 midnight
6pm
March
6am
12 midnight
6pm
April
12 midnight
6am
6pm
May
12 midnight
6pm
6am
June
12 midnight
6pm
6am

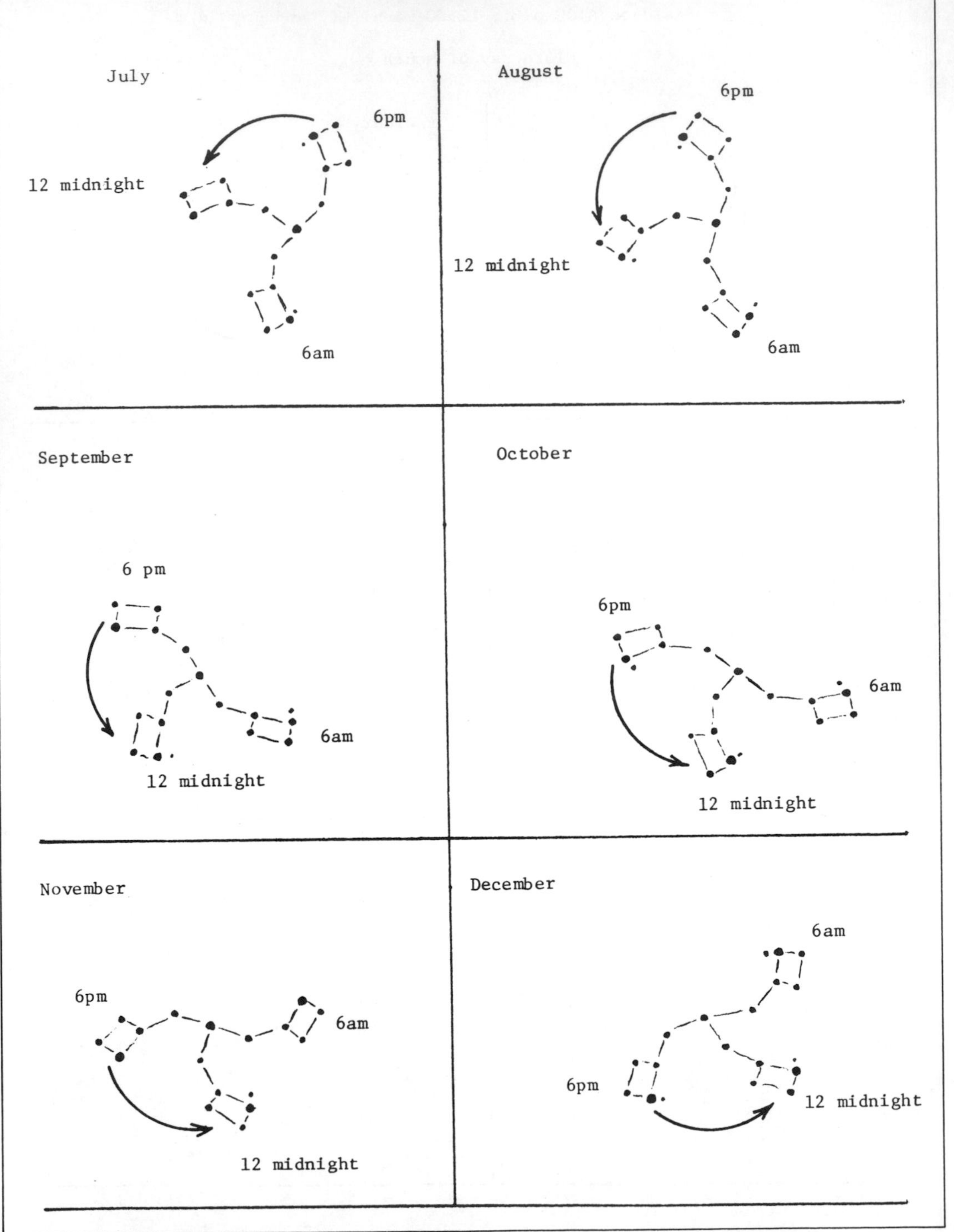
A Celestial Time Clock using the Orientation of the Little Dipper
-- continued --
July
6pm
12 midnight
6am
August
6pm
12 midnight
6am
September
6 pm
6am
12 midnight
October
6pm
6am
12 midnight
November
6pm
6am
12 midnight
December
6am
6pm
12 midnight

Circumpolar/Spring URSA MAJOR

STAR NAMES:

α	-"Dubhe"	ζ	-"Mizar/Alcor"
β	-"Merak"	ϵ	-"Alioth"
γ	-"Phecda	η	-"Alcaid"
δ	-"Megrez"	ι	-"Dnoces"

CONSTELLATION NOTES: Try to forget the impression that Ursa Major is a "Big Dipper", and picture it as a great Bear--the pointer stars, Dubhe and Merak form a line to Polaris, an easy way to identify the north. Be sure to examine Alcor and Mizar; they can be seen with the naked eye, but a low-power telescope reveals that Mizar (the brightest component) is also a double star. The galaxies given below can easily be seen in binoculars on a dark night. M-51 is the Whirlpool.

DATE OBSERVED:__________ NAME:__________ TIME:________ U.T.

OPTICAL EQUIPMENT:__________ __________

Visual Impressions:

M51__________

M81__________

M82__________

M106__________

Stars:

__________ Color:__________

__________ Color:__________

Alcor/Mizar__________ Colors:__________

COMMENTS:

LEO

Early Spring

STAR NAMES: α -"Regulus" R.A. 10h 06m; DEC. $+12^{\circ}13'$
β -"Denebola" 11h 46m; DEC. $+14^{\circ}51'$

CONSTELLATION DATA: Leo the Lion is one of the few constellation which actually looks like what its legendary figure represents. The figure is void of bright star fields but does contain some interesting galaxies for the amateur.

DATE OBSERVED:____________TIME:______U.T. OBSERVER:_______________

OPTICAL AID:____________________

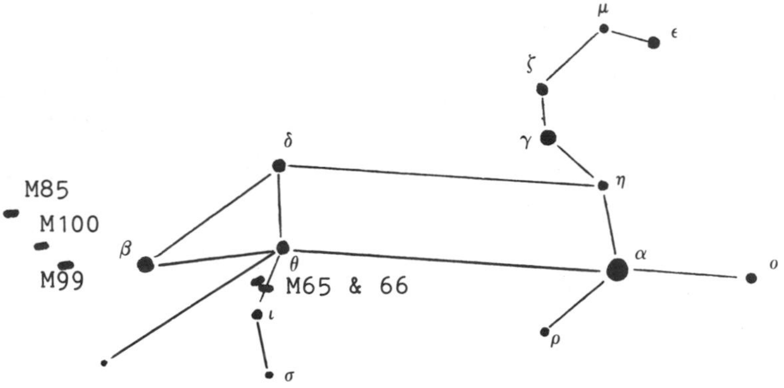

Visual Impressions:

M 85 (12h 23 m ; $18^{\circ}28$m)________________________________

M 100 (12h 20 m ; $16^{\circ}06$m)________________________________

M 99 (12h 16 m ; $14^{\circ}42$m)________________________________

M 65 (11h 16 m ; $13^{\circ}23$m)________________________________

M 66 (11h 17 m ; $13^{\circ}17$m)________________________________

Stars:

α ____________________________________ Colors:__________

β ____________________________________ Color:__________

OBSERVE:

Using a telescope, look at γ Leonis; record the following data:

1) Telescope used_______________________
2) Lowest power needed to separate stars_________________
3) Color of both stars_____________________________
4) Compare the magnitudes of both stars__________________

Look at every other star at high powers in the telescope and record what you see.

Spring/Early Summer

CANES VENATICI
COMA BERENICES

STAR NAMES: α C. Venatici - "Cor Caroli"

CONSTELLATION NOTES: The famous Realm of the Galaxies lies between Coma Berenices and Leo, to the west. The observer should scan this area with binoculars on the first very dark night, searching for faint patches of light which most likely will be galaxies. A 10" scope reveals hundreds of galaxies in this immediate area.

DATE OBSERVED:________________NAME:________________________

OPTICAL EQUIPMENT:____________________ TIME:____________U.T.

M94 β α C. Venatici

M 3 β γ M64 M53 α C. Berenices

Visual impressions:

M 94____

M 3____

M 53____

M 64____

α C. Ven.____________________________ Color:________

β C. Ven____________________________ Color:________

α C. Ber.____________________________ Color:________

β C. Ber.____________________________ Color:________

C. Ber. Star Cluster________________________________

COMMENTS:

Spring

CORVUS AND CRATER

STAR NAMES: α Cor -"Alchiba"
γ Cor -"Gienah"

Constellation Data: These two constellations are inconspicuous, mainly because of their low southerly declination. Corvus is the most easily-recognized, appearing as a lop-sided square just WEST of bright Spica in Virgo; Crater is the next group of stars WEST of Corvus, appearing as a faint circlet of stars. Look up some historical data on these two interesting and ignored constellations.

DATE OBSERVED________________TIME:______U.T. OBSERVER:__________

OPTICAL EQUIPMENT:__________________________

M104

CRATER

θ ϵ δ γ Hydrae η ζ γ α β

η δ γ CORVUS β ϵ α

Visual Impressions:

M104____

Stars:

α Cor____________________________ Color:__________

η Cor____________________________ Color:__________

α Cra____________________________ Color:__________

θ Cra____________________________ Color:__________

TRANSLATE: Look up the legend and/or history behind these two little constellations and write a summary below:

COMMENTS:

BOÖTES

Spring/Early Summer

STAR NAMES: α -"Arcturus"
ϵ -"Izar"

CONSTELLATION NOTES: Boötes, with its bright star Arcturus is a dominant spring constellation. The striking color (determine for yourself) of Arcturus is exaggerated with optical aid.

DATE OBSERVED:________________ TIME:______________U.T.

NAME:______________________ OPTICAL AID:______________

Visual Impressions:

M 3____

M 5____

Stars:

α ______________________________ Color:__________

ϵ ______________________________ Color:__________

β ______________________________ Color:__________

δ ______________________________ Color:__________

COMMENTS:

TRANSLATION: "Arcturus" -

"Izar" -

VIRGO

Early Spring

STAR NAMES: α -"Spica"
γ -"Porrima"
ϵ -"Vindemiatrix"

CONSTELLATION DATA: This constellation is dominated by bright Spica. Since it is crossed by the ecliptic, occassional lunar occultations of Spica may occur. Virgo is famous for containing a portion of the realm of the galaxies, primarily the Coma-Virgo group. Virgo is easily found by first locating Spica which lies just west and south of bright Arcturus in Bötes.

OBSERVATION DATE:__________TIME:________U.T. OBSERVER:__________

Optical Aid:____________________

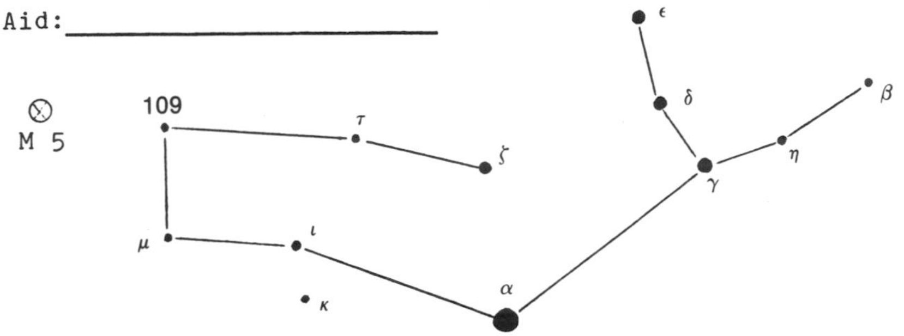

M 5
109
τ
ζ
ϵ
δ
β
η
γ
μ
ι
α
κ

Visual Impressions:

M-5____

Stars:

α ____________________________ Color:________

γ ____________________________ Color:________

ϵ ____________________________ Color:________

DISCOVER:

Look up some data on the star SPICA. Write a summary, including distance, spectral class, size, age, and any unusual characteristics it might have.

FACTS: Virgo is one of the ZODIAC constellations. Can you name the other eleven?

Why was the Zodiac and its constellations important to the people of medieval times?

What significance does it have today for astronomers?

LIBRA

Late Spring

STAR NAMES: α -"Zubenelgenubi"
β -"Zubeneschamali"

CONSTELLATION DATA: This star pattern is perhaps the least-defined of all in the northern sky. There are no striking star patterns, nor are there any excessively bright stars in Libra. However, this constellation has the distinction of possessing two of the most unusual star names in the entire Universe!

OBSERVATION DATE:________________TIME:______________U.T.

OBSERVER:__________________OPTICAL AID:_________________________

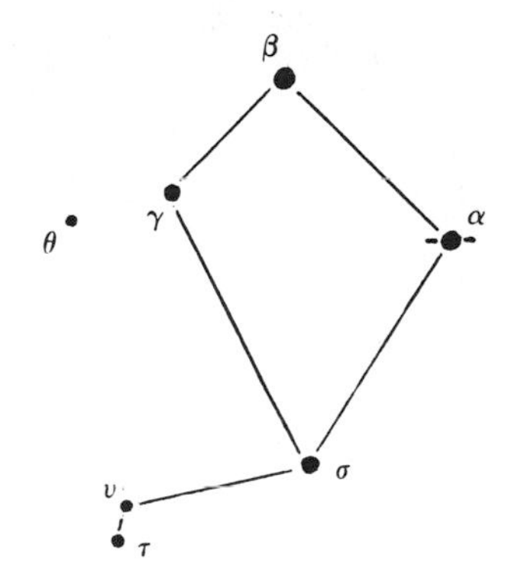

Visual Impressions:

Stars:

α ____________________________________ Color:________

β ____________________________________ Color:________

DISCOVER:
Survey the constellation with a telescope, looking particularly at the stars 7th magnitude and brighter. Note any unusual colors, or multiple star systems that you might encounter. Put your observations below:

TRANSLATE:
Find the Origin and Meaning of the following two star names

"Zubenelgenubi" -

"Zubeneschamali"-

Spring

SERPENS CAPUT

CONSTELLATION DATA: This largely ignored constellation lies just West of Ophiuchus. The head (top) of Serpens is very rich in background stars. Serpens is a very nice double for all telescopes, although it is easily resolved in even a 2" scope. A famous variable, R Serpenis, can be found just south of the head of this constellation, varying from magnitude 5.6 to less than 13 in about 357 days.

DATE OBSERVED:____________TIME:_________U.T. OBSERVER______________

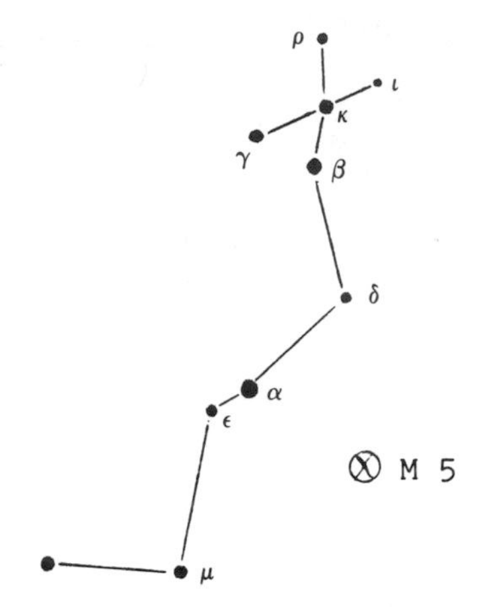

Visual Impressions:

M 5 (15h 16' ; +2°16')______________________________

Stars:

μ ______________________________ Color: __________

α ______________________________ Color: __________

OBSERVE:
In the circle below, sketch the field at low power of δ Serpens, being careful to orient the drawing to match the sky directions.

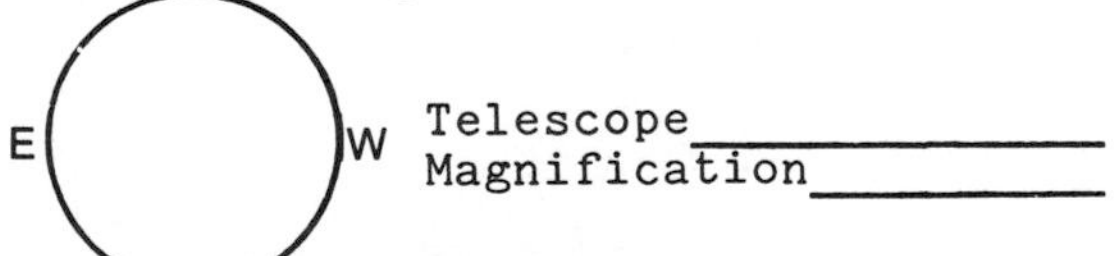

Telescope______________
Magnification__________

DISCOVER: Locate some information on Serpens Caput, and find out what the name means, and explain the difference between it and SERPENS CAUDA, which lies far to the east of it.

Find out the total number of stars contained in M 5.

Late Spring

OPHIUCHUS

STAR NAMES: α -"Ras Alhague"
η -"Sabik"

CONSTELLATION DATA: Ophiuchus is easily one of the largest constellations, but particularly void of bright stars. The north "point" of Ophiuchus is nearly midway between bright Arcturus of Böotes, and Altair of Aquila. Several bright globular clusters are found within Ophiuchus.

DATE OBSERVED:__________TIME:________U.T. OBSERVER________________

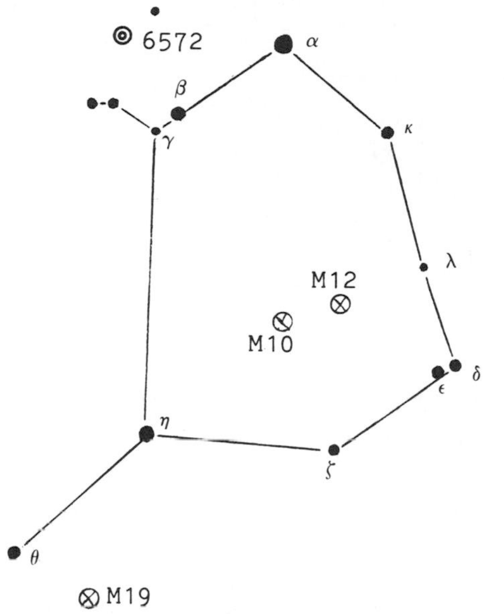

6572 α β γ κ λ M12 M10 δ ε η ζ θ M19

Visual Impressions:

M-10____

M-12____

M-19____

Of the three globular clusters listed above, which one appears:

Largest________ Brightest_________?

NGC 6572____

Stars:

α ______________________________ Color: ____________________

η ______________________________ Color: ____________________

TRANSLATE: "Ophiuchus" -
"Ras Alhague" -

Late Spring

CORONA BOREALIS

STAR NAMES: α -"Alphecca" (Gemma)

CONSTELLATION DATA: Corona Borealis is a very neat circlet of stars just west of Hercules. Look up the translation of "Corona Borealis" and write in the appropriate space below. Perhaps the most famous feature of this constellation is the variable star "R Corona Borealis", a highly irregular and unusual star observed easily by amateur astronomers.

DATE OBSERVED:______________________TIME:____________U.T.

OBSERVER'S NAME:___________________OPTICAL AID:

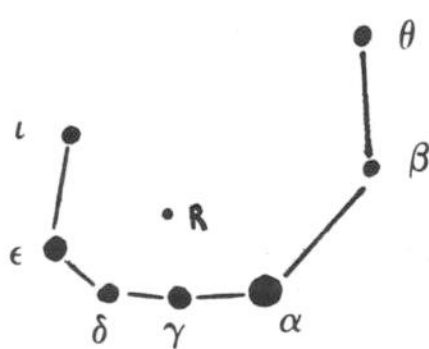

Visual Impressions:

R Cor Bor_________________________ Color:____________

Stars:

α _________________________ Color:____________

ϵ _________________________ Color:____________

DISCOVER:

Use source material to write a short summary of the variable R Cor Bor, including the reason given for the unusual light changes.

Draw a rough light curve of R Cor Bor:

TRANSLATE: What does Corona Borealis mean?

Write a short report on another constellation, CORONA AUSTRALIS:

COMMENTS:

Late Spring/Summer

HERCULES

STAR NAMES: α -"Ras Algethi"

CONSTELLATION NOTES: Hercules is a large constellation marked with many stars. Several novae have been discovered within this constellation's boundaries. The most notable deep-sky object here is M-13, the brightest globular cluster seen from the northern hemisphere.

DATE OBSERVED:______________________________TIME:____________U.T.

OBSERVER'S NAME:_________________________ OPTICAL AID:_______________

ι θ M92 τ ϕ σ χ ρ π η M-13 o ξ μ ϵ ζ λ δ 6210 β γ α

Visual Impressions:

M-13____

NGC 6210____

M-92____

Stars:

α ____________________________________ Color:_____

β ____________________________________ Color:_____

COMMENTS:

Translation: "Ras Algethi" -

Facts: About Messier 13: Distance________L.Y.; Size________

Estimated # of stars__________; Age__________years

What spectral type stars?_______;

Summer

LYRA

STAR NAMES: α -"Vega"

CONSTELLATION DATA: Because of its compact size, bright stars, and the dazzling blue brightness of Vega, Lyra is an easily recognized constellation in the early summer months, resembling a small harp of ancient design. Beta Lyrae is a variable star which one can monitor with the naked eye. Epsilon (ϵ) Lyrae is the famous "double-double" star. M-57 is the famous Ring Nebula, a bright planetary nebula visible in even the smallest telescopes.

DATE OBSERVED:________________TIME:______U.T. OBSERVER:____________

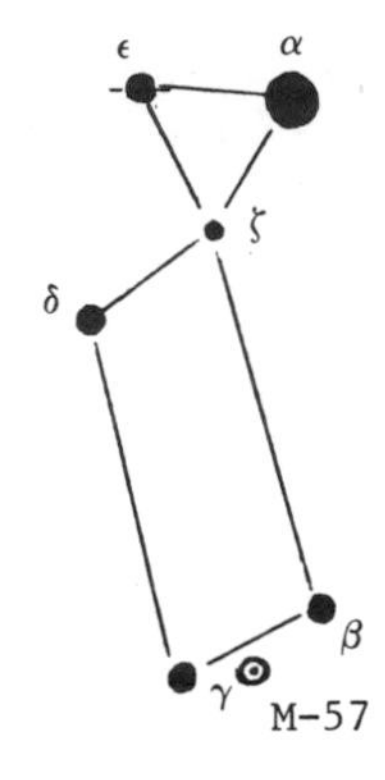

Visual Impressions:

M-57____

Stars:

α ______________________________ Color ____________

β ______________________________ Color ____________

δ ______________________________ Color ____________

Observe Epsilon (ϵ) Lyrae with the following instruments and record what you see:

With Naked Eye_______________________________________

With Binoculars______________________________________

With 6" Scope____

In the circle at left, sketch exactly what you see with high power in the 6" telescope — be sure to record what power was used.

DISCOVER: Look up some data about planetary nebulae and record on the back of this sheet the origin, number, etc. of these strange-appearing objects.

CYGNUS

STAR NAMES: α -"Deneb" R.A.20h 40m; DEC. +45°06'
γ -"Sadr" 20h 20m; DEC. +40°06'

CONSTELLATION DATA: Be sure to scan under a dark sky with binoculars through this area - the Milky Way view is beyond description. Take careful notes on what is seen and record them on the next page.

DATE OBSERVED:________________TIME:______U.T. OBSERVER:___________

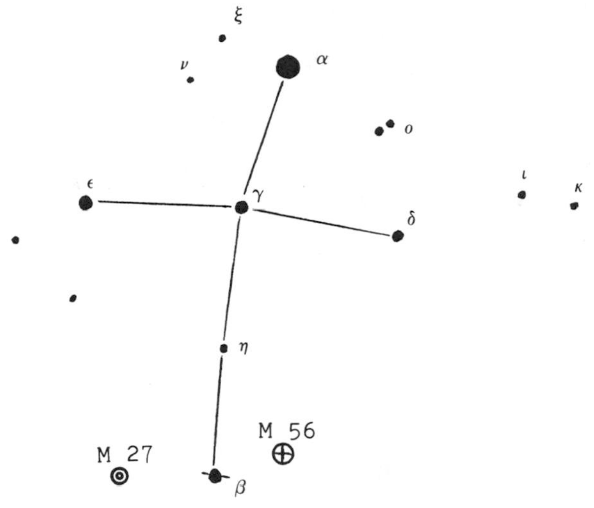

Visual Impressions:

M 27 (19h 57m; 22°35m)____

M 56 (19h 15m; 30°05m)____

Stars:

α ______________________________________ Color:__________

γ ______________________________________ Color:__________

OBSERVE:

With low power, observe Albireo (β Cyg) and record the color of the brighter component and the fainter component:

Brighter Star:_______Fainter Star:_________

On the following sheet, draw in faint stars, clusters, and any object of interest seen in low-power binoculars. Be sure to record the outlines of major star clouds of the Milky Way. This MUST be done under very dark sky conditions with no moon present.

DISCOVER:

Look up the legend behind Cygnus, the Swan and write a short summary. Can you picture a swan from the star outline? List below other constellations which to YOU actually look like what they are supposed to represent. (AT LEAST 10)

VISUAL STUDY OF
CYGNUS

OBSERVER:______________________LOCATION:__________________________

DATE:____________TIME:_____________U.T. BINOCULARS:____________

Be sure to sketch both naked-eye and binocular impressions in the above star field. USE PENCIL, because you will probably have to erase many times. Do the chart as neatly and accurately as possible.

NOTES:

Summer

SCORPIUS

STAR NAMES: α -"Antares" R.A. 16h 26m; DEC. -26° 19 m
δ -"Dschubba" 15h 57 m -22° 30 m
λ -"Shaula" 17h 30m; -37° 04 m

CONSTELLATION DATA: The scorpion is one of the most well-defined constellations in the sky. You should scan the entire length of this constellation to appreciate the richness of its star clouds.

DATE OBSERVED:________________TIME:_______U.T.

OBSERVER:_____________________________

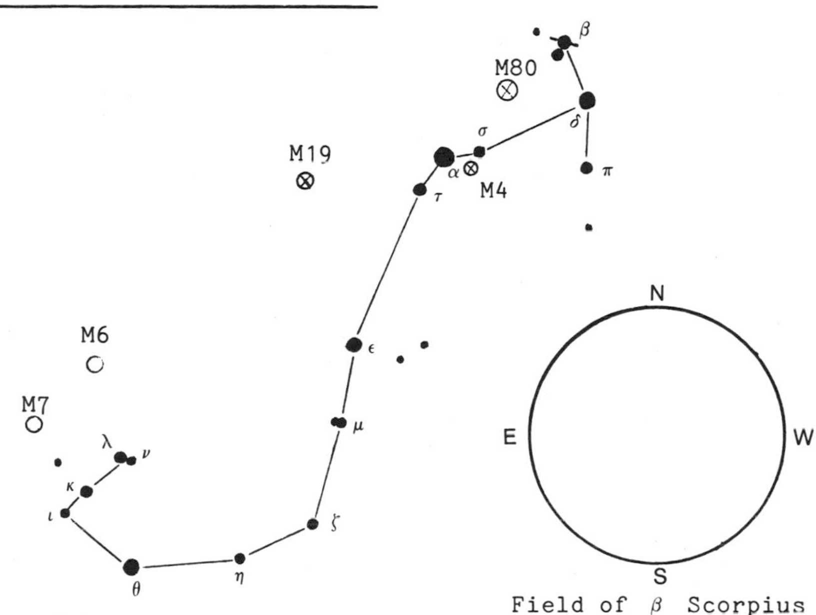

Field of β Scorpius

power:____

Visual Impressions:

M 6 (17h 37m; -32° 11m)________________________________

M 7 (17h 51m; -34°47m)________________________________

M 19 (16h 59m; -26°12m)________________________________

M 80 (16h 14m; -22°51m)________________________________

Stars:

α ______________________________ Color:__________

λ ______________________________ Color:__________

OBSERVE:

Examine β Sco with a telescope at low power to resolve into its main components; in the small circle above, record the field of view. Now change to high power (it may take very high power) and look at the brighter component of the two. Record what you see (you are looking for a faint companion).

INDEX